Process and Method in Canadian Geography

Geomorphology

Process and Method
in Canadian Geography

title: # Geomorphology

Selected Readings

edited by

J. G. Nelson

M. J. Chambers

Methuen

Toronto   London   Sydney   Wellington

Library of Congress Card Number 75-83146

SBN 458 90240 3 (Hardbound)

SBN 458 90250 0 (Paperback)

Design by Hiller Rinaldo

Printed and Bound in Canada

73 72 71 70 69 1 2 3 4 5 6

Preface

This volume is part of a series on Canadian Geography. Companion works include *Weather and Climate, Vegetation, Soils, and Wildlife,* and *Water.* Volumes on other topics are being planned.

Each volume comprises a collection of recent articles gathered, for the most part, from learned journals. The series is Canadian in the sense that the papers deal with Canadian topics or have been prepared by geographers working in Canada; it is geographic in the sense that they are concerned with the landscape, and the physical and cultural processes at work on it. For purposes of general definition the landscape can be thought of as *that zone at or near the surface of the earth that is perceived, used or affected by man.* This zone obviously includes surface phenomena such as vegetation, animals, and soils. The zone changes, however, with changes in knowledge and technology, so that more and more contemporary geographers are becoming concerned about human penetration into the oceans and into space. A process can be thought of as *a change that varies in space and time.* A process can be considered as cultural when it is *totally or primarily a result of the activities of man.* Examples are: urbanization with its cities and towns; agriculture with its fields and crops. A process can be considered as physical when it is *totally or primarily a result of forces independent of man.* Examples are ocean currents and hurricanes. As the articles in these volumes will show, however, the distinction between cultural and physical processes is, in many cases, increasingly difficult to draw. Floods are influenced by cultural activities, such as lumbering, as well as by physical processes, such as thunderstorms. Climate is affected by cities as well as by the general circulation of the atmosphere.

In recent years geographers have placed increasing emphasis on the study of processes. While this is partly a result of improved instrumentation, research methods, and technique, it also reflects a general trend in geography toward a more analytical approach. In this and companion volumes, stress has therefore been placed on articles illustrating research techniques and the study of processes. An attempt, however has also been made to add to the usefulness of the volumes by including studies that are concerned with many parts of Canada in order to provide knowledge and appreciation of the Canadian landscape.

In our initial thinking about the volumes, we recognised that it would be necessary and desirable to include relevant research undertaken by biologists, geologists, and members of other landscape-oriented disciplines. Geography, as with other disciplines, interrelates with many fields of study, and a considerable amount of cross-fertilization of thought and technique occurs. Now that a number of volumes have been formulated and organized, we find that some of them contain more work by practitioners in related disciplines than we had originally anticipated. Our concern, however, is to deal with those processes, techniques, problems, and topics that we believe to be of interest and importance to geographers in the wide sense. Moreover, the broader content of some of the volumes may make them extremely useful to members of related disciplines, as well as to those many scholars and students whose work and interest carry them into the so-called interdisciplinary areas where so many important contemporary problems lie.

Aside from their possible interest and value to some members of the public, the various volumes in this series should be useful at a variety of academic levels. Many of the articles in each volume will be of interest to the beginning student. They would also seem to be suitable for more advanced undergraduate courses, for example, the first courses exclusively on biogeography, geomorphology or climatology, or for broader courses in resources or conservation. The volumes could also be valuable in certain graduate courses as they describe important recent developments in research fields. They are a guide to the state of the art—or the science—in Canada. For those wishing to investigate a topic further, all the original bibliographies have been retained. These references are considered to be of greater value to the reader than other material that might be introduced, such as comments and reviews interjected between articles by the editors. Most of the maps, diagrams, and photographs have been retained in each article.

Contents

Contributors

Andrews, J. T.
Formerly at the Geographical Branch, Department of Energy,
Mines and Resources; now at the Institute of Arctic and Alpine
Research, University of Colorado, Boulder, Colorado.

Ashwell, I. Y.
Department of Geography, University of Calgary.

Bird, J. B.
Department of Geography, McGill University.

Brunger, A. G.
Department of Geography, University of Western Ontario.

Cameron, H. L.
Late of the Department of Geology, Acadia University.

Chepil, W. S.
Late of United States Department of Agriculture,
Manhattan, Kansas.

Farvolden, R. N.
Department of Geology, University of Western Ontario.

Gardner, J. S.
Formerly Department of Geography, McGill University.

Ives, J. D.
Formerly Director, Geographical Branch, Department of Energy,
Mines and Resources, Ottawa; now Director, Institute of Arctic
and Alpine Research, University of Colorado, Boulder, Colorado.

Jungerius, P. D.
Department of Physical Geography and Soils,
University of Amsterdam, Holland.

Harris, S. A.
Department of Geography and Meteorology, University of Kansas,
Lawrence, Kansas.

Loken, O. H.
Inland Waters Branch, Department of Energy, Mines and Resources.

Mackay, J. R.
Department of Geography, University of British Columbia.

Mathews, W. H.
Department of Geology, University of British Columbia.

McPherson, H. T.
Department of Geography, Queen's University.

Morgan, A. V.
Department of Geology, University of Birmingham, England.

Nelson, J. G.
Department of Geography, University of Calgary.

Parry, J. T.
Department of Geography, McGill University, Montreal.

Pissart, A.
Department of Geology and Physical Geography,
Liege University, Belgium.

Radforth, N. W.
Formerly Chairman of the Organic and Associated Terrain Research
Unit, McMaster University; now at the Department of Biology,
University of New Brunswick, Fredericton.

Schaerer, P. A.
Division of Building Research, National Research Council, Ottawa.

Straw, A.
Formerly Department of Geography, McMaster University; now
Department of Geography, University of Sheffield, England.

Webber, P. J.
Department of Botany, Queen's University, Kingston, Ontario.

Wood, H. A.
Department of Geography, McMaster University.

Introduction

In gathering together this series of readings in geomorphology, the objective has been to prepare a representative sample of modern Canadian research in this field. It may therefore appear unbalanced in favor of glacial and periglacial topics: in fact this is a true reflection of the interests of the majority of Canadian geomorphologists. As so much of the landscape bears the imprint of glacial activity, it is not unreasonable that those involved in the study of this landscape should concentrate in this area of the discipline. One of the great advantages, which stimulates this research, is the presence in Canada of living examples of the processes that have played such a significant role in moulding the landscape of the more temperate regions to the south. Thus studies of the Barnes Ice Cap on Baffin Island, permafrost features in the McKenzie delta, and valley glaciers in the Rockies show that these areas are being used as vast laboratories in order to increase our understanding of processes that have a far broader significance and application than is at first apparent.

This abundance of potential research topics in glacial and periglacial landforms has attracted many workers — almost to the exclusion of many other significant fields. Thus, coasts and beaches, rivers, and aeolian features, together with their respective processes, have received comparatively little attention; yet the development of drainage systems and the action of wind have been two of the most important factors in modifying the post-glacial landscape. The extent to which changes have taken place since deglaciation is by no means fully understood, and more detailed and painstaking research is necessary before any precise statement of the degree and rate of erosion, transport, and deposition can be made.

In the past few years, however, there has been a noticeable trend in Canadian geomorphology away from static and descriptive studies toward a quantitative, dynamic approach to the subject. This is partly the result of improved instrumentation, which allows sufficient accuracy for short-term or small-scale measurements to be recorded; it also reflects a growing interest in the processes of landscape development.

It is this progress toward a quantitative analysis of the rates of change observable today, together with improved technique in the absolute dating of past events, that has provided the theme of this volume. The aim has been to emphasize the variables in the complexity of a compound system and to illustrate the various techniques that have been developed to study these variables.

In his article, J. T. Parry traces the history of geomorphology in Canada by means of a fascinating study of the subject from its early — and often misguided — attempts at broad generalizations within a vast and uncharted land, through the periods of regional air-photograph mapping, and analysis right up to the present trend. It is only because this foundation has been so thoroughly laid during the post-war period that it is now possible to focus attention on the details that allow the processes operating on the landscape to be examined.

It is important to note the influence that the Canadian Government has had on the growth of the subject to its present stature through directing research within its own departments, and through sponsored research within the universities. The government has also been partly responsible for the concentration of research in glacial and periglacial topics, particularly in arctic and alpine environments. It was while Ives was director of the Geographical Branch that his introduction to the topic of glaciers was written. Under his direction the great volume of research on and around Baffin Island (see articles in this volume by Løken and Andrews) was carried out, which increased considerably our knowledge of arctic Canada. This brief general discussion of glaciers sets the scene for a series of more detailed papers that examine specific problems using a variety of techniques. The article by Nelson *et al.* shows what can be done with a series of historic photographs of the Drummond Glacier, where an account of the recession during the past eighty years has been carefully documented. This pattern of retreat has been integrated with a set of modern ablation and recession measurements, providing a comprehensive picture of the recent history and current activity of the glacier.

Another technique of plotting recent glacial recession by lichonometry is explained in detail by Andrews and Webber. This approach obviously has more potential in arctic areas than that described by Nelson *et al.*, as the photographic record in the north is limited. It is stressed that lichens can be used in deglaciation studies by those who have not specialized in botany. The same applies to the use of dendrochronology, which is here used to supplement the lichenometry, and serves to validate the accuracy of the recent lichen growth-rate curve.

Another problem commonly encountered in studies of mountain glaciation is the determination of maximum ice-height relative to

the peaks. Løken presents a carefully investigated case for suggesting that the mountains of northeastern Labrador-Ungava were
never covered by ice, or if they were, it was at a very early stage
of the Pleistocene epoch. Evidence is drawn from a variety of observations ranging from an X-ray analysis of clay minerals to the
presence or absence of glacial erratics. The discussion by Ives,
which follows, introduces a significant problem into the sphere of
geomorphological research. Here, two specialists, both experts in
the field, having the same evidence, come to different conclusions.
It is possible to explain this difference by the paucity of evidence;
but, as conclusions are often based on a similarly sparse amount
of information, it is important to take the subjective element into
account when dealing with interpretive studies such as this.

The analysis of material left behind by retreating ice has always
been of major importance in the study of glacial processes, and a
series of four papers has been selected for this volume in order to
illustrate a variety of approaches to the study of till, erratics, and
moraines. Those by Andrews and Mackay deal with a series of
small irregular ridges, which, from a superficial examination, appear similar. Examination of the composition of the ridges, together
with an assessment of field relationships with other glacial features
leads the respective workers to differing conclusions as to the origin
of these features.

The contribution by Harris presents a detailed critique of the
technique that is probably the most widely used and least understood of all — macro-fabric analysis. The difficulties in sampling,
the assumptions concerning till deposition, and the problems of
interpretation are all clearly outlined, with a warning relating to
the many possible errors that may be incurred. Morgan selects
the lithology variations in till sheets as a criterion for determining
the source of interacting ice-masses. The validity of this method
rests upon the assumption that the different ice-masses bear material of unique lithological characteristics. In the area where he
worked, Morgan found that three distinct ice-masses could be
identified merely by the lithology of the till-stones. A surprising
amount of information was gathered by this relatively simple
technique.

In the realm of periglacial studies in Canada, a series of articles
has been chosen with a view to illustrating the wide variety of
processes involved in this general field. Straw describes the effect of
freeze-thaw activity and mass movement on an outcrop of dolomite
in Southern Ontario. Gardner records a detailed inventory of all
rapid mass movements over a period of two months in a high alpine

zone. His analysis of these data shows some of the problems that must be solved before the geomorphic significance of such activity can be assessed. Matthews and Mackay discuss the possible ways of instrumenting a study of snow-creep, the problems involved, and the geomorphic significance of this almost ubiquitous process in the mountain environments of Canada.

Schaerer's article also deals with mass movements of snow; but in this instance he is concerned with the interaction of man with the natural process of snow avalanches. He shows that only by a comprehensive study of snow types, weather conditions, and valley forms, together with experimentation into practical methods of protection, can a long-term economical defence program be operated. Ives, in his paper on iron-mining, is also concerned with man's interaction with an inhospitable environment. Again, it is the economics of the situation that causes most concern, not the mere technical challenge. A cogent argument is made for more fundamental research into the nature and distribution of permafrost. The economic and other man-related facets of these natural phenomena largely account for the extensive research of this type that has been sponsored by the Canadian Government. Permafrost has perhaps the broadest implications of all, and articles relating it to climate, soils, and vegetation will be included in the subsequent volumes of this series.

The article by Farvolden is not easy to classify, as it deals with pre-glacial river courses, the development of the modern drainage on top of glacial drift, the degree of integration of these two systems, and the implications of this with respect to ground water. The reconstruction, stage by stage, of the process of superimposition of the present-day drainage system helps to explain some of the problems that face the geomorphologist in Alberta. McPherson's study of the Red Deer River demonstrates how several techniques can be applied to one problem, each contributing a certain amount of information. The synthesis of information presents a comprehensive picture of the river in its complex relationships with channel and floodplain.

The articles by Cameron and Wood make an interesting comparison in approach to the analysis of coastal deposition. The former employs old records and charts, and a modern series of air photographs to illustrate the changes in outline that have occurred on the Sable Island sands off Nova Scotia. By this means the development of the bar over a period of two hundred years has been mapped. Although Wood makes use of old records to demonstrate that change has been taking place, his main approach is quantitative. He uses meteorological and hydrographic data to calculate the energy available for erosion transport and deposition along a Lake Erie spit. In this way, a direct relationship between wind and beach processes is established, and predictions can be made about future changes. Mackay's report on coastal recession in the Yukon reveals a remarkably rapid rate of erosion. His suggestion that massive ground-ice may be partly responsible for this recession introduces another variable into the complex processes that are so difficult to analyse and quantify.

The problem of understanding erosion processes does not merely apply to water, and the article by Chepil outlines some basic research into the mechanics of wind erosion. Although much work has been done in this field since he wrote his article, the fundamental principles and factors that he considers serve as a valuable introduction. Only when this basic material is fully understood can a geomorphologist move on to study the significance of wind in landscape modification. Bird outlines the aeolian processes that are currently observable in the arctic regions, and an attempt is made to assess their geomorphic significance, but there is clearly a great need for data. Pissart examines deflation in some detail, concluding that it is of minor importance in the area where he worked—mainly because of relatively light winds. The most important influence, apparently, is the irregular distribution of snow by wind, which gives rise to numerous nivation landforms.

The lack of detailed research in limestone areas of Canada is reflected in a dearth of published material. Some broad descriptive accounts have been produced, and some more specific work is currently being undertaken in the eastern Rockies, but this is not yet available. In the arctic, Bird has compiled a comprehensive account of the geomorphic processes active in limestone terrains that illustrates the unique nature of this rock type and the importance of solution at approximately 0° Centigrade. Again the need for research is stressed.

Jungerius brings another valuable technique to the geomorphologist's repertoire in his precise analysis of soils in the Cypress Hills. He uses this as a tool in the interpretation of the geomorphic history of the area. Mechanical, clay mineral, and heavy mineral analysis all contribute elements that, when synthesized, form a coherent picture of past processes and events.

In the final article of this collection, Radforth discusses the interrelationships between terrain features and vegetation. The potential value of this in-landscape interpretation is considerable, and quantitative data on colonization and growth rates of muskeg may well

provide as useful a dating technique as dendrochronology or lichenometry. This paper raises many other interesting topics that
stretch beyond the scope of this volume into the realm dealt with
in the book of readings entitled *Vegetation, Soils and Wildlife*.

J. G. NELSON
M. J. CHAMBERS

1

Geomorphology in Canada

J. T. Parry

In the Beginning

The beginnings of geomorphology in Canada are difficult to discover, but as in other countries the roots are intergrown with those of geology and scientific exploration. Undoubtedly the first writer to demonstrate a clear appreciation of the variations in regional character from one part of Canada to another was David Thompson. Unfortunately, he did not prepare the materials in his journals for publication until after 1840, and they were not finally published until 1916,[1] by which time their significance as pioneer documents on the geography of Canada was overlooked. It is apparent from his journals that Thompson frequently prepared a complete geographical synthesis of the areas traversed, in which was included a detailed description of the topography.[2] It is in these descriptions that the first contributions to Canadian geomorphology are to be found.

Throughout the nineteenth century there were very few natural scientists in Canada, and fewer still were concerned with problems of the Canadian landscape. Almost inevitably foreign visitors provided the first topographic descriptions, and European viewpoints provided the theoretical basis for landscape study. Catastrophism,

An abridged version of a paper first published in *Canadian Geographer*, Vol. XI, no. 4 (1967), pp. 280-311.

[1] J. B. Tyrrell (ed.), "David Thompson's Narrative of his Explorations in Western America, 1784-1812," (Champlain Society, 1916).
[2] P. Warkentin (ed.), "The Western Interior of Canada: a record of geographical discovery, 1912-1917," (McClelland and Stewart, 1964), pp. 91-105.

with its special elaboration, diluvialism, was invoked to explain many features in Upper Canada and in the west. For example, in 1823 the sediments of what came to be known as Lake Agassiz were ascribed by W. H. Keating to a lacustrine episode, which was terminated by a spectacular lake burst,[3] the high terraces of Lake Athabaska were cited by Samuel Black as evidence of the great Flood,[4] Niagara Falls and the Niagara gorge were considered to be the result of the drainage of the waters of the Flood,[5] while the Great Lakes, and the myriad other lakes of the Shield, were interpreted as the remnants of this great deluge.

In Europe, in the 1840s, there was a shift in the climate of opinion, so that uniformitarianism penetrated where catastrophism had previously prevailed. This change was largely the result of the work of Charles Lyell, and his visits to North America in 1841 and 1845 ensured successful transplantation of the new doctrine. Unfortunately, Lyell's uniformitarianism was marred by his overemphasis of the role of marine processes in landscape sculpture, and his insistence on the iceberg-drift origin of glacial deposits. Thus, while he accepted a fluvial origin for the Niagara gorge,[6] and correctly ascribed the features on Mount Royal and the Scarboro bluffs to wave action,[7] he also believed that intensive erosion could result from the mass movement of ocean water, and so the overdeepened basins of the Great Lakes were considered to be the result of the swirling action of estuarine currents.[8] The reader of Lyell's *Travels in North America* is presented with a vision of a land half-sunk beneath Arctic seas, in which icebergs ground against one another, and became stranded along the coastlines.[9] The simpler land-ice theory of Louis Agassiz[10] was eclipsed for nearly a quarter of a century.

---

3 W. H. Keating, "Narrative of an Expedition to the Source of St. Peter's River, Lake Winnepeg, Lake of the Woods, etc. 1823," (Ross and Haines, 1959), Vol. II, pp. 98-100.

4 E. E. Rich (ed.), "A Journal of a voyage from Rocky Mountain Portage in Peace River to the sources of Finlays Branch and North West Ward, 1824," (Hudson's Bay Record Society, 1955), pp. 209-10.

5 H. D. Rogers, "On the falls of Niagara and the reasoning of some authors respecting them," *Amer. J. Sci.*, 27 (1835), pp. 329-30.

6 C. Lyell, "A memoir on the recession of the falls of Niagara," *Proc. Geol. Soc.*, 3 (1842), pp. 595-602.

7 C. Lyell, "On the ridges, elevated beaches, inland cliffs and boulder formations of the Canadian Lakes and the valley of St. Lawrence," *Proc. Geol. Soc.*, 4 (1842-3), pp. 19-22.

8 *ibid.*

9 C. Lyell, "Travels in North America," (London, 1845), Vol. II, pp. 82-3.

10 L. J. Agassiz, "Etudes sur les glaciers," (Neuchâtel, 1840).

One of Lyell's converts, William Logan, was appointed to the newly created post of provincial geologist in 1842.[11] This appointment marked the beginning of the work of the Geological Survey in Canada, an important event in the history of landform study because, although the immediate concern of Logan and his assistants was the examination of Canada's coal and mineral resources, their long-term project was nothing less than a field survey of all the provinces. Involved in this was not only geological mapping, but topographic survey as well, since virtually no hypsometric data were available at that time.

In the period 1850–1900, Canada, like the United States, had its scientific frontiersmen. It was the period of safari-like reconnaissances in which a few scientists, usually assisted by local guides, prepared field notes on what they could observe along the rivers and portages, and from the vantage points of high summits. The reports of Logan and his small group of surveyors provided the first reliable statements about the geology and physiography of the settled parts of Canada, while the reports and maps of the Palliser expedition, 1857–59, which was sponsored by the Royal Geographical Society, and the Hind-Dawson expeditions, 1857–58, supported by the Canadian government, made available the first accurate information about the western prairies and the adjacent cordilleras. A significant contribution to regional physiography was made by James Hector, chief scientist of the Palliser expedition in identifying the three prairie steps, and in distinguishing the major vegetation zones of the area,[12] while to H. Y. Hind must be given the credit of reviving the land-ice theory, and proposing that an ice sheet of continental proportions had covered the whole interior of Canada, prior to a marine incursion which supposedly spread the drift materials.[13]

In 1870, the responsibilities of the Geological Survey of Canada were increased to include the western territories. During the next thirty years members of the Survey carried out a series of magnificent exploratory traverses, and almost everything then known of the remote areas of the Rocky Mountains and the Shield was derived from their journals. The reports of A. R. C. Selwyn, G. M. Dawson,

---

[11] B. J. Harrington, "Sir W. E. Logan," *Geol. and Nat. Hist Survey of Canada, Rept. of Progress 1875-76* (1877), pp. 8-21.

[12] J. Hector, "Physical Features of the Central Part of British North America, with special reference to the Botanical Physiognomy," *Edinburgh New Philosophical Journal* (n.s.), Vol. XIV, no. 2 (1861), pp. 263-68.

[13] H. Y. Hind, "Observations on Supposed Glacial Drift in the Labrador Peninsula, Western Canada, and on the South Branch of the Saskatchewan," *Q. J. Geol. Soc. London,* Vol. XX (1864), pp. 122-30.

R. Bell, R. G. McConnell, J. Macoun, J. B. Tyrrell, J. Richardson,
R. W. Ells, J. W. Spencer, and A. P. Low[14] may be less well known
than those of J. W. Powell, G. K. Gilbert, C. E. Dutton, A. R. Mar-
vine, and F. V. Hayden, but their writing shows the same freshness
of approach and freedom of opinion, and the same appreciation of
the significance of landforms in interpreting the geological history
of an area.

The first major achievement of the Survey was the publication
of the *Geology of Canada* which appeared, appropriately enough, in
1863,[15] twenty-one years after the establishment of the Survey, and
provided the first comprehensive treatment of Canada's structure
and scenery. In later years, the explorations of members of the
Survey in different parts of Canada led to the investigation of several basic problems in geomorphology; for example, in the writings
of G. M. Dawson one finds a review of the relative merits of the
land-ice and iceberg hypotheses for the origin of glacial drift.[16]
R. G. McConnell provided the first pieces of evidence for estimating
the size and depth of the continental ice mass.[17] A. P. Low correctly
identified abandoned shorelines and high level deltas in Labrador
and around James Bay, and demonstrated their significance in the
study of postglacial isostatic response.[18] W. Upham outlined the
successive stages of Lake Agassiz, and provided evidence of its proglacial character.[19] J. B. Tyrrell was the first to comment on the
nature and significance of certain glacial landforms—"ispatinows"

14 Two of these scientists, G. M. Dawson and J. Macoun, carried out important work before joining the Geological Survey. Dawson was attached to
the North American Boundary Commission Survey of 1872-74, and mapped
the natural features along the 49th parallel for a distance of over 800 miles,
while Macoun was responsible for the assessment of settlement possibilities
on the prairies, first for the C.P.R. and then for the Canadian government.
A useful summary of the work of the Geological Survey is given in Alcock,
F. J., "A century in the History of the Geological Survey of Canada," *Nat.
Mus. Can., Special Contribution* no. 47-1 (1947).
15 "Geology of Canada, 1863," with a preface by W. E. Logan, Geol. Survey
of Canada (1863).
16 G. M. Dawson, "Report on the country in the vicinity of the Bow and
Belly Rivers, North West Territory," *Geol. and Nat. Hist. Survey of Canada.
Report of Progress, 1882-83-84* (1885), C. pp. 149-52.
17 R. G. McConnell, "Report on an Exploration in the Yukon and Mackenzie
Basins, N. W. T.," *Geol. Surv. of Canada Annual Report, 1888-89* (n.s.), Vol.
IV (1890), D. pp. 24-29.
18 A. P. Low, "On explorations in James Bay and the country East of Hudson Bay drained by the Big, Great Whale, and Clearwater Rivers," *Geol.
and Nat. Hist. Survey of Canada. Annual Rept. 1887-88* (n.s.), Vol IV (1889),
J. pp. 26-33, 61-62.
19 W. Upham, "Report of Exploration of the Glacial Lake Agassiz in Manitoba," *Geol. and Nat. Hist. Survey of Canada, Annual Report, 1888-89* (n.s.),
Vol. IV (1890), 1. pp. 1-110.

or drumlins—which, along with striae, provided the bases of his theories of the "Labradorean," "Keewatin," and "Patrician" centres of continental ice dispersion.

It must be remembered that treatment of landforms in the early geological survey reports was more or less incidental. At the turn of the century, Canadian geology was following the European lead, giving little encouragement to geomorphology, in spite of the successful debut of this discipline in the United States. The studies of regional physiography, inspired by W. M. Davis, which form such an impressive part of the American geological literature, are almost completely lacking from the Canadian; the only significant contribution of the period is the *Physiography of Nova Scotia* by J. W. Goldthwait,[20] which deserves recognition as the first Canadian monograph to provide a geomorphological approach to landscape study. But Goldthwait's interest in landforms was not shared by his contemporaries; at the universities, geologists concentrated on minerals and oil, while the members of the Survey proceeded with the immense task of mapping the Dominion. Geography was still not established as an academic discipline, and its recognition required strenuous endeavour over nearly half a century. Fortunately, Canadian geographers, unlike their American colleagues, maintained the holistic view of their chorological task, and geomorphology was accepted as part of the curriculum once geography had gained university status.

The Founding Fathers

The first university course in geography was given as long ago as 1906 at the University of Toronto,[21] to be followed by another at l'Ecole des hautes études, Montreal, in 1910.[22] Both of these, however, were courses in economic geography, and it was the newly established University of British Columbia which took the first significant step in the promotion of physical geography. J. S. Schofield gave a course on physical geography to students in the Department of Geology and Mineralogy at the first session of the univer-

[20] J. W. Goldthwait, "Physiography of Nova Scotia," *Canada, Dept. of Mines, Geol. Surv. Mem.* 140, 1924.
[21] The curriculum of the College of New Brunswick included geography as a separate study at the time of its foundation in 1800, but the course appears to have been given for only a few years, and so this early development can only be considered abortive. A brief treatment of this episode is provided by M. V. Williams, "Geology and Geography," *Trans. Roy Soc. Can.*, Vol. XL, sect. 4 (1946), pp. 125-39.
[22] M. R. Dobson, "Geography in Canadian Universities," *Misc. Paper No. 2, Geographical Branch, Dept. Mines and Tech. Surveys*, 1950, p. 1.

sity in 1915–16.[23] The course was given for many years, and in 1922 the Department changed its name to Geology and Geography, thus becoming the first department in a Canadian university to use "geography" in its title. In a sense, this was a false dawn, because geography at the University of British Columbia was subordinate to geology, and no immediate advances were made in geomorphology.

As in the nineteenth century, it was external stimuli which led to further progress in Canada. The new wave of European teaching after the First World War was introduced to Canada very largely through the endeavours of Raoul Blanchard, who made Canada his country of adoption. In 1927, Blanchard, who was then professor of geography at Grenoble, and already well known for his research work on many Alpine topics, was invited to the University of Chicago to give a course of lectures. This provided the occasion for his first visit to Canada, and so commenced a long and close association with French Canada, the true significance of which can be judged from the deep personal affection for Blanchard, and the high opinion of his work, which is expressed by so many of the contributors to the *Mélanges géographiques canadiens offerts à Raoul Blanchard.*[24] Blanchard visited Quebec regularly each year from 1929 until the outbreak of the Second World War.[25] From 1933 to 1938 he gave conferences at the Université de Montréal, and his field work covered all parts of the province except Ungava, from the Gaspé and the Côte nord in the early years, to the Laurentians and the Eastern Townships later on. After the war, he returned to become the first director of the newly established Institut de géographie at the Université de Montréal. In addition, he gave courses at Université Laval in 1952 and again in 1958, and so, in the words of Pierre Dagenais, "Blanchard fut le principal artisan de la création d'un enseignement de la géographie dans nos universités."[26]

The manifesto of geography which Blanchard presented in Canada was based on the teachings of Paul Vidal de la Blache, the founder of the French school, and so it is not surprising to find that regional geography is given great emphasis. However, Blanchard had a proper appreciation of the significance of the physical setting. The importance of geology and structure as a basis for regional

23 J. L. Robinson, "Geography at the University of British Columbia," *Can. Geogr.*, 13 (1959), pp. 46-47.
24 L. E. Hamelin (ed.), "Mélanges géographiques canadiens offerts à Raoul Blanchard" (Les Presses universitaires Laval, 1959).
25 L. E. Hamelin (ed.), "La géographie de Raoul Blanchard," *Can. Geogr.*, Vol. V, 1 (1961).
26 P. Dagenais, "Monsieur Raoul Blanchard," *Rev. Can. de Geog.*, Vol VIII, 1-2 (1959), pp. 80-83.

differentiation is frequently underlined, and his writings contain many penetrating observations on landforms and the sculpturing processes. In his study of the Laurentides, for example, Blanchard notes[27] the general accordance of height, and reviews the evidence for considering the summits as parts of a former erosion surface. In the same way, his treatment of the valleys leads to a consideration of valley form, and the efficacy of glacial erosion under a continental ice mass. Geomorphology was always subordinated to geography in Blanchard's writing, but its relevance in helping to explain the geographical personality of an area was never overlooked. Fernand Grenier, in his obituary of Blanchard, noted that his work provided the inspiration for the first generation of French Canadian geographers: "il a ouvert la voi a des nombreuses recherches qui se poursuivent toujours."[28] This is true for geomorphology, and many other branches of geography.

Blanchard's achievement in Quebec had its counterpart in Ontario, where Griffith Taylor assumed a pioneer role in accepting the invitation of President Cody to establish a department of geography at the University of Toronto in 1935. Taylor found in Canada the same sort of challenge he had met in Australia, and he devoted himself over a period of eighteen years to the building of the first independent geography department.[29]

Taylor's undergraduate training was in geology with a strong emphasis on mining geology, and his first university appointment was as a demonstrator in geology. However, he also developed a keen interest in physiography, and had the unique opportunity, while still a research student at Cambridge, of accompanying W. M. Davis on a field excursion through the Alps. In 1910, he returned to Australia to take up an appointment with the Weather Service, and this led to his being attached to Scott's expedition to the Antarctic as meteorological observer. During his year in the Antarctic, Taylor was able to collect climatic and glaciological information which provided the basis for his doctoral thesis on Antarctic glaciation.[30]

It is not surprising that Taylor placed great emphasis on the physical aspects of geography. He was more convinced than Blan-

---

[27] R. Blanchard, "Etudes canadiennes: Les Laurentides," *Rev. de Geog. Alpine*, Vol. XXVI, 1 (1938), pp. 6-18.

[28] F. Grenier, "Raoul Blanchard, 1877-1965," *Can. Geogr.* Vol. IX, 2 (1965), pp. 101-3.

[29] G. T. Taylor, "Geography at the University of Toronto," *Can. Geogr. J.*, Vol. XXIII (1941), pp. 152-4.

[30] Details of Taylor's life are given in his autobiography, *Journeyman Taylor* (Educational Books Ltd., 1958), and in the obituary article by D. F. Putnam "Griffith Taylor, 1880-1963," *Can. Geogr.*, Vol. VII, 4 (1963).

chard of the value of geomorphology for its own sake, and not simply as the preliminary to regional synthesis. In his second year at Toronto, Taylor instituted laboratory studies to demonstrate the physical basis of geography, and teach students cartographic techniques and the use of surveying and draughting instruments.[31] Much of the laboratory work was based on the local area so that students could benefit from field studies, and when the honours programme was established in 1940, provision was made for more intensive field work.

Blanchard and Taylor, the first two honorary presidents of the Canadian Association of Geographers, founded geography in Canadian universities, and by the same token, they established geomorphology as part of the geographical curriculum. Progress was slow but at the outbreak of the Second World War, geography, with geomorphology as an intrinsic part, was being taught at five universities. The number of professional geographers could be counted on one hand, but already some research in geomorphology was being done, for example, Taylor's first publications on Canadian topics dealt with topographic control in the Toronto area,[32] and the structural basis of Canadian geography,[33] while Blanchard's Etudes canadiennes, which appeared in eight parts between 1930 and 1939,[34] provided the first physiographic treatment of southern and eastern Quebec.

Even so, research in physical geography was very limited, as can be judged from a simple statistical analysis of the geographical periodical literature published between 1930 and 1940. The total numbers in the different columns are derived from counts made in bibliographies of periodical literature on Canadian geography published between 1930 and 1963,[35] and it can be seen that physical geography ranks fourth, with geomorphological studies contributing only 4.7 per cent of the total.

The geomorphological literature of the 1930s has been subdivided into ten major categories according to topic. Regional physiography

---

31 G. T. Taylor, "The Geographical Laboratory," (University of Toronto Press, 1938).
32 G. T. Taylor, "Topographic control in the Toronto region," *C.J.E.P.S.*, Vol. IV, 4 (1936), pp. 493-511.
33 G. T. Taylor, "The structurral basis of Canadian geography," *Can. Geog. J.*, Vol. XIX, 5 (1937), pp. 297-303.
34 A complete list of Blanchard's works is given in L. E. Hamelin (ed.), "Mélanges géographiques canadiens offerts à Raoul Blanchard," pp. 33-45.
35 *Bibliography of Periodical literature on Canadian Geography 1930 to 1958*, Bibliographic Series no. 22, parts 1-6 (Geog. Branch, Dept. of Mines and Tech. Surveys, 1959). Other issues in the series covering the period 1956-1963 are nos. 23, 26, 28, 29, 32, 33.

received most attention, followed by studies of glaciers and rivers. But, it must be remembered that only a very small proportion of the geographical literature of the period was actually written by geographers, while in geomorphology the proportion was even smaller. The writers about geomorphology included geologists, naturalists, mountain climbers, conservationists, and even engineers. Many of their contributions were valuable in focussing attention on particular problems or crucial areas; for example, H. C. Cooke's treatment of the physiography of the Shield,[36] E. Antev's analysis of the late glacial stages in eastern Canada and Manitoba,[37] G. M. Stanley's examination of the raised beaches around Lake Huron[38] and A. P. Coleman's numerous studies of different aspects of Pleistocene geology.[39]

The New Frontier

The decade 1940–50 was most significant, both for geography and geomorphology in Canada. A. H. Clark, in commenting on the period, noted that Canadian geographers had been engaged in a serious struggle for recognition, indeed actual survival.[40] However, by 1946 the battle was won, geography had emerged in six universities,[41] and the federal government, having been convinced of the value of geography in developing national resources, had established a Geographical Bureau (1947), which was elevated to the status of a Branch in 1950.[42] It is interesting to note that this same battle for recognition had been fought by European geographers more than half a century before. Once accepted, the results were the same; geography became respectable, and by the same token geomorphology was also accepted into the halls of learning as a suitable topic for teaching and research. Although it is important to recog-

---

[36] H. S. Cooke, "Studies of the physiography of the Canadian Shield," *Trans. Roy. Soc. Can.*, 23, ser. 3, sect. 4 (1929), pp. 91-120; 24, ser. 3, sect. 4 (1930), pp. 51-87; and 25, ser. 3, sect. 4 (1931), pp. 127-80.

[37] E. Antevs, "Retreat of the last ice sheet in eastern Canada," *Canada Geol. Surv.*, Mem. 146 (1925); "Late glacial correlations and ice recession in Manitoba," *Can. Geol. Surv.*, Mem. 168 (1931).

[38] G. M. Stanley, "Lower Algonquin beaches of Penetanguishene peninsula," *Bull. Geol. Soc. Amer.*, 47 (1936), pp. 1933-60; "Lower Algonquin beaches of Cape Rich, Georgian Bay," *Bull. Geol. Soc. Amer.*, 48 (1937), pp. 1665-86.

[39] A. P. Coleman, "The Last Million Years," (University of Toronto Press, 1941).

[40] A. H. Clark, "Contributions to geographical knowledge of Canada since 1945," *Geog. R.*, Vol. XL (1950), pp. 285-312.

[41] G. H. T. Kimble, "Geography in Canadian Universities," *Geog. J.* 108 (1946), pp. 114-15.

[42] T. Lloyd, "The Geographical Bureau," *Can. Geog. J.*, Vol. XXXVI (1948), pp. 39-41.

nize that geographers have not been the only contributors in the development of geomorphology, nevertheless, it is apparent that it was only with the establishment of geography that a way was opened for further progress.

Geomorphological research work in the 1940s was limited, and as Clark has indicated, professional geographers were primarily concerned with the study of economic activities. There is, in fact, an actual decrease in the geomorphological contribution of the 1940s, compared with the previous decade: whereas between 1930 and 1940 articles on geomorphology made up 4.7 per cent of the over-all total, and 50 per cent of the total in physical geography, they amounted to only 4.0 per cent of the total, and 32 per cent of the physical geography in the period 1941–50.

The analysis of the publications in geomorphology according to topic shows that in the 1940s studies of glaciers and regional physiography were most common. However, the most significant aspect of the geomorphological research work is not revealed, and this is the seminal nature of several of the studies undertaken at this time, studies which were continued in the next decade, and which provided the signposts for the future development of Canadian geomorphology. Among the published works of this type which could be cited are the reports on glaciology and glacial geomorphology resulting from P. D. Baird's expedition to Baffin Island,[43] R. F. Flint's analysis of the growth of the Labrador and Keewatin ice sheets,[44] J. L. Jeness's review of permafrost in Canada,[45] Chapman's and Putnam's series of articles which were expanded into a book on the physiography of southern Ontario,[46] the investigations of patterned ground in the St. Elias Range, Yukon Territory and in Victoria Island by R. P. Sharp[47] and A. L. Washburn,[48] and V. Tanner's study of Labrador.[49]

The great significance of the work of Tanner, and Chapman and Putnam can be judged from the fact that every subsequent publication dealing with either Labrador or southern Ontario draws upon

43 P. D. Baird et al., "The Glaciological Studies of the Baffin Island Expedition 1950," J. Glaciology, Vol. II, 2 (1952), pp. 2-9.
44 R. F. Flint, "Growth of the North American ice sheet during the Wisconsin age," Bull. Amer. Geol. Soc., 54 (1943), pp. 325-62.
45 J. L. Jeness, "Permafrost in Canada," Arctic, 2 (1949), pp. 13-27.
46 L. J. Chapman and D. F. Putnam, "The Physiography of Southern Ontario," (University of Toronto Press, 1951).
47 R. P. Sharp, "Soil structures in the St. Elias Range, Yukon Territory," J. Geomorph., Vol. V, 4 (1942), pp. 274-301.
48 A. L. Washburn, "Patterned ground," Rev. can. géog., 4 (1950), pp. 5-59.
49 V. Tanner, "Outline of Geography, Life and Customs of Newfoundland-Labrador," Acta Geografica, 8 (1944), pp. 1-906.

their material. The *Physiography of Southern Ontario*, now revised in its second edition, is the result of painstaking and detailed field work over an area of nearly 30,000 square miles, involving the mapping of landforms and surface materials at a scale of one inch to the mile. The traversing was supplemented by air-photo interpretation, and the printed maps are at a scale of 1:250,000. The text provides a comprehensive treatment of the geology and the main structural features, the glacial stages and the glacial landforms, and in addition there is a subdivision of the area into physiographic regions on the basis of topography and the predominant surface material. In the fifteen years since its first publication, the book has become a standard reference on the landscape and surface geology of southern Ontario.

In much the same way, Tanner's work has become a standard reference on Labrador. It was based on observations made during the Finland-Labrador expedition, 1937, and the Tanner-Labrador expedition, 1939, and so the treatment is less detailed than that of southern Ontario. However, a complete regional physiography is attempted, with the systematic examination of the geology and structure of Labrador, the denudation chronology in late Cenozoic time, and the effects of the Pleistocene glaciation with its attendant isostatic response.

One of the noteworthy features of the periodical literature of the 1940s is the increasing proportion of articles devoted to northern Canada—23.2 per cent compared with only 11.2 per cent in the 1930s. An important stimulus to northern research was provided in 1945 with the establishment of the Arctic Institute of North America, an international agency with representation from the United States, Canada, and Greenland on its board of governors. The Institute has promoted research in the arctic and subarctic through its library facilities, its publication programme, and its generous grants for field work. In its first ten years of operation the Institute helped to finance 177 projects, of which 21 were of a geomorphological nature.[50]

The interest in the arctic and subarctic reflected in the geomorphological literature of the 1940s is indicative of the very real concern of the governments of Canada and the United States in obtaining information about all aspects of the environment of the north. This sudden awareness of the northland is understandable in view of the strategic significance of the area in the days of the cold war,

---

[50] A full list of the field research projects sponsored by the Arctic Institute of North America between 1945 and 1955 is given in *Arctic*, Vol. VII, 3-4 (1955), pp. 354-66.

before the development of intercontinental missiles, when the Soviet Union and the western world faced each other over the pole, and it appeared that the most likely theatre of military operations would be the arctic. It was this situation which led to the construction of northern airfields, and the aerial photography of vast stretches of Labrador-Ungava and Keewatin. Using the trimetrogon method, the photo squadrons of the R.C.A.F. achieved some remarkable results, no less than 950,000 square miles were photographed during one summer season alone, and by the middle fifties the trimet coverage of northern areas was complete. Progress with vertical photography from 30,000 and 35,000 feet was somewhat slower, but the coverage was half completed by the late fifties. The research value of this essentially military operation was enhanced by the filing and index-ing of all these photographs in the National Air Photo Library, Ottawa, where they are available for study. The Library was estab-lished as early as 1925 as a centralized reference unit, and by 1954 it contained 2,500,000 prints.

The possibilities of using this unrivalled source of information in physiographic investigations were realized almost immediately; indeed some pioneer work using air photographs (mainly high- and low-level obliques) had already taken place in the 1930s,[51] for example, J. T. Wilson's examination of the drumlins of Nova Scotia[52] and the eskers northeast of Great Slave Lake,[53] and N. E. Odell's study of the regional geomorphology of northern Labrador.[54] The availability of an extensive air-photo coverage produced a revolution in geomorphological thinking in Canada, which marked the departure from the European tradition, with its emphasis on detailed studies of small areas and its overtones of nineteenth-century positivism; an outlook quite unsuited to the Canadian situa-tion, where the vast areas to be studied, the difficult field conditions, and the small numbers of researchers made a mockery of such a small-scale approach. Air-photo interpretation provided a new per-spective, ideally suited to regional physiography in general, and glacial geomorphology in particular. The spectacular results that could be achieved with "aerogeomorphology" were demonstrated in the late forties and early fifties.

[51] A. C. T. Sheppard, "What Canada is doing with the aid of camera and aeroplane," *Photogram, Eng.*, Vol. IV, 1 (1938), pp. 30-41.
[52] J. T. Wilson, "Drumlins of Southwest Nova Scotia," *Trans. Roy. Soc. Canada*, 32, ser. 3, sec. 4 (1938), pp. 41-47.
[53] J. T. Wilson, "Eskers north-east of Great Slave Lake," *Trans. Roy. Soc. Canada*, 33, ser. 3, sect. 4 (1939), pp. 119-29.
[54] N. E. Odell, in A. Forbes, *Northernmost Labrador mapped from the Air* (American Geographical Society, 1938).

The first projects of this type were jointly sponsored by the Arctic Institute and the Defence Research Board of Canada, which had been concerned with all aspects of northern research during the war, and was convinced of the value of air-photo interpretation as a means of acquiring information about surface conditions. This early work, undertaken by G. V. Douglas of Dalhousie University and his daughter, M. V. Douglas, was never published, but it deserves to be better known. The initial study covered the coasts of Labrador,[55] and it was followed by a reconnaissance treatment of the whole of Ungava and the interior of Labrador.[56] At the same time, a project was under way in the Geography Department at McGill University, involving the use of air photographs for the study of ice conditions in Hudson Bay.[57] These three investigations were very significant, because in a sense they were a test of the aerial photograph as a geomorphological tool and their success set the pattern for the development of Canadian geomorphology in the next decade.

Go North Young Man

There are several commentaries on the status of geography in Canada in the early fifties,[58] and it is apparent that there was an increasing interest in physical geography at both the university and the government level. M. R. Dobson's analysis of geography in Canadian universities indicates that in 1950, general courses in physical geography were offered at seven universities: British Columbia, Laval, McGill, McMaster, Montreal, Toronto, and Western

[55] G. V. Douglas, "The Coast-Line of Labrador," 5 vols. (unpublished report prepared for the Defence Research Board of Canada, 1949).
[56] G. V. and M. C. V. Douglas, "Ungava (New Quebec) and Interior of Labrador," 3 vols., (unpublished report prepared for the Defence Research Board of Canada and the Arctic Institute of North America, 1949).
[57] F. K. Hare, and M. R. Montgomery, "Ice, open water, and winter climate in the Eastern Arctic of N. America," Arctic, Vol. II, 2 (1949), pp. 79-89, and II, 3 (1949), pp. 149-64.
[58] These commentaries include the following: M. R. Dobson, Geography in Canadian Universities (Misc. Paper No. 2, Geog. Branch Dept. of Mines and Technical Surveys, (1950); L. D. Stamp, Geography in Canadian Universities: a Survey. Report of a survey under the auspices of the Canadian Social Science Research Council, 1951; J. L. Robinson, "Geography in Canada," in The status of geography in countries adhering to the Int. Geog. Union (Publication No. 7, Vol. XCII Int. Geog. Cong., Washington, 1952), pp. 10-3 and "The development and status of geography in universities and government in Canada," Year Book of the Association of Pacific Coast Geographers, 13 (1951), pp. 1-13; J. W. Watson, "A report on the status of geography in Canada, 1950-52," Third Pan-American Consultation on Geography (Washington, 1952), pp. 1-3; and "Geography in Canada," Scottish Geogr. Mag., 66, 3-4 (1950), pp. 170-72; P. Dagenais, "Status and tendencies of geography in Canada," Can. Geog., 3 (1953), pp. 1-15.

Ontario, as part of an honours or concentrated programme in geography. However, specialization in geomorphology at the graduate level was only possible at four universities: British Columbia, McGill, Montreal, and Toronto. It is interesting to note that four of these universities also offered specialized courses on northern Canada or the Arctic.

McGill University was able to take the lead in this development as the result of two important circumstances, firstly, the allocation of funds for research on the surface characteristics of northern Canada by the Defence Research Board which permitted the organization of the McGill University Research Group, and secondly, the establishment of the Subarctic Research Laboratory at Knob Lake.

The work of the McGill Research Group was the direct outcome of the new scale of operations permitted by air-photo coverage. Their task was the investigation of the surface characteristics of Ungava-Labrador; an area of 500,000 square miles. This involved field reconnaissance to establish appropriate classification systems for vegetation and landforms, and the elaboration of air-photo interpretation keys. Trimetrogon photography taken from 20,000 feet, and vertical photography at scales between 1:14,000 and 1:58,000 were used, and two series of maps were compiled at a scale of 1:500,000, one of vegetation-cover type, the other of surface type.[59] A parallel project, sponsored by the Arctic Institute of North America, involving some members of the McGill Group, was the preparation of reconnaissance maps of surficial deposits and structural features for the whole of Ungava-Labrador.[60] The initial mapping was at a scale of approximately 1:500,000, and from the 37 sheets covered at this scale, five separate maps were compiled showing drumlins, eskers, rock and drift distribution, faults, foliations, etc., and the physiographic regions, at a scale of approximately 1:5 million.[61]

The McGill Subarctic Research Laboratory was constructed in the summer of 1954, with the assistance of mining and construction companies involved in the development of the Knob Lake iron-ore deposits. The Laboratory has served as a centre for the field study

[59] F. K. Hare, *A photo-reconnaissance survey of Labrador-Ungava* (Geog. Branch Memoir 6, 1959).
[60] M. C. V. Douglas and R. N. Drummond, "Air Photograph Interpretation of Quebec-Labrador," 35 vols. (unpublished reports arranged by N. T. S. map areas prepared for the Defench Research Board of Canada, 1953).
[61] M. C. V. Douglas and R. N. Drummond, "Glacial features of Ungava from air photos," *Trans. Roy. Soc. Can.*, 47, ser. 3, sect. 4 (1953), pp. 11-16; "Map of the physiographic regions of Ungava-Labrador," *Can. Geog.* 5 (1955), pp. 9-16.

of a wide range of subarctic topics. Even in the early years, the research programme covered eight different fields ranging from limnology to ionospheric observations, but it was apparent from the first that the Station provided magnificent opportunities for geomorphological research, particularly glacial geomorphology.[62]

McGill University was not alone in its emphasis on northern research. Although none of the research projects at other universities approached the size of the McGill programme, there were several similar applications of aerogeomorphological techniques; for example, at the University of Toronto, W. G. Dean's investigations of the drumlinoid landforms of the Barren Grounds[63] and J. T. Wilson's study of the glacial features between the Mackenzie River and Hudson Bay.[64]

In addition, there were several individuals who took advantage of the opportunities afforded by the Arctic Institute, the Geographical Branch, the Geological Survey, and the Quebec Department of Mines, to undertake summer field work in the Canadian north. P. D. Baird, R. D. Goldthwait, and H. R. Thompson spent several field seasons examining the glaciers and glacial landforms of Baffin Island. J. B. Bird embarked on a field programme which was eventually to cover nearly the whole of the central arctic and subarctic. R. Mackay, after initial studies on Cornwallis Island, concentrated on the Mackenzie delta with its unique assemblage of permafrost and periglacial features. J. K. Stager began field work in the same area, and gave particular attention to pingoes. W. A. Wood and his colleagues established a research station on the Seaward Icefield lying along the Alaska-Yukon boundary, while P. Gadbois, J. K. Fraser, M. Marsden, R. Bergeron, W. F. Fahrig, and R. N. Drummond undertook regional surveys in different parts of the north.[65]

Besides encouraging independent research on the geomorphology of northern Canada of the type discussed above, several departments of the federal government were responsible for major programmes dealing with particular aspects of the arctic and subarctic environ-

[62] R. N. Drummond, "Research at the McGill Sub-arctic Research Laboratory," *Can. Geog.* 11 (1958), pp. 45-6; "The first three years," *McGill Subarctic Research Papers*, 22 (in press), pp. 3-8.

[63] W. G. Dean, "The drumlinoid landforms of the Barren Grounds, N. W. T.," *Can. Geogr.*, 3 (1953), pp. 19-30.

[64] J. T. Wilson *et al.*, "Glacial features between the Mackenzie River and Hudson Bay plotted from air photographs, *Bull. Am. Geol. Soc.*, 64 (1953), pp. 1413-14 (abstract only).

[65] Much of this research was published and a complete reference list is given in J. B. Bird "Recent conrtibutions to the physiography of northern Canada," *Zeit für Geomorph.*, 3 (1959), pp. 151-74.

ment. At the Geological Survey of Canada, which had increased the scale of its operations in the north after the Second World War, it was found that conventional methods were quite inadequate for even exploratory level mapping over the vast areas involved, and so a new approach was tried in 1952. Two helicopters were used by a party of geologists for aerial traverses and spot landings, with the result that 57,000 square miles of southern Keewatin were covered at the exploratory level (allowing mapping at the scale of 1:500,000) in a single summer.[66] Operation Keewatin was so successful that the same methods were adopted in Operation Franklin in 1955. Although not specifically geomorphological in intent, much valuable information on surficial deposits and landforms was obtained from these surveys.

The Division of Building Research, National Research Council, became involved in northern research because of the peculiar problems associated with construction in the arctic and subarctic. The wartime Canol project, and the activities of the U.S. Army Corps of Engineers in the late forties, provided the first information about permafrost, and after a general survey of permafrost conditions in northern Canada, the National Research Council decided to establish a northern research station at Norman Wells in 1952 to serve as a base for field and laboratory studies. Particular attention was given to the distribution of permafrost, and the location of sporadic and discontinuous permafrost bodies in the southern fringe area. In addition, the factors influencing thermal conditions in the ground were studied.[67]

The Defence Research Board maintained its interest in northern Canada, and became involved in sea-ice studies in the Arctic Ocean. In 1946, the discovery of large, stable ice masses, later known as ice islands, floating amid the pack ice, gave rise to speculation as to their origin, and their possible strategic value. Their tracks were plotted from observations made by the U.S.A.F. and the R.C.A.F., and in 1952 the first landing was made on one of the ice islands known as T3, where a semi-permanent base was established.[68] Aerial reconnaissance and air-photo interpretation indicated that the most likely source area for these ice masses was the ice shelf along the northern coast of Ellesmere Island, and so in 1953, the Defence

66 C. S. Lord, "Operation Keewatin, 1952: a geological reconnaissance by helicopter," *Precambrian*, 26, 4 (1953), pp. 26-30.
67 A general review of research prior to 1955 is provided in R. F. Legget "Permafrost Research," *Arctic*, 7, 3-4 (1955), pp. 153-58. The National Research Council maintains a complete bibliography: *List of Publications on Permafrost and Building in the North* (1966).
68 I. S. Koenig *et al.*, "Arctic Ice Islands," (Arctic, 5, 2 (1952), pp. 67-103.

Research Board and the Geological Survey of Canada sponsored a preliminary field survey in preparation for a joint U.S.-Canadian expedition in 1954.[69]

During the same period the Geographical Branch, which had become the chief employer of geographers in Canada, was also involved in the study of ice in northern Canadian waters.[70] The survey was concerned with all aspects of floating ice, the different types of ice, the distribution and movement, the factors governing formation and decay, and the associated navigational problems. All the existing documentary information on floating ice was collated, and the distribution of sea-ice in the Canadian arctic was examined on all available air photographs taken between 1947 and 1952.[71] Initially only Canadian arctic waters were included, but the survey was later extended to include the Gulf of St. Lawrence and eastern Canadian waters.

In addition, the Branch also became involved in terrain studies. Starting in 1952, a series of reports on surface conditions in northern Canada was prepared. Existing information was supplemented by air-photo interpretation, the aim being to describe all aspects of the surface: deposits and landforms, vegetation, and water bodies including their freeze and break-up characteristics.

The greater emphasis placed upon geomorphology in the early fifties is reflected to only a limited extent in the periodical literature; thus only 6.8 per cent of the total is devoted to geomorphology. However, the concentration on northern studies is very apparent. It can be seen that 47.7 per cent of the total number of geomorphological articles dealt with arctic or subarctic topics compared with only 23.2 per cent in the 1940s. The largest number of articles dealt with glacial geomorphology followed by sea-ice survey and related matters (appearing in the miscellaneous column), regional physiography, and periglacial phenomena.

An analysis of the papers presented at the annual meetings of the newly formed Canadian Association of Geographers (Table I) reveals the same emphasis. In 1955, for example, nearly half the programme was devoted to geomorphology, and all the papers in this section dealt with some aspect of northern research, with regional physiography and periglacial topics receiving the most attention.

---

[69] G. Hatterseby-Smith, "Northern Ellesmere Island, 1953 and 1954," *Arctic*, 8, 1 (1955), pp. 3-36.

[70] J. K. Fraser, "Canadian Ice Distribution Survey," *Arctic Circular*, 5, 5 (1952), p. 56.

[71] This part of the survey was undertaken by Miss M. C. V. Douglas and Miss J. McCarthy of the Defence Research Board.

TABLE I.

	'51	'52	'53	'54	'55	'56	'57	'58	'59	'60	'61	'62	'63	'64	'65	'66	Total
Coastal Geomorphology	1						1				3	1			1		7
Fluvial Geomorphology						3		2									5
Glacial Geomorphology		1	2	1		1		3	2	1	2	1	1	4		1	20
Karst Studies									1			2			1		4
Periglacial Studies			1		2	1	1	1	1		2		2	1		2	14
Non-glacial Pleistocene					1											1	2
Regional Physiography		2		2	3	1		1	3	1	2		2	2	1		20
Miscellaneous												2		1	1	1	5
Total	7	11	12	22	14	13	17	30	27	22	21	23	26	42	32	21	
Geomorphological Topics	1	1	1			2		3	4	1	1	2	2	3	3	5	
Arctic and Sub-arctic Papers		2	2	3	6	4	1	5	3	1	5	6	4	5		1	

Table I. Geomorphological presentations at the annual meetings of the Canadian Association of Geographers. C: coastal geomorphology; F: fluvial geomorphology; G: glacial geomorphology; K: karst studies; M: miscellaneous; P: periglacial studies; PL: non-glacial aspects of the Pleistocene; R: regional physiography.

The north became the new frontier in Canada, and the acceptance of this challenge by Canadian geomorphologists obviously provides one of the highlights in any review of the development of geomorphology in the twentieth century, in much the same way that the surveys of the western frontier created the most dramatic episodes in the nineteenth century. However, it must be remembered that equally valuable research was carried out in southern Canada. These studies are more difficult to analyse because there is a greater diversity of topic, and there are comparatively few large projects. However, mention should be made of R. J. Lougee's contribution to the study of isostatic response in eastern Canada,[72] the summary of the literature on erosion surfaces in eastern Canada compiled by J. B. Bird and F. K. Hare,[73] the Saskatchewan Glacier Project,[74] and C. P. Gravenor's work on glacial landforms.[75]

A Decade of Research at the Universities and the Institutes

Many new graduate schools of geography were established during the late fifties and early sixties. In 1955, there were only seven Canadian universities offering advanced degrees in geography, but this total increased to eleven in 1960, and twenty in 1966. Inevitably, this development allowed increasing scope for research in geomorphology, and prompted the elaboration of new methods, and the investigation of a wide range of topics.

McGill University confirmed its position as a centre for geomorphological work, with continued emphasis on its arctic programme. At the Subarctic Research Laboratory, studies of glacial geomorphology soon became predominant. Of the 32 theses which have been prepared by graduate students working at the Laboratory prior to 1966, exactly half deal with topics in glacial geomorphology, and of the ninety or so papers published in different journals as the result of work carried out at the Laboratory in the same period, nearly half deal with some aspect of glacial geomorphology.[76] The work of J. D. Ives, who was director from 1957 to 1960, had a con-

[72] R. J. Lougee, "The role of upwarping in the post-glacial history of Canada," *Rev. can. geog.*, 7, 1-2 (1953), pp. 3-14, and 8, 1-2 (1954), pp. 3-51.
[73] J. B. Bird and F. K. Hare, "Upland Surfaces in Eastern Canada," *Commission for the study and correlation of erosion surfaces around the Atlantic, Eighth Report*, Part 4 (Int. Geog. Union, 1956), pp. 54-64.
[74] M. F. Mehr *et al.*, "Preliminary data from Saskatchewan Glacier, Alberta, Canada," *Arctic*, 7, 1 (1954), pp. 3-26.
[75] C. P. Gravenor, "The origin of drumlins," *Am. J. Sci.*, 251, 9 (1953), pp. 674-81; "The origin and significance of prairie mounds," *Am. J. Sci.*, 253, 8 (1955), pp. 715-28.
[76] P. W. Adams, "The Laboratory in 1964," *McGill Sub-arctic Research Papers*, 22 (in press), pp. 9-13.

siderable influence on the pattern of research. The basic problems of
the deglaciation (including the location of the last ice centre),
which had been studied initially in the heart of the peninsula, were
investigated farther afield, and related problems, such as the late-
glacial marine transgression, the proglacial lakes, and the isostatic
response to deglaciation were also examined. In addition, there have
been studies of organic terrain, periglacial features, fluvioglacial
morphology and mass wasting, as well as snow and ice surveys in
the immediate vicinity of Schefferville.[77] At the present time, there
is increasing emphasis on more detailed investigations, and it is
very likely that the next twelve years in the history of the Laboratory may be even more productive than the last.

Continued interest in the Arctic is reflected in the two major
research programmes undertaken at the Department of Geography
of McGill University in the late fifties. The first of these, directed
by J. B. Bird, was supported by the R.A.N.D. Corporation, a private
American agency undertaking research for the United States Air
Force. A research group was formed in 1955, and their task was the
production of a series of reports and maps on the physiography of
the southern Canadian arctic between Baffin Island and Banks
Island. The work involved the compilation of information from
existing sources, and the interpretation of air photographs in order
to provide an accurate assessment of the terrain conditions. The
group was in existence for seven years, and during that time seven
reports were prepared, each one covering an area of approximately
500,000 square miles.[78] These reports are equally valuable as reference sources, and as monographs on regional physiography.

The second project, the Jacobsen-McGill expedition to Axel
Heiberg Island (1959–1962), led by F. Muller, received support from
the National Research Council and from private sources. All aspects
of the physical geography of the central part of Axel Heiberg Island
were investigated,[79] and special attention was given to glaciology,
and the techniques of glacier mapping. General maps at a scale of
1:50,000 were prepared to show the glaciers and landforms of the

[77] Appendices C. and D, *McGill Sub-arctic Research Papers*, 22 (in press),
pp. 272-82.
[78] R. A. N. D. Corp., *A report on the physical environment of the Great
Bear River area N. W. T.* (RM-2122-1-PR, 1963). The other reports have the
same general title and cover northern, central and southern Baffin Island,
the Quoich River area, the Thelon River area, and Victoria Island respectively.
[79] Reports on the activities over the four years were published in the *Axel
Heiberg Island Research Reports McGill University: Preliminary Report,
1959-1960* and *Preliminary Report 1961-1962*.

expedition area. In addition, special maps of the lower parts of the White and Thompson glaciers were produced at a scale of 1:5,000 with a 5-metre contour interval, to show small-scale features of the ice and the adjacent moraines, such as drainage ways, moulins, crevasses, ablation mounds, and patterned ground. The sheets, which are printed in six colours, represent a significant contribution in the field of morphological mapping.[80]

During the sixties, two other programmes at McGill have contributed to geomorphological research. The Planetary Surface Interpretation Project, directed by J. B. Bird and A. Morrison, has investigated the feasibility of distinguishing physiographic regions and major topographic features on photographs taken from space craft,[81] and the Terrain Evaluation Project, directed by J. T. Parry, has been concerned with the interpretation and classification of surface morphology, considered as one aspect of applied geomorphology in the study of land locomotion or terramechanics.

All the French-speaking universities in Canada now offer specialized courses in geomorphology, but only two of them have developed strong research programmes in this field. At l'Université de Montréal, the main emphasis has been on regional physiography and morphological mapping. In the latter field, G. Ritchot has been one of the first to apply the techniques developed in France to sample areas in southern Quebec. At l'Université Laval, where the Institut de géographie became independent in 1958, considerable attention has been given to regional physiography, but far more significant has been the contribution to la géomorphologie froide (glaciare et periglaciare) inspired by L. E. Hamelin.[82] One of the first pieces of research was the collection of information for the *Carte préliminaire de phénomènes periglaciaires au Canada*, first published in 1960.[83] A comprehensive study of periglacial processes and features in Canada was published in the following year,[84] and it will be supplemented in the near future by an illustrated glossary of periglacial phenomena prepared by L. E. Hamelin and the late

[80] F. Muller, "Large scale maps for glaciological research in the Canadian High Arctic," *Revista Cartografica*, 12, 12 (1963), pp. 315-24.
[81] J. B. Bird and A. Morrison, "World Atlas of Photography from Tiros Satellites I to IV," *N.A.S.A. Contract Report*, CR 98 (McGill University).
[82] L. E. Hamelin, "Bilan vicennal de géomorphologie a l'Institut de Geographie de Quebec," *Bulletin de l'Association des géographes de l'Amérique francaise*, 10 (1966).
[83] J. C. Dube and L. E. Hamelin, "Carte préliminaire de phénomènes périglaciarires du Canada," (1960).
[84] L. E. Hamelin, "Périglaciaire du Canada: idées nouvelles et perspectives globales," *Cahiers de géog. Québec*, 10 (1961), pp. 141-203.

F. A. Cook.[85] In addition, the problems involved in the preparation of large-scale maps of periglacial features have been investigated, and a set of symbols has been devised allowing the differentiation of periglacial features on the basis of genesis, form, and size.[86]

The establishment of the Centre d'études nordiques in 1961 gave an additional stimulus to the geomorphology of cold regions, which was felt both at Laval and farther afield. Preliminary discussion of such a centre took place as early as 1954, but it was another seven years before the necessary funds were available for its foundation, and the inauguration of a multi-disciplinary research programme devoted to northern Quebec.[87] Twenty-five per cent of the 70 or so projects supported by the C.E.N. during the past five years have been in geomorphology, and of these, 15 have been presented as theses for higher degrees at Laval and other universities. Nearly a quarter of the 23 publications of the C.E.N. deal with geomorphology, particular attention being given to regional physiography. In 1962, facilities at Fort Chimo were made available by the Ministère des richesses naturelles, and this station has undoubtedly stimulated interest in the Ungava Bay area.

At the Ontario universities during the last ten years, research in geomorphology has been comparatively limited; however, mention must be made of some recent developments which augur well for the future. The University of Western Ontario has recently established a field station on the shores of Lake Erie, near Port Bruce, which will serve as a centre for the study of the shoreline and mass-wasting processes which affect the high bluffs along the lake margin. At McMaster University a programme of detailed work on the glacial drifts and periglacial phenomena of southeastern Ontario has been commenced by A. Straw, while D. C. Ford has formed a research group for the study of karst landforms and solution processes in Ontario and British Columbia, with particular emphasis on cavern genesis and sedimentology. An inventory of Ontario caverns has been prepared, and research is continuing on karst developments along the Niagara escarpment, doline formation in Ontario, cavern systems in the Selkirk Range, B.C., and the isotope dating of drip-stone features. In the Department of Geography of the University of Toronto A. V. Jopling is largely responsible for the

85 L. E. Hamelin, and F. A. Cook, "Illustrated Glossary of periglacial Phenomena," (Centre des études nordiques, in press).
86 L. E. Hamelin, "Cartographie géomorphologique appliquée au périglaciaire," Cahiers de géog. Québec, Vol. VII, 14 (1964), pp. 193-209.
87 L. E. Hamelin, "Le Centre d'Etudes nordiques de l'Université Laval, (C.E.N. Publication 475, 1966).

development of substantial hydrological laboratory facilities and the design of a 5-acre outdoor geomorphological laboratory.

In the west of Canada, the University of British Columbia has offered graduate work in geomorphology since 1948, and although there have been few theses presented in this field, there has been no lack of research. The studies of the Mackenzie delta initiated by J. R. Mackay and J. K. Stager in the fifties, and continued through to the present have been particularly outstanding because of their comprehensive nature. Over the years almost every aspect of the geomorphology and hydrology of the delta has been investigated: the landforms as such—oriented lakes, river channels, and pingos—the geomorphic history, with particular emphasis on the late-glacial episode, and finally the contemporary processes—solifluction, shore-line recession, the break-up pattern of river ice, and freezing and thawing in the active layer.[88]

At other universities in the west, geomorphology has only been introduced very recently, and so it is not surprising that research has been somewhat limited. At the University of Calgary, interest has centred on what might be termed alpine and foothills geomorphology, with the main emphasis on glacial and periglacial studies in a mountain environment. At the University of Alberta, little attention has been given to geomorphology as such, but mention may be made of the programme of research conducted by A. H. Laycock on the patterns of water deficiency and supply in the prairie provinces: a study in applied geography, involving several aspects of geomorphology. At the University of Saskatchewan the Institute of Northern Studies, which was founded in 1960, has supported several projects in geomorphology.[89]

The expansion in geomorphological research at the universities during the last decade has been paralleled and indeed, in several instances, complemented by research supported by private institutions. The Arctic Institute of North America has continued its support of arctic and subarctic projects and between 1956 and 1966 31 awards, representing 10 per cent of the total, were made for geomorphological research. Another organization, the Geological Association of Canada, made an important contribution to geomorphology with the publication of the Glacial Map of Canada in 1958.[90]

[88] J. R. Mackay has published many articles on the Mackenzie delta, but his most recent comprehensive treatment of the area is "The Mackenzie Delta Area," *Geogr. Branch Canada*, Mem. 8 (1963).

[89] *Institute for Northern Studies, University of Saskatchewan, Sixth Annual Report* (1965-66).

[90] "Glacial Map of Canada," (Geological Association of Canada, 1958).

This map is, in a sense, the outcome of all previous research on the glacial geomorphology of Canada, but it owes most to the studies in aerogeomorphology undertaken in the late forties and fifties, which have been mentioned above. The map scale, 1:3,801,600, obviously restricts the amount of information that can be included; however, by skillful use of colour and a variety of symbols, it has been possible to show glacial landforms such as moraines, eskers, and drumlins, the directions of ice movement, the extent of the major proglacial lakes, the limits of the post-glacial marine transgression, and the distribution of present-day glaciers. This last topic was treated in another important publication, which appeared in 1959, the *Geographic Study of Mountain Glaciation in the Northern Hemisphere* prepared by the American Geographical Society.[91]

A Decade of Research in the Federal and Provincial Governments

In several government departments there has been an increasing interest in both pure and applied geomorphology in recent years. The National Research Council has supported several pioneer programmes of a multi-disciplinary type including various aspects of geomorphology. For example, the Photogrammetric Research Section provided the ground control and produced the maps of Salmon Glacier for the University of Toronto Expedition (1956–57),[92] and was responsible for the glacier maps of the Jacobsen-McGill Axel Heiberg expedition, which was discussed above. In the Division of Building Research, investigations into the nature and distribution of permafrost have continued. Field studies of permafrost and the site indicators, such as vegetation, drainage, and soil type, have been made in several parts of northern Canada, notably by J. A. Pihlainen and R. J. E. Brown. In 1960, Brown prepared a provisional map showing the southern limits of the different types of permafrost,[93] and a more detailed map was published in 1967 as the result of field work in the Mackenzie District and the prairie provinces.[94] The National Research Council has also supported research on organic terrain and mass wasting through its Committee on Soil

---

[91] "Geographic study of mountain glaciation in the Northern Hemisphere," (American Geographical Society, Department of Exploration and Field Research, 1959).

[92] D. Haumann, "Photogrammetric and Glaciological Studies of Salmon Glacier," *Arctic*, 13 2 (1960), pp. 75-110.

[93] R. J. E. Brown, "The distribution of permafrost and its relation to air temperature in Canada and the U. S. S. R.," *Arctic*, 13, 3 (1960), pp. 163-77.

[94] G. Robinson, "Permafrost investigations on the Mackenzie Highway in Alberta and Mackenzie District," *National Research Council, Division of Bldg. Research Tech.*, Paper 175 (1964), and Map 1246A, *Geol. Surv. Can.* (1967).

and Snow Mechanics, and its sponsorship of the Muskeg Research conferences and the Canadian Soil Mechanics conferences. N. W. Radforth's investigation of organic terrain[95] and the studies of eastern Canadian flow slides by N. R. Gadd and others[96] provide instances of the value of such interdisciplinary research.

The Defense Research Board continued to support research on northern Canada during the late fifties and early sixties, with particular emphasis on air-photo interpretation techniques and ground studies of the polar continental ice shelves. In the former field, the main contributions to geomorphology were the collection and interpretation of a representative sample of air photographs of arctic Canada by M. Dunbar and K. R. Greenaway,[97] and the preparation of a series of air-photo interpretation keys for organic terrain by N. W. Radforth.[98] The ice-shelf studies made no direct contribution to geomorphology except for Operation Hazen, which included the examination of the physiography of the Lake Hazen area, Ellesmere Island.[99]

The Geological Survey of Canada has paid considerably more attention to the study of Pleistocene stratigraphy and landforms during the last decade. Facilities for radiocarbon dating and palynological work have been established, and investigations of unusual structures such as cryoturbation features and glacio-tectonic forms have been included in an expanded programme for the mapping of surficial deposits in the settled parts of Canada. Unfortunately these studies are too numerous to discuss separately, but mention may be made of the work of N. R. Gadd in the St. Lawrence and Ottawa valleys, H. A. Lee, E. P. Henderson, and O. L. Hughes in the Shield, A. M. Stalker in Alberta, V. K. Prest in the St. Lawrence Valley and Ontario, and D. M. Baird in the national parks. In addition, it should be noted that several provincial departments of mines and research councils have supported work in surficial geology, for example, that of C. P. Gravenor and L. A. Bayrock in Alberta, E. A. Christiansen

---

95 N. W. Radforth, "Organic terrain," in "Soils in Canada," (*Roy. Soc. Can.*, Special Publication, no. 3, 1961).
96 Papers on flow slides by P.M. Bilodeau, W. J. Eden, and N. R. Gadd, were presented at the Tenth Canadian Soil Mechanics Conference, 1956, and published by the Associate Committee on Soil and Snow Mechanics: *Technical Memorandum*, No. 46 (1957).
97 M. Dunbar and K. R. Greenaway, "Arctic Canada from the Air," (Ottawa, 1956).
98 N. W. Radforth, "Organic Terrain Organization from the Air (altitudes 1,000-5,000 feet), (Handbook No. 2, D. R. no. 124, D.R.B. Ottawa, 1958).
99 D. I. Smith, "Operation Hazen: the geomorphology of the Lake Hazen region, N. W. T.," *McGill Univ. Geog. Dept.* (Miscellaneous Papers, no. 2, 1961).

in Saskatchewan, W. H. Matthews and J. E. Armstrong in British Columbia, A. Dreimanis and P. F. Karrow in Ontario.[100]

The most substantial contribution to geomorphological research in the last decade has been made by the Geographical Branch until its dissolution in 1967; the Branch employed the largest group of geomorphologists in Canada, and the programme of research centred around the interests of the director, J. D. Ives who, after specializing in geomorphology, became director of the McGill Subarctic Research Laboratory, and devoted his attention to the study of the glacial and post-glacial history of Labrador-Ungava. The Branch was able to develop several major research programmes, which were beyond the scope of universities or private organizations; some of these were continued from the early fifties,[101] others were begun more recently.[102]

The programme of arctic and subarctic terrain studies was continued until 1959, by which time air-photo interpretation keys had been prepared for fourteen areas, ranging from the Albany River to the Boothia Isthmus and Cornwallis Island.[103] Another programme, the ice distribution survey, is one which continued through the fifties and is still in progress. Besides the collection of data from bibliographic sources the Branch also arranged for actual ice observations, from United States and Canadian icebreakers in the Arctic, and from R.C.A.F. aircraft, R.C.N. vessels and shore stations in the Gulf of St. Lawrence. Reports on ice conditions in Ungava Bay, Hudson Bay, the Gulf of St. Lawrence, and eastern Arctic waters were prepared by W. A. Black and C. N. Forward,[104] and published by the Branch in its series of Geographical Papers;[105] in addition, several special reports were produced such as those by M. Brochu on drift ice in the Gulf of St. Lawrence.[106]

[100] A comprehensive bibliography covering surficial geology is given in R. F. Ligget (ed.), "Soils in Canada," (*Roy. Soc. Can.*, Special Publication no. 3, 1961).
[101] N. L. Nicholson, "The Geographical Branch, 1947-57," *Can. Geogr.*, 10 (1957), pp. 61-68.
[102] A useful summary of the activities of the Geographical Branch has appeared in the *Newsletter* of the Canadian Association of Geographers from 1960 onwards.
[103] J. K. Fraser, "Activities of the Geographical Branch in Northern Canada, 1947-57," *Arctic*, 10, 4 (1957), pp. 246-50.
[104] W. A. Black, "Geographical Branch Program of Ice Surveys of the Gulf of St. Lawrence 1956 to 1962," *Cahiers de géog.*, 11 (1962), pp. 65-74.
[105] Numerous reports on ice conditions were prepared and the reader is referred to the *Bibliographic Series* no. 5: 18, 19, 23 *et seq.* (Geog. Branch).
[106] M. Brochu, "Dynamiques et caractéristiques de dérive de l'estuaire et de la partie nord-est du golfe St. Laurent," *Direction de la géographie, Ministère des mines et des relevés techniques, Etude géog.*, 24 (1960).

Several new research programmes, such as the study of periglacial phenomena were begun in the late fifties. The most important contribution to this programme was made by F. A. Cook. His early work included studies of the types of patterned ground on Cornwallis Island, and an investigation of the thermal regime of the active layer in the same area.[107] A three-stage programme was started in 1958, involving the compilation of information on periglacial phenomena from printed sources,[108] the preparation of a glossary of periglacial terms,[109] and finally the elaboration of techniques for the continuing study of periglacial processes. In addition, several experimental maps of periglacial phenomena were prepared, for example, B. Robitaille's map of the Mould Bay area, Prince Patrick Island (1:72,000),[110] D. A. St. Onge's maps of Ellef Ringnes Island (1:250,000, seven colours), and Isachsen (1:30,000, also in seven colours) which provides sets of symbols and shading for structural, fluvial, marine, nivial, glacial, solifluction, and cryoturbation features.[111]

The Geographical Branch had also been associated with the recently completed Polar Continental Shelf Project which was begun in 1958. Besides ice distribution surveys and periglacial studies of the type discussed above, the terrain conditions, particularly the glaciers and ice caps of the islands bordering the shelf, have been examined. Meighen, Ellef Ringnes, and Borden islands were studied in the first few years and they were followed by Prince Patrick and the Queen Elizabeth Islands.

In 1961, the Branch embarked upon an even more specialized programme in the eastern Arctic. After a preliminary reconnaissance, the Penny and Barnes icecaps in central Baffin Island were selected for long-term studies of glaciology, glaciometeorology and glacio-fluvial processes. Pit studies, gravity and seismic traverses, and mass balance studies have been carried out on the ice caps, and test studies of some ice-cored moraines have been completed. The glacial landforms adjacent to the Barnes Icecap have been examined in detail, and till fabric and lichenometric techniques

---

[107] F. A. Cook, "Geographical Branch Studies in Periglacial Geomorphology," *Cahiers de géog.*, 7 (1959), 209-10.

[108] F. A. Cook "A selected bibliography on periglacial phenomena in Canada," *Geog. Branch, Bibliographical Series*, no. 24 (1959).

[109] With the death of F. A. Cook, the responsibility for completing the glossary has been taken over by the co-author L. E. Hamelin.

[110] B. Robitaille, "Présentation d'une carte géomorphologique de la région de Mould Bay, Ile-du-Prince-Patrick, Territoires du Nord-ouest," *Can. Geog.*, 15 (1960), pp. 39-43.

[111] D. A. Onge, "Géomorphologie de l'Ile Ellef Ringness, T. du Nord-ouest," *Geog. Branch*, Memoir II (1964).

have been used to unravel the recent geomorphological history. In addition, sections of both the east and west coasts have been studied, partly for the evidence they provide on the post-glacial isostatic readjustment, and partly in search of information on present day coastal processes. The Baffin Island programme was conceived as part of a long-range scheme for the detailed study of a representative cross section of the eastern arctic.[112]

Finally, mention must be made of the inventory of Canadian glaciers, begun in 1960, and the Prairie Physiography Programme, begun in 1961. The latter has included the preparation of experimental morphological maps for several areas in southern Saskatchewan at scales of 1:50,000 and 1:250,000, and the investigation of minor landforms such as prairie mounds. Studies of structural geomorphology, river terraces, and soil-landform relationships have also been carried out in the Cypress Hills.

It is apparent from this review that at both government and university levels the amount of geomorphological research has increased rapidly in the last decade. This is reflected in the proportion of papers on geomorphological topics presented at meetings of the Canadian Association of Geographers (Table I), which has increased from a quarter of the total in the first seven years of the Association's history to a third in the last seven years. It is also reflected in the steady increase in the amount of geomorphological literature appearing in the last decade, accounting for 8.4 per cent and 9.1 per cent of the total periodical literature in the periods 1956–60, and 1961–63 respectively. The interest in northern Canada has intensified to such an extent that it now dominates geomorphological research in this country, accounting for 53.3 per cent of the published geomorphological literature. This emphasis is in part the result of concentration on certain topics, such as glacial geomorphology and studies of existing glaciers, surveys of sea and river ice (which account for the greater part of the miscellaneous column), and periglacial studies, all of which, by their very nature, can only be carried on in an arctic or subarctic environment. However, the concentration on northern Canada is also apparent in more general fields such as regional physiography, and the study of non-glacial Pleistocene phenomena.

Two significant aspects of the geomorphological research undertaken in the last few years are not revealed by the diagrams. There has been a relative decline in the numbers of reconnaissance type

---

[112] O. Løken, "Science on Baffin Island," *Can. Geog. J.*, Vol. LXXII, 2 (1966), pp. 38-47.

studies, and an increase in the numbers of detailed field investigations. In addition, there are indications of a move toward the inclusion of ancillary techniques such as lichenometry, palynology, fabric analysis, and geochronology (particularly radiocarbon dating methods) with standard geomorphological procedures such as the study of morphology and stratigraphy. In a few instances, the projects have been truly interdisciplinary, involving scientists in cognate fields.

Canadian Geomorphology: The Prospect

From this inventory of studies in Canadian geomorphology several interesting features have emerged, and it is valuable to have seen these in their historical context, because it provides the proper perspective for the future. If progress in geomorphology is to continue at a steady pace, it is important that some attention be given at the present time to the direction of future development. This is obviously a complex question, but an approach can be made by considering some of the component parts, such as the choice of topics for investigation, the methods of data collection and storage, the procedures of interpretation and analysis, and the relations of geomorphology with its nearest neighbours, geography and geology.

The first of these factors, the choice of topic, in fact involves several choices, including the field of study and the study area. It is apparent that Canada with its range of conditions from the muskeg lowlands of the Shield to the high mountain chains of the Rockies, and from the ice-clad islands of the Arctic archipelago to the sagebrush flats of the dry prairie offers a tremendous variety of terrain types and geomorphic processes. Canada's natural laboratory has been richly furnished. The review of investigations that has been presented in the last few pages indicates that some parts of the laboratory have been most effectively used, resulting in important contributions to glacial and periglacial geomorphology; however, into other parts of the laboratory we have scarcely ventured. Many branches of geomorphology are practically untouched in Canada, and remarkable opportunities exist for both landform and process studies in weathering and mass wasting, coastal, structural, and karst geomorphology.

The field of study obviously determines the study area to a considerable extent; however, because of their vast territory, Canadian geomorphologists suffer less in this regard than most other national groups. Equally valuable results could be obtained in many branches of geomorphology from research conducted in either northern or southern Canada, and yet the main contribution, so far, has not been in the south but in the north. The publication of J. B. Bird's book on the Physiography of Arctic Canada[113] produces the extraordinary situation of our having a comprehensive and substantial treatment of northern Canada, while its counterpart on southern Canada is little more than a gleam in some geomorphologist's eye. There is a critical need for sound geomorphological work in the settled parts of Canada as a basis for economic and regional studies, and if one accepts the view that the researcher should be concerned to some extent with aiding national development, there is good reason to advocate the rerouting of some of our research effort from northern to southern Canada.

The methods of data collection and the procedures of interpretation and analysis hinge directly one upon the other, since the type of programme determines the form in which field evidence has to be presented. During the last decade, increased emphasis has been placed on experimental and analytic techniques demanding precise information about the amounts of material and energy involved in geomorphic processes, and the expression of surface form in suitably stratified arrays of morphometric data. Developments in other fields such as engineering and physics are directly responsible for this trend in geomorphology, and it is now apparent that it is more than a passing fancy. Canadian geomorphologists have been remarkably reticent about methodological issues, and it seems that the quantitative revolution has been largely ignored. This is unfortunate, because "there is reason to believe that a new era of discovery is underway."[114] However, it is important to see the new regime in proper perspective, lest its apparent benefits prove to be of no more lasting value than the gifts of Pandora's box. Quantification in geomorphology has two aspects which it is well to keep distinct. Firstly, there is precise measurement: the monitoring of geomorphic processes and the morphometry of landforms; secondly, model building: the establishment of geomorphological relationships by statistical and mathematical methods.

Careful observation and attempts at precise measurement have been part of geomorphology since its beginning in the seventeenth century, and additional efforts in this direction are to be welcomed. Until quite recent times the problem of landform quantification presented a serious obstacle to further progress. Accurate contoured maps provided the first comprehensive quantative statements about

---

[113] J. B. Bird, "The Physiography of Arctic Canada," (Johns Hopkins University Press, 1967).

[114] J. P. Miller, "Geomorphology in North America," *Przeglad Geograficzny*, Vol. XXXI, 3-4, (1959), p. 577.

landforms, and geomorphologists were quick to use them for simpler procedures such as profiling and altimetric frequency analysis, but it is only in the last few years that the possibility of translating cartographic data into morphometric data has been demonstrated. For example, the work of R. E. Horton[115] and A. N. Strahler[116] has rationalized the geometry of drainage basins, thus allowing the direct comparison of one basin with another, and providing a means of relating morphometric, kinematic, and dynamic properties. The full potential of this approach has not yet been realized, perhaps because of a distrust for data derived from topographic maps. However, there is no reason why morphometric data cannot be obtained in other ways, for example, from multiple profiles provided by airborne radar altimeters or gas lasers, or from measurements made on a stereomodel with a sophisticated plotting instrument, such as the Wild orthograph linked to a digital read-out. All discrete landforms, whether positive or negative, can be expressed in morphometric terms such as length, breadth, elongation, peakedness, profile area, volume, while assemblages can be expressed in terms of spacing, density, parallelism, local relief, etc., thus allowing direct comparison of one individual in a group with another, or the group as a whole with another group. Once a sufficient population has been sampled, it should prove possible to devise suitable class-intervals on a naturalistic basis, so that the geomorphologist can talk about a second-order drumlin with third-class peakedness and fourth-class elongation, in the same manner in which a meteorologist speaks of a Force 2 wind gusting to Force 3, or a seismologist speaks of an earthquake of intensity 4 at normal focus. There is no reason why Canadian geomorphology with its tradition of air-photo interpretation should not be able to make a substantial contribution along these lines by adapting photogrammetric methods to the particular problems of landform quantification.

By providing precise data about landforms, morphometric procedures can go a long way toward remedying a basic weakness in geomorphology—the prolixity and lack of precision in the nomenclature. There has never been any real systematization in geomorphic terminology as a whole, and this is particularly obvious in landform nomenclature, which is duplicative, multilingual, indefinite, and lacking any precise rules of application. There is a serious

---

[115] R. E. Horton, "Erosional development of streams and their drainage basins," *Geol. Soc. Am. Bull.*, 56 (1945), pp. 275-370.
[116] A. N. Strahler, "Hypsometric analysis of erosional topography." *Geol. Soc. Am. Bull.*, 63 (1952), pp. 1117-42.

need for the construction of a fabric or matrix in which known landforms can be placed and which will accommodate new forms as they are discovered.

In the same manner, possibilities exist for improving our knowledge of geomorphic processes by the collection and analysis of numerical data. The current trend in geomorphology involves increasing use of basic physical and chemical approaches to investigate the inner workings of particular processes and it is apparent that such problems demand a much greater use of mathematics, statistics, and computer science, than has been customary in the past. Some geomorphological process are more amenable to careful monitoring than others, and Canadian geomorphologists are fortunate in having within their territory a variety of subaerial processes, many of which are sufficiently rapid to yield significant results after a few years of study. The prime opportunities lie in the study of fluvial, glacial, coastal, and certain of the more rapid mass-wasting processes. In many cases, study sites lie close to universities, or within reach of moderately financed expeditions.

The final aspect of field data collection is that of storage. So far, the only way in which geomorphic data has been stored is in publication, manuscript, and map form, which has meant that retrieval has been more or less determined by the thoroughness of the bibliographic search. The literature becomes more voluminous every year, and it is apparent that Canadian geomorphology would benefit from the establishment of a national data-storage system. The advantages of such a system are obvious enough.[117] It would allow for the automatic storage of current data, thus providing a rapid retrieval system. It would permit the immediate comparison of different sets of data, and by accumulating information it should lead to the discovery of the actual frequency distributions of particular geomorphic attributes, and allow the recognition of trends, associations, and groupings. In the same way, hypotheses which appear to be true in a particular locality could be tested with a larger population, and conceptual or theoretical frameworks could be checked against actual data. Geomorphology is unlike the more exact sciences such as physics or chemistry, where careful observations have been accumulated over a long period of time, and the hypotheses and laws are of universal application. It has only reached a rudimentary stage, real knowledge is very limited, and

---

117 A convincing case for a national storage system for geological data has been produced recently. "A National System for Storage and Retrieval of Geological Data in Canada," (National Advisory Committee on Research in the Geological Sciences, 1967).

by virtue of its subject matter it is a detective science—the evidence is incomplete, isolated in space and time, and often inconclusive. Modern data-storage systems and computer techniques offer one of the most promising lines of advance for a science of this type, by providing for the retention of each piece of information and its statistical testing.

The possibilities for the development and testing of theoretical models are very considerable; however, it must be remembered that we are dealing with a variable mixture of solids, liquids, and gases, and the parameters in the equations will be extremely difficult to establish. The discovery of cause-effect relationship between process and landform is complicated by interactions between processes and by human activity. Once the time factor is added, it becomes even more complex, because of the difficulty of determining how much each landform owes to present processes, and how much to processes operating in the past, when climatic conditions may have been very different. Another complicating factor is introduced by the diastrophic processes.

Perhaps the best prospects at the present time are for studies of scale models, restricted to a particular situation where the conditions can be set up as required, thus eliminating the complex variables such as diastrophic and climatic history, and allowing the systematic investigation of the process and response factors. In many situations the sealing laws still have to be worked out, and it is difficult to maintain similarity between the model and the actual situation, particularly with regard to the dynamic dimensions. However, sufficiently impressive results have been obtained from model studies of wave action, tidal action, glacier flow, turbidity currents, wind action, and fluvial action to demonstrate the tremendous possibilities for scale models in geomorphological research.[118] The establishment of a geomorphological laboratory for research on scale models is well within the capabilities of many universities in Canada; however, for long-term projects it might be more effective for such a facility to be sponsored by a government organization such as the National Research Council.

The more complex relationships are outside the scope of scale models, because there is no possibility of testing the model performance against the actual conditions. For such situations, conceptual or theoretical mathematical models are available,[119] but

[118] A good selection of model studies in different fields is given by C. A. M. King, "Techniques in Geomorphology," (Edward Arnold, 1966).
[119] A comprehensive treatment of theoretical models has been provided by A. E. Schneidegger, "Theoretical Geomorphology," (Prentice-Hall, 1961).

it is apparent that in many cases they cannot be validated. Thus, if one were to describe the relationship between isostatic uplift and marine regression or valley deepening it would be only an imagined relationship, not only unproved, but incapable of proof at the present time. Models of this type, such as Davis' cycle of erosion, or Penck's theory of piedmont-treppen have most frequently been stated in a non-mathematical form following the rules of deductive logic. However, mathematics can serve as a language into which verbal reasoning can be effectively translated, and conceptual models should benefit from formulation in both the logical and the physico-mathematical modes. It should be remembered that the mathematical formulation is no more precise than the deductive, because there is no way of verifying the initial premises on which it is based.

A final factor, which is of some importance for Canadian geomorphology at the present time, is the relation of geomorphology to geography and geology. In the review of the historical development of geomorphology in Canada presented in the last few pages it has been shown how tenuous the link with geology has become, and how geomorphology has grown as the foster child of geography. It is unlikely that geology will seek to re-establish its authority; the separation has been too lengthy. More of a problem for the future of geomorphology is the nature of the tie with geography.

A well-argued case for an autonomous geomorphology has been presented quite recently in Canada,[120] and it has been apparent for some time that geomorphology cannot develop to its full extent if it is obliged to relate its work at all points to the general field of geography. Davis himself distinguished between geomorphography and geomorphogeny, and although he emphasized the former, it is now generally accepted that Davisian geomorphology is far from being a complete system. It was formulated in terms of the prevailing geographical theory; the regional synthesis with its emphasis on explanatory description. Most of the recent demands for a geographical geomorphology have come from the United States,[121] although some echoes have been heard in Canada, and it becomes increasingly apparent that to accede to such demands would reduce geomorphology to the status of a branch of applied geography, its field restricted to the provision of landform description suitably pre-

120 L. E. Hamelin, "Géomorphologie – géographie globale – géographie totale associations internationales," *Cahiers de géog. Québec*, 16 (1964), pp. 199-218.
121 G. Robinson, "A consideration of the relations of geomorphology and geography," *The Professional Geographer*, Vol. XV, 2 (1963), pp. 13-17.

digested to allow for the socio-economic treatment of one area after another. Such a future is quite unacceptable—*ex nihilo nihil fit.*

Canadian geomorphology has come of age, and it must be independent of geographical control to achieve its full potential. However, a self-conscious and defensive isolation would be equally damaging. Modern science is becoming increasing synoptic, and geomorphology can both contribute to, and benefit from, an interdisciplinary dialogue among the natural sciences. By accepting rigorous scientific standards in its procedures, and encouraging collaboration with researchers in allied fields, Canadian geomorphology can make an even more effective contribution to earth science in the future.

Nature does not reveal all her secrets at once.
We may imagine ourselves initiated in her mysteries;
We are as yet but loitering in her outer courts.   *Seneca*

2

Glaciers

J. D. Ives

Figure 1.

First published in *Canadian Geographical Journal*, Vol. LXXIV, no. 4 (1967), pp. 110-17.

Glaciers are natural phenomena of great beauty and of intense scientific interest. The mountaineer, skier, and tourist, in many of the world's mountain regions, is familiar with them. Over the past million years, glacial pulsations have repeatedly overwhelmed most of Canada and, whilst major changes in the broad elements of the landscape have been effected in some areas, it is the myriad of repeated events on a small scale that has helped to give Canada its distinctive physical personality. Glaciers are interesting objects of study in themselves; they also represent vast natural reservoirs of fresh water in those western and arctic sections of the country where they are abundant today; and knowledge of their past fluctuations, their erosive and depositional characteristics, their indirect and direct impact upon relative movements of land and sea level, and the massive energy of their meltwaters, is vital to an understanding of Canada's physique.

Figure 2. Permanent ice occurs in all shapes and sizes. This view shows parts of two valley glaciers, plateau ice patches and ice carapaces which cap the mountain tops nearly 6,000 feet above sea level. The bare rock precipice in shadow at the top left is 1,800 feet high, the valley glacier in the foreground lies more than 4,000 feet beneath the camera. The large view, far left, shows where these two glaciers meet very close to sea level in Inugsuin Fiord.

Figure 3. A river of ice plunges down toward the fiord from the upland snow accumulation grounds in northeast Baffin Island.

Scientists and mountain folk have known for several hundred years that glaciers move. Early observations in areas such as the Alps showed that rocks on the surface of glaciers moved progressively down-valley year by year. Icelandic farmers, more than 500 years ago, learned to appreciate this motion when the great outlet glaciers of the Icelandic ice caps advanced to overwhelm field and farm and when their meltwaters spread a ruin of boulders and gravel in their paths. Later, more detailed inspection showed that, if a line of rocks were placed across a glacier surface, the central ones moved more quickly that those near the sides. It was also seen that there were some similarities between glacier movement and the flow of a river. The rate of movement not only varies with the supply of ice from above, but with ice thickness and with variation in rainfall; changes occur over very short periods, as well as day-by-day and with the seasons.

In high mountains and high latitudes the warmth of summer is not sufficient to melt all the snow that fell during the previous winter. Thus, at the end of each summer some snow is left over; this accumulates from year to year to form ice caps and glaciers. There is a theoretical line, or zone, on a high mountain where the amount of snow melt in summer just balances the amount of snow accumulation in winter. This is called the *snowline*; above it permanent snow is found, generally increasing in amount with height; below it, except with unusually snowy winters or cool summers, all the previous winter's snow melts by the end of summer. Where winter snowfall is very heavy, such as in the Alps, Iceland, and the Coast Range of British Columbia, the summers may be quite warm, but still not warm enough to melt all the snow. "Temperate" glaciers frequently form under these conditions where the entire ice mass is at the melting/freezing point. In high latitudes relatively little snow falls in winter, but the summers are so cool that some snow may remain unmelted in favourable areas. This results in the formation of cold, or arctic type, glaciers and ice caps, where the temperature of most of the ice remains below the freezing point throughout the year. Very high mountains in middle latitudes produce a similarly "cold" ice cover.

Once created, an ice cap or glacier may persist even though the climate may become warmer or dryer. A large ice cap will tend to create its own climate; if it could be removed artificially, as, for instance, the Greenland ice sheet, the Icelandic glaciers, or the Barnes Ice Cap on Baffin Island, no ice cap would redevelop under present climatic conditions. Because of the subtle interplay between variations in weather and climate and the behaviour of glaciers, they may be regarded as giant natural thermometers, although it requires much patience and ingenuity to read them correctly.

"Cold" glaciers move much more slowly than "temperate" glaciers and meltwater is largely confined to the surface or the moat between the glacier margin and the hillside; "temperate" glaciers are frequently "afloat", their meltwater percolating right through the ice mass to the glacier bed and issuing from the snout as a prematurely born river. From a topographic point of view the two main sub-classes are the ice caps and ice sheets that mantle flat mountain tops, plateaux, or even entire continents (Antarctica) and the valley glaciers that are more tightly restricted in their flow patterns by the confining valley walls.

Valley glaciers may be simply described as rivers of ice that flow slowly down steep-walled valleys. They originate in the permanent snowfields above the snowline and their rate of movement depends upon the amount of snow supply, the steepness of the valley floor and the temperature of the ice mass. They range greatly in size, from less than a mile to more than 50 miles long; some are as much as 20 miles wide.

Ice caps sometimes mantle entire mountain ranges or cover the tops of plateaux, and spread out in all directions down the steepest slopes. The ice tongues, as outlet glaciers, or a more even perimeter of the ice cap itself, depending upon the topography, push to lower and lower levels until the amount of ice melt, normally increasing with decreasing elevation, balances the rate of supply from above. The Barnes Ice Cap sits on top of the extensive inland plateau of Baffin Island. Its greatest dimensions are 90 miles long, 45 miles wide and 2,000 feet thick: in this respect it is its own mountain and raison d'être, as the underlying land is no higher than the surround-

Figure 4. This glacier was left "hanging" above Itirbilung Fiord after the major ice stream in the main valley melted away about 7,000 years ago. Today it tumbles nearly 3,000 feet from its feeding grounds on the high plateau between fiords.

ing rolling plateau. It is an accident of nature, a relict of the once great inland ice sheet that covered most of Canada 11,000 to 18,000 years ago. The surface slopes are very gentle except along the margins where cliffs up to 120 feet high occur, fronting a series of ice-dammed lakes. These lakes give further indications of the accidental nature of the ice cap as they occupy the upper reaches of broad valleys which would drain to the southwest in Foxe Basin but for the inert ice mass that blocks their normal paths.

The Antarctic and Greenland ice caps are the two largest in the world and together they contain more than 95 per cent of the world's ice by volume. If the Antarctic ice caps were to melt, the world's oceans would rise between 120 and 180 feet above their present level. This would inundate most of the world's largest cities. The size of the large ice sheets is difficult to envisage; the Barnes Ice Cap is about the size of Prince Edward Island, whereas the Antarctic ice sheet has an area nearly 50 per cent greater than that of the United States.

The exact nature of glacier movement is not yet fully understood. It is believed to be a combination of flowing, melting, refreezing, shearing and slipping. Perhaps as much as 90 per cent of the movement of a temperate glacier is slippage of the entire ice mass over its bed. The thicker, central parts of a glacier move faster than the thinner margins that are restricted by friction with the valley sides. This differential movement within the ice mass causes shearing and fracturing of the surface ice, and particular crevasse patterns form, trending obliquely down-glacier.

Crevasses also form above ice falls: as the gradient increases, the otherwise uninterrupted slope of the ice surface is bent convexly upward, causing a crescentic crevasse pattern. Where a glacier extends onto a plain beyond the immediate confines of its valley walls, the ice spreads out into a broad lobe. This expansion causes surface tensions and creates a radial crevasse pattern with individual crevasses forming approximately perpendicular to the ice margin.

Crevasses are essentially cracks in the upper layers of glaciers. At depth, ice approaches a plastic state and spreads out under its own weight. Thus extra large surface tensions, which tend to cause the larger and deeper crevasses, are compensated for at depth by a form of plastic flow within the ice mass. This limiting tendency becomes effective at depths of between 150 and 200 feet, thus controlling the size of crevasses.

Most glaciers and ice caps in Canada have been getting thinner, and their fronts have been retreating over the past 60 or 70 years.

Figure 5. The Barnes Ice Cap mantles a section of the interior Baffin Island plateau over a length of 90 miles and a maximum breadth of 45 miles. Although it is late August the ice-dammed lake in the right centre is still largely ice-covered.

This glacier recession is a reflection of a significant warming trend in the climate of wide areas. The signs of thinning and retreat are very conspicuous. Ground newly uncovered by the retreating ice front is raw, light-coloured, unvegetated, in contrast to the ground beyond the trimline which was the position reached by the glacier at its greatest recent extent. Many small glaciers have completely disappeared since the turn of the century, larger ones have become 300 to 500 feet thinner and in some instances their snouts have retreated several miles, frequently leaving loops of morainic ridges to mark former positions. The general cause of thinning and retreat is higher summer temperatures that result in greater melt, a longer period of melt, and a greater proportion of precipitation falling as rain rather than as snow. Lessening of snowfall in the winters would have the same general effect.

Were the reverse to occur — increased winter snowfall, lower summer temperatures — and it is believed that such a trend may be underway in some areas today, then the glacier tongues would receive an increased supply of ice from their feeding snowfields, would lose less by melt, would first thicken and then advance. Glacier responses to climatic change are by no means immediate and simple, however, and a time lag between climatic change and glacier response will occur and will vary with the topographic setting and the size of each glacier. It is not uncommon to find neighbouring glaciers advancing while the majority continue to retreat.

Figure 6. Detail of crevasse pattern on the inside bend of a valley glacier. The sharp crested ridge of boulders, called the lateral moraine, is actually composed largely of ice which the covering mantle of debris has protected from melting.

Another feature of glaciers, at least in summer, is the abundant meltwater usually associated with them. Glaciers may be regarded as a somewhat unusual phase in the normal hydrological cycle in that precipitation in frozen form is locked up for a period and tapped over an extended interval. Thus where hydroelectric plants draw their water supplies from watersheds containing glaciers, the ice masses help to regulate the run-off pattern. During a warm, dry summer, for instance, when streams normally run low and lakes fall in level, the increased ice melt serves as a counterbalance. Thus an understanding of the nature of glaciers can be a most important factor in assessing waterpower potential in glacierized areas. This is one of the reasons why Canada is slowly beginning to pay increasing attention to its frozen reservoirs of fresh water.

When glaciers end in the sea their snouts may be afloat or aground in shallow water, depending on the thickness and rate of supply of the ice and the depth of the water. The waves erode the glacier front even where it is aground in shallow water and a line of ice cliffs may form. Pieces of ice which then break off become small icebergs. Where the glacier tongue is afloat, or very nearly so in deeper water, the buoyancy effect so created tends to cause large chunks of ice to break off. The size and shape of icebergs is related to the glacier thickness, its crevasse pattern and the depth of water. Extremely large tabular bergs break from the floating margins of the Antarctic ice sheet. The outlet glaciers of the Greenland ice cap are frequently very productive and produce vast icebergs that may project several hundred feet above the water. Many of these drift southward via the East Baffin and Labrador currents into the more frequented parts of the North Atlantic, so becoming a menace to shipping.

Although most of the world's glaciers have been thinning and retreating for the past seventy years, 12,000 years ago much of Canada and adjacent parts of the United States were inundated with glacier ice. On at least four occasions during the last million years, ice sheets grew to cover much of North America, northwest Europe, central Asia and South America. As each successive ice age tends to obscure the effects of the previous one, most is known about the final ice age. In North America it is known as the Wisconsin Ice Age and it lasted about 70,000 years, ending about 8,000 years ago. Many large fluctuations of the southern margin occurred during this time, involving retreats and advances across southern Canada and northern United States of several hundreds of miles. The northern margins in the Canadian Arctic probably also experienced fluctuations although little is known about them. The Laurentide ice sheet was the main centre of the Wisconsin Glaciation, originating on the high uplands of Labrador-Ungava and Baffin Island and slowly growing to fill in Hudson Bay and to cover most of Canada. The western mountains maintained their own centre, as did Greenland. Less is known about the High Arctic islands, although they probably maintained independent ice caps of their own in partial contact with the ice caps on neighbouring islands. Land in the extreme northwest may have remained ice-free. In its central parts the Laurentide ice sheet probably exceeded 12,000 feet in thickness; it thinned progressively toward the margins, and was not sufficiently thick along the Baffin Island and Labrador east coasts to prevent the higher mountains from sticking out as nunataks. New evidence is being accumulated indicating that even low-lying coastal strips, where backed by high mountains along these coasts, remained ice-free.

Figure 7. One hundred and fifty years ago this glacier at the head of Itirbilung Fiord extended across the floor of the valley to dam up a large lake. Today, the small, wasting ice tongue lies more than three miles behind the outermost moraines which mark its previous extent.

Figure 8 shows the relationship between the location of present-day glaciers and the outer limit of the continental ice sheet at the height of the Ice Ages. No present-day glacier occurs outside this outer limit. With a prolonged deterioration in climate, the existing glaciers would expand and begin to coalesce to form ice caps. This certainly happened at the beginning of each Ice Age.

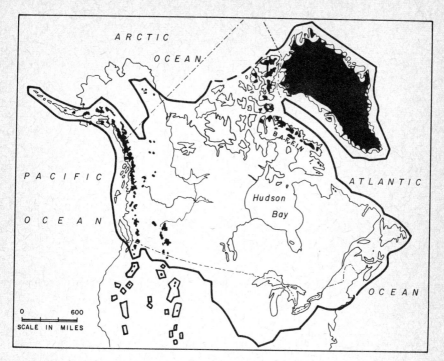

Figure 8. The heavy line shows the maximum area of North America that was covered during the Ice Ages. The black area is the present-day distribution of glaciers.

What is probably more important, however, is that, as the regional snowline lowered with deteriorating climate, large areas of the upland country of Labrador-Ungava, Baffin Island and Keewatin would become vast snow accumulation grounds simultaneously. Thus large ice sheets would grow in these areas and play a much more significant role than the mountains of the east coast. Similarly, at the close of each Ice Age some of the final ice sheet remnants were located in the central tracts of these immense upland areas: this occurred in Labrador-Ungava and Keewatin about 7,000 years ago, whereas the disintegration of the Baffin Island ice cap proceeded much more slowly, and the process is still incomplete — witness the slowly shrinking Barnes Ice Cap of today.

The study of glaciers has its own special kind of interest and excitement; one is not only pitting one's mind against secretive Nature, but her sheer physical difficulties can form half the struggle. The scientific approach to glaciology is best pursued along several

5

Figure 9. Revoir Pass is a great glacier trough that runs from the head of Eglington Fiord to Sam Ford Fiord. The original trough glacier has long since melted and today only small mountain glaciers tumble down its walls to the trough floor.

fronts: while much remains to be learned about the straight-forward geography of Canada's ice and snow, glaciology is multi-disciplinary in scope and the physicist, engineer, mathematician, botanist, geologist and geographer are all required. One very fruitful approach at the present stage of relatively slight knowledge of the characteristics of Canada's glaciers, is a combined study including geomorphology, glaciology and climatology. The more that can be learned about the characteristics of present-day glaciers and their relationship with climate, the more readily we can interpret the landforms left behind by the great ice sheets of the past. The approach should be reversible, for the greater our knowledge of the history of past glaciation, the more extensive will be our insight to interpret present-day glacier characteristics. This parallel approach, however, has sometimes been followed too rigidly. Most of the Canadian glaciers are of the mountain and valley glacier type, whereas the more typical feature of the Ice Ages was a great continental ice sheet. This is one of the reasons why plateau ice masses, such as the Barnes Ice Cap, are particularly suitable for long-range detailed investigation.

3

Recession of the Drummond Glacier, Alberta

J. G. Nelson, I. Y. Ashwell, A. G. Brunger

Since 1962, a programme of geomorphological, glaciological, and climatological studies has been conducted in the Red Deer River valley, Banff National Park (Figure 1 inset). The work began with the reconnaissance mapping of deposits and landforms in the valley and has progressed to include detailed geomorphological

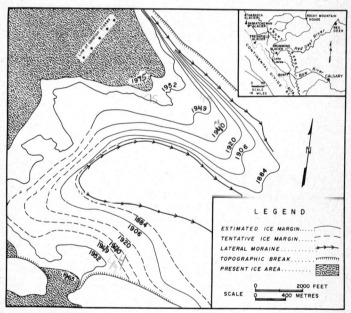

Figure 1. Recession of the Drummond Glacier, 1884-1965. Inset map shows the Drummond and other glaciers in the Rocky Mountains of Alberta, Canada.

First published in *Canadian Geographer*, Vol. X, no. 2 (1966), pp. 71-81.

[1] In November, 1964, Dr. G. Hattersley-Smith of the Defence Research Board pointed out to the authors that the name "Drummond Glacier" was not officially in use and required the approval of the Canadian Permanent Committee on Geographical Names. The name was therefore submitted as being suitable for the glacier, particularly as it had been in common usage for some time, and it was subsequently approved by the Committee on Geographical Names, Department of Mines and Technical Surveys.

mapping, fluvial studies, measurements of glacier movement, and the summertime operation of a weather station in the Drummond Creek area, one of the headwater valleys of the Red Deer. Results of the work have been presented in several reports in the last two years,[2] and only the amount and rate of glacial recession will be discussed in this article. Conclusions are based on photographs, some tree-ring data, radio-carbon dates, and recent measurements of ice-wastage.

Location

The Drummond Glacier is situated in the headwaters of the Red Deer River in the eastern ranges of the Canadian Rockies. The icefield covers approximately ten square miles and lies at an altitude of between 8,000 and 10,000 feet above sea level. The icefield is dome-shaped, fairly uniform at the surface and remarkably free of crevasses. It is enclosed in an amphitheatre of peaks, one of which is Mt. Drummond and at least six of which exceed 10,000 feet in altitude. By tourist standards the Drummond area seems to have been very little visited; all roads end fifteen miles or more from the glacier.

Photographic Record

Although fur traders and others are known to have been active in the Canadian Rocky Mountains prior to 1800, little reference is made to the Red Deer valley. The latter does not seem to have been used much as a communication route, probably because the Bow River valley, in close proximity to the south, provided much better access to and through the mountains.

One of the first recorded trips made up the Red Deer Valley by a white man was that of George W. Dawson in 1884. A geologist, he travelled alone except for Indian guides. He took many photos including two excellent shots of the Drummond Glacier from a point about two miles south of the ice near the junction of the Drummond Creek and the main Red Deer River (Figure 2). The quality of the photographs is good, and it is possible to plot the position of the southern lobe of the ice with considerable accuracy.

[2] J. J. Gardner, G. Nelson and I. Y. Ashwell, "Alpine Studies in the Upper Red Deer River Valley," *Can. Alpine J.*, XLVII (1964), pp. 137-44.

H. Macpherson, "Glacial geomorphology of the Upper Red Deer River Valley," Unpub. M.Sc. thesis, University of Alberta at Calgary (1963).

G. Hattersley-Smith, "Report on Glacier Research in Canada," (1964), *Can. Geophys. Bull.*, XVII, sec. 2 (1964).

A. G. Brunger, J. G. Nelson and I. Y. Ashwell, "Reports on Geomorphological Studies in the Upper Red Deer River Valley," unpub. paper presented at the Canadian Association of Geographers annual meeting (May 1965).

Figure 2. Drummond Glacier as photographed by G. W. Dawson in 1884. Note the extent of the main tongue of ice on right side of picture. Comparison with later photos reveals the large amount of ice-wastage. Nothing is said about the photographic evidence for vertical ablation in the text, primarily because of difficulties in estimating the amount. However, the vertical ablation has obviously been appreciable since 1884. The Dawson photo is provided by courtesy of the Canadian Geological Survey.

In the early twentieth century increased interest in glaciers accompanied the development of the Banff area as a tourist resort. Mountaineers began to note the positions of glaciers, and the Canadian Alpine Club, founded in 1907 by A. O. Wheeler, had glacier observation as one of its objectives. Predictably, the observations that were made generally came from the most accessible areas of the Rockies, the upper Red Deer apparently having been visited by only a few climbers, none of whom are known to have referred to the position of the Drummond. However, in the first decade of the century, Wheeler surveyed much of the area, and travelled throughout the front ranges of the Rockies. In 1906 he moved eastward from the Bow valley and took numerous photographic panoramas from mountain peaks — including several of the Drummond Glacier which permit very precise estimation of its position. The next known record of the Drummond Glacier is a photograph taken by M. P. Bridgland, who was surveying in the area of the foothills and front ranges in the period 1917-20. The photo is not good in detail, however, largely because it was taken from a mountain some distance to the south of the glacier.

The photographs so far described are derived from government-sponsored surveys and were taken by persons who travelled the area for purposes of mapping or research and who were undoubtedly among the first white people in the upper Red Deer valley. It is difficult to estimate the number of other travellers in the area. A number of mountaineers and packers undoubtedly were there during the early part of the century but their interest was apparently mainly non-scientific, and few of them returned with photographs.

In 1930 Mr. Leonard Leacock travelled into the Red Deer valley and photographed the Drummond ice from Dawson's viewpoint. His two photographs show that the glacier had retreated still further and are good enough to permit plotting of the glacier margin with reasonable accuracy. Another photograph was taken in 1939 by Colin Wyatt.[3] Shot from a more distant vantage point on the southern side of the Red Deer valley, about five miles away from the ice, the view is better than Dawson's in that the ice is completely visible and the margin can be mapped fairly accurately. The photo can be compared with one which was taken from a similar position by J. Gardner in 1963, and which shows considerable recession from the 1939 position.

For the period 1939-62 no photographs of the Drummond Glacier are known to the authors, except for the aerial photographs taken by federal and Alberta government agencies on September 3, 1951, and July 31, 1952; the provincial photos, by the Department of Lands and Forests (no. 160-5114X, 2426-49; no. 160-5113X, 2426-15), provide a very clear picture of the glacier and environs and have been used as the base for mapping retreat stages. Since the 1962 field season a large number of photographs of the Drummond have been taken by personnel involved in Red Deer valley studies; Figure 3 gives one example. Figure 1 shows six

Figure 3. A view of the Drummond Glacier as it was in early July 1964. Comparison of this with earlier photos reveals changes not only in the glacier but in the position of the river, fluvial deposits, mass wastage deposits, and so forth. Photographs by M. Sundstrom.

3 Colin Wyatt, *The Call of the Mountains*, (London, 1952).

positions of the ice-margin which have been plotted using all the previous photographic evidence; the symbols for estimated and tentative ice-margin are intended to suggest relative difficulty in plotting.

There are a number of long-term records from other glaciers in the nearby area. Among the best known are those by Field and Heusser,[4] whose data are considerable and are derived from dating of moraines by botanical and other techniques, from photographs, and more recently from direct measurement of recession. Only their records from the Saskatchewan and Freshfield glaciers are included in this report, primarily because the time periods for which retreat data are available for these two glaciers make them most suitable for long-term comparison with the record of recession from the Drummond. The data available for all three glaciers are tabulated in Table I. The measurements of average annual recession show that during the period 1884-1965 the Drummond receded at varying rates, increasing through about the first fifty years to a maximum average annual recession in the

TABLE I. RECESSION OF DRUMMOND, SASKATCHEWAN, AND FRESHFIELD GLACIERS.

Time periods	Frontal recession in feet (and metres) Total amount	Annual amount
	DRUMMOND GLACIER	
1884–1906	1080 (330)	49 (15.0)
1906–1920	850 (260)	61 (18.6)
1920–1930	850 (260)	85 (26.0)
1930–1939	1115 (340)	124 (37.9)
1939–1952	1115 (340)	86 (26.2)
1952–1965	1062 (325)	82 (25.0)
	SASKATCHEWAN GLACIER	
1893–1912	400 (122)	21 (6.5)
1912–1924	600 (184)*	50 (15.3)
1924–1948	2600 (80)	108 (32.5)
1948–1964	2026 (69)	126 (38.5)
	FRESHFIELD GLACIER	
1869–1902	1050 (321)	32 (9.8)
1902–1930	1508 (463)	54 (16.5)
1930–1953	2817 (862)	123 (37.5)
1952–1954	150 (46)	75 (23.0)

*Taken from W. O. Field, "Glacier Observations in the Canadian Rockies, 1948." *Can. Alpine J.*, XXXII (1949), 99-114: maximum recession figure is used.

4 W. O. Field and C. J. Heusser, "Glacier and Botanical Studies in the Canadian Rockies," *Can. Alpine J.*, XXXVII (1954), pp. 128-40.
5 W. O. Field, "Glacier Observations in the Canadian Rockies," (1948), *Can. Alpine J.*, XXXII (1949), pp. 99-114.

1930-39 period, and decreasing since then. Generally comparable patterns are recorded from the Saskatchewan and Freshfield glaciers, although the similarity is not readily apparent for the Saskatchewan because the period 1924-48 shows an average annual recession of 108 feet, while the period 1948-64 shows a higher figure of 126 feet. However, the earlier of the two periods includes about six years of records (1924-30) in which the rates of recession were probably closer to the average of 50 feet for 1912-24 (Table I) than to the average of 108 feet. In other words, the average annual rate of recession of the Saskatchewan Glacier is considered to have been appreciably higher during the period 1930-39 than it was during the pre-1930 and post-1939 periods.

Because of differences in periods of measurement and other problems, the authors are not prepared at this stage to make any further statements about regional rates of retreat, or about possible variations in climate or other controls. It should be noted, however, that a relatively large number of observations of the Drummond are available and that the retreat record for this glacier seems to be the most accurate of any in the eastern Rockies for the period 1884 to the present.

Recession Estimates Based on Dendrochronological and Carbon 14 Evidence

Tree-ring samples have been collected from Englemann Spruce located at several points in the Drummond valley. Several trees within a half-mile or so of the prominent lateral moraine surrounding the present ice were found to be about 300 years old. A study of areas shown to be covered by ice in Dawson's photo of 1884, and since deglaciated, reveals little or no tree-growth as yet. Although it is recognized that an eighty-year period without re-growth is a long one, perhaps influenced by the high altitude, this incomplete and unsatisfactory dendrochronological analysis still indicates a period of at least 400 years since the ice penetrated beyond the lateral moraine which contains the present ice-field.

C_{14} dates have been obtained for two samples recovered about fifteen miles[6] and two-and-one-half miles[7] respectively from the Drummond Glacier. The first date was from a small amount of charcoal recovered from a till-like deposit located on the north side of the Red Deer valley near Scotch Cabin. The result, which

---

[6] Personal communication with K. J. McCallum, Department of Chemistry, University of Saskatchewan (December 4, 1964).

[7] Personal communication with A. Stalker, Geological Survey of Canada (March 19, 1965).

was greater than 33,000 years, is quite out of line with dates which are available for nearby areas. The second date was obtained from a sample of wood and organic debris located about nine feet from the surface of a large mudflow on the east side of the Drummond valley. Several layers of wood and organic debris buried by former mudflows were exposed during the digging of a cross-section along the stream bank. The C_{14} date for the lowest of these was $2,930\pm150$ years B.P., a minimum date for deglaciation inasmuch as an unknown amount of colluvium, and perhaps other debris, underlie the sample horizon.

Recent Measures of Glacier Recession and Vertical Ablation

In recent years the retreat of the glacier has been observed from both frontal and vertical stakes sunk into the ice. The lateral retreat of the glacier was first measured in 1962. In the following year, frontal recession was measured along a line of five stakes which has been the basis for the subsequent observations expressed in Table II.

TABLE II. CUMULATIVE FRONTAL RECESSION OF THE DRUMMOND GLACIER.

Distance in feet (and metres) of stakes from ice at periodic observations

Stake	July 6, 1962	Sept. 6, 1962	July 23, 1963	Aug. 3, 1963	July 15, 1964	Aug. 10, 1964	July 14, 1965	Sept. 1, 1965
A 1962	Placed	45* (12.2)	Adjusted to	68 (20.8)				
			#2 1963					
#1 1963	—	—	Placed	—	32.4 (10.0)	39.8 (12.2)	54.9 (16.8)	93.6 (28.6)
#2 1963	—	—	Placed	—	45.4 (13.8)	54.4 (16.6)	59.9 (18.3)	98.1 (30.0)
#3 1963	—	—	Placed	—	48.4 (14.8)	61.3 (18.8)	64.3 (19.7)	81.1 (24.7)
#4 1963	—	—	Placed	—	77.5 (23.7)	90.8 (27.8)	98.1 (30.0)	Disappeared
#5 1963	—	—	Placed 18 ft. away	—	70.4 (21.6)	101.2 (30.9)	116.4 (35.6)	157.3 (48.0)

*Snow on the ice probably affected measurement.

The vertical ablation of the glacier was first measured in 1963. On a line across the glacier approximately half a mile from the tongue, ten stakes were inserted to depths of about ten feet, and an aluminum tube consisting of ten-foot sections and intended for ice-movement studies, was also inserted into the glacier. These stakes, which required re-setting in 1964, four being inserted to depths of about thirty feet, have provided ablation measurements since 1963 and have been measured as often as practicable in each field season. Although a check was not possible during the summer of 1965 because of bad weather, the re-surveying during 1964 indicated a glacial advance of about twenty-five feet during the period from early August 1963 to early July 1964. The cumulative results are shown in Table III.

TABLE III. CUMULATIVE VERTICAL ABLATION OF THE DRUMMOND GLACIER

Length in inches (and centimetres and metres) of exposed stake at periodic observations

Stake	Aug. 7, 1963	July 16, 1964	Aug. 10, 1964	July 26, 1965	Aug. 24, 1965	Sept. 1, 1965
1963	Placed	79.0 (200.0)	105.8 (260.8)	123.5 (312.8)	—	194.0 (493.0)
#1 1964	—	Placed	8.9 (22.7)	29.4 (74.9)	85.5 (217.0)	90.0 (228.0)
#2 1964	—	Placed	21.8 (55.8)	37.4 (95.0)	90.0 (228.0)	93.5 (237.0)
#3 1964	—	Placed	28.4 (72.3)	46.2 (117.5)	106.5 (270.5)	109.5 (278.0)
#4 1964	—	Placed	26.0 (66.0)	42.2 (107.0)	100.0 (254.0)	107.0 (256.0)

Calculated from the cumulative amounts of recent frontal recession of the Drummond Glacier (Table II), the approximate annual amounts of recession were as follows: July 6, 1962–August 3, 1963, 68 feet (20.8 metres); July 23, 1963–July 15, 1964, 51 feet (17.0 metres); July 15, 1964–July 14, 1965, 27 feet (7.3 metres). The

1962-63 measurements are not very useful because the recession was measured at only one stake and because there was a time over-lap with part of the following recession period. The measurements for the 1963-64 and 1964-65 periods are more reliable, being the average for the five stakes located approximately parallel to the southern edge of the Drummond Glacier. Although the recession for 1963-64 was considerably greater than that for the next year, the rate may well increase significantly in 1965-66 because 35.3 feet (10.8 metres) have already been recorded for the period July 14–September 1, 1965. All recent measurements from frontal stakes are smaller than the average rate of recession estimated by the use of photos for the period 1952-65 (Table I).

The amounts of vertical ablation measured on the Drummond Glacier were: August 7, 1963–July 16, 1964, 79 inches (200 centi-metres); July 16, 1964–July 26, 1965, 40 inches (101 centimetres). Also the ablation recorded between July 26 and September 1, 1965, was 71 inches (180 centimetres), clearly greater than the 1964-65 amount and thus comparable with the frontal observations. In fact, the over-all vertical measurements reflect the pattern of frontal recession; in this important sense the two sets of measurements reinforce one another.

Although the above recent measurements of recession of the Drummond Glacier cover only a very short period of time, it is interesting to compare them to similar measurements from nearby ice-sheets. Of the surveys which the Water Resources Board has made of the tongues of several glaciers in the eastern Rockies since 1945,[8] unfortunately only two were continued until 1964. Never-theless, these do provide relatively precise data on the recession of the Saskatchewan and Athabaska glaciers. The average rate of annual retreat of the Saskatchewan in the period 1960-62 was 133 feet (41 metres), and in 1962-64, 88 feet (27 metres). The Atha-baska Glacier advanced 1.5 feet (0.46 metres) in 1960-62, and in 1962-64 receded at an annual rate of 56 feet (17 metres). In sum then, the Drummond Glacier seems to be similar to the Saskat-chewan and the Athabaska glaciers in that the rate of recession has decreased in very recent years, the Athabaska actually advanc-ing for one short period.

---

[8] Canada, Dept. of Northern Affairs and National Resources, Water Resources Branch, *Survey of Glaciers on the Eastern Slope of the Rocky Mountains in the Banff and Jasper National Parks*, prepared by K. F. Davies (Calgary, Alberta, September 10, 1964).

Conclusion

The need for increased glacier observations was outlined in 1964 by Hattersley-Smith[9] when he listed a number of glaciers which were suitable for study because they possessed advantages such as ease of access or a history of observation. He stated that "it seemed best to choose glaciers representative of the different ranges in Western Canada."[10] Those chosen for the Rocky Mountains were all to the west of the Bow valley, and included none representative of the front ranges east of the Bow valley. It would therefore appear that the present report on the Drummond provides the first details on recession in the eastern ranges. The authors plan to continue their measurements for several years, and also, in the near future, to publish estimates of recession of the Bonnet and Hector glaciers, located near the Drummond.

[9] G. Hattersley-Smith, *et al.*, "Proposed Programme for Recording the Variations of Existing Glaciers in Canada," *Can. Alpine J.*, XLVI (1963), pp. 128-34.
[10] *Ibid.*, p. 132.

4

A Lichenometrical Study of the Northwestern Margin of Barnes Ice Cap: A Geomorphological Technique

J. T. Andrews and P. J. Webber

Introduction

The Barnes Ice Cap lies on a subdued plateau-like surface between 400 and 700 meters above sea level on north-central Baffin Island (70°N, 74°W). In 1950, an expedition from the Arctic Institute of North America studied its southeastern lobe, and Hale (in Ward, 1952) made botanical transects from the ice-cap margin to the area of mature lichen colonization. On the basis of an assumed growth of 1 millimeter a year for *Alectoria minuscula* Nyl., he concluded that recession following a readvance began about 1860. In 1961 the Geographical Branch began a long-term research program in north-central Baffin Island, and from 1961 to 1963 the research effort was concentrated about the northwestern margin of the Barnes Ice Cap. During the first summer, large differences in lichen diameters were noted above and below the shoreline of Glacial Lake Lewis, a former ice-dammed lake, and were measured by Andrews (Ives, 1962). In 1962 systematic observations based on a standard method and the use of more species of lichens resulted in two important conclusions: (1) that the maximum lichen diameters increase in size away from the ice cap; (2) that the maximum diameters are similar both on contemporaneous moraines and within former glacial-lake basins. These conclusions led to a detailed consideration of lichenometry, which makes it possible to date substrates when historical records are inadequate.

Reprinted from *Geographical Bulletin*, no. 22 (1964), pp. 80-104, by permission of the Department of Energy, Mines and Resources.

The aims of this paper, which presents the results of three field seasons, are (1) to assess the usefulness of lichenometry for glacial geomorphological studies when its basis is the work of observers whose botanical training is limited and (2) to provide a relative time scale for the glacial fluctuations that occur about the margin of the Barnes Ice Cap. This paper is therefore complementary to Beschel's basic analysis of lichenometry as applied to glaciology and physiography (Beschel, 1961a).

Concept and Method of Lichenometry

The colonization of a new surface by lichens and their growth on it depend upon substrate type, climate and microenvironment. This study concerns only epipetric lichens. The rock types of central Baffin Island belong to the typical Archaean basement complex, the predominant type being granite gneiss. As no species in this area was seen to have a clear preference for a rock type, the preference factor receives no further consideration. The factors of temperature and the availability of liquid water are of prime importance in the growth of lichens. Beschel (1961a) found proof of this when he observed a decrease in the growth rate of *Rhizocarpon tinei* (Tornab.) Runem. (*R. geographicum* s.l.) along a gradient from the wet coastal to the dry continental regions of West Greeland.

The climate of central Baffin Island is severe, having a precipitation of about 37 centimeters water equivalent a year, an annual mean temperature of $-10°C$ and a continentality index of perhaps

TABLE I. "LICHEN FACTOR" OF R. GEOGRAPHICUM (L.) DC OF DIFFERENT REGIONS.

Region	Author	"Lichen factor" (mm per century)
Greenland (Søndre Strømfiord area)	Beschel (1961a)	2 to 45
Baffin Island	Andrews and Webber	5.4
Greenland (Disko area)	Beschel (1963a)	15
Axel Heiberg	Beschel (1963b)	4 to 15
Italy (Gran Paradiso)	Beschel (1957)	13 to 25
North Sweden	Stork (1963)	20
Austria	Beschel (1957)	21 to 93
South Norway	Stork (1963)	46
Switzerland (Steingletscher)	Beschel (1957)	60

50 per cent (Mackay and Cook, 1963). The index for Disko, West Greenland, is 23 per cent. Table I compares growth rates of *R. geographicum* for various regions and shows the necessity of establishing a series of growth rates for each species in each region. The effect of microenvironment on lichen growth, combined with the possibility of successive colonizations by more thalli, would perhaps preclude the usefulness of lichenometry for dating, but using a species' maximum diameter reduces these disadvantages to a minimum (Beschel, 1957). The assumption is that lichens of maximum diameter are both the oldest and the optimally growing. It can be applied, however, only to the most common and successful species in an area and to those species whose thalli retain their identity.

The rate of growth of crustose and foliose lichens is not constant over their life span. It is initially sigmoidal and then becomes generally linear until the onset of senescence, when growth decreases or even stops. During the linear phase, the length of which varies with species, the diameter of the thallus may be considered directly proportional to its age, providing there has been no great change of climate during the life of the lichen. (For an account of lichen growth-curve patterns, see Beschel, 1961a, pages 1045-1047). Diameter measurement rather than the more accurate but time-consuming planimetric method of Hale (1959) is used at this stage as a means of rapidly determining lichen size.

A basic assumption of lichenometry as an indication of time since deglaciation is that the substrate is devoid of all living lichens immediately before deglaciation. No evidence was seen of the survival of lichens under the Barnes Ice Cap, and it is unlikely that any survivors were measured in the present study. Lichen thalli in association with ice bodies have been reported by Goldthwait (1960), Beschel (1961b) and Falconer (1963: personal communication), but as yet there is no positive proof that these thalli were alive. Similarly, no evidence was noted of lichen survival in the former snowbank areas, although in the differences of diameter and density there were variations that may reflect differences in the duration of lichen survival. Survival beneath ice requires rapid freezing and thawing and absence of movement within the ice. Stork (1963) suggests that lichens may begin to grow on boulders that have been carried along on the glacial surface, but the nature of the ice cap is such that this possibility is remote.

As 10 observers contributed to the sampling at 290 lichen stations, the method used had to be simple and as close as possible to uniformity. A lichen station was established at a point along a transect or on a geomorphological feature. Sampling at each station covered a circular area of 10-meter radius. Beschel (1961a) recommended at least 100 square meters, and Stork (1963) five separate 5 x 5 meter quadrats. When a station was situated on a slope, sampling was usually restricted to the contour. Whenever a special feature was sampled, care was taken to limit the sampling to some specific and uniform part of the feature, for example, to the proximal side of a moraine. Wet and snowpatch sites were not sampled. Each station was diligently searched and the diameters of all large thalli of the common lichens were measured and recorded to the nearest half millimeter. For each species only the 10 largest diameters were permanently recorded, and in this study only the maximum value is used. The act of measuring and recording other large thalli is a means of encouraging observers to be careful and adhere to some degree of consistency. At each lichen station a record was made of site aspect, slope and substrate type, and additional notes on vegetation and geomorphological features were taken as required. When sampling was carried out in recently deglaciated areas, willow and soil samples were occasionaly collected. The data from each site were later transferred in the field to printed cards.

The greatest density of sampling was around the Lewis Glacier. The station locations within the whole study area are shown in Figure 1. Isophyses (Greek: *isos*, equal; *physis*, a growth) of two lichens were plotted by interpolation between sample points.

Lichen Determinations

Not all lichens are suitable for dating; only those epipetric lichens with distinct and almost circular thalli were used. If a thallus was elliptical, the shorter diameter was recorded. Measurements were restricted to about 10 easily recognizable species; frequently, only four were ever abundant enough to provide reliable data. The lichen species described in the following paragraphs were introduced to the observers. Their usefulness in the present study is assessed, but in each new area a reassessment must be undertaken. The principal references for the determinations were Dahl (1950), Hale (1954) and Runemark (1956). The genera in this list of lichens are arranged in the sequence of the growth form of the thalli—fruticose-foliose-crustose.

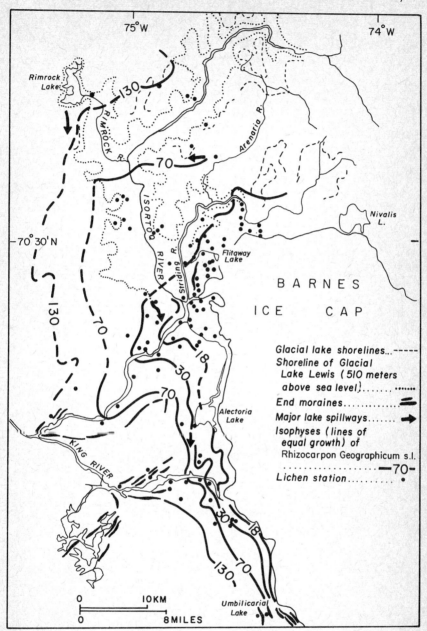

Figure 1. Rhizocarpon geographicum isophyses for the northwestern margin of the Barnes Ice Cap. Also shown are the principal end moraines and former glacial-lake shorelines. The isophyses present a consistent picture of the glacial retreat and make possible a firm determination of relative chronology and a tentative decision on absolute chronology. The figure also gives an idea of the sampling network around the ice cap.

Alectoria minuscula Nyl., a black, fibrous lichen, is not easily separated from the similar *Alectoria pubescens* (L.) Howe. *A. minuscula* is usually the most abundant, but no evidence could be found of a difference in growth rates between the two species. Thus the measurements were grouped, and all data used are those for *A. minuscula* s.l. In the remaining sections of the paper the two species are grouped as *A. minuscula*. These species are among the first lichens to colonize newly exposed rock surfaces in the field area, and, further, are easily recognized by the observer. *A. minuscula* is very common, and diameters of up to 140 millimeters have been found. Larger diameters have also been measured, but these thalli are often incomplete, their margins having eroded away. The senescent phase of the growth curve probably comes at 120 millimeters, there being at this point an apparent decrease in the size ratio in relation to *R. geographicum*.

Umbilicaria proboscidea (L.) Schrad. is easily recognized and presents no taxonomic problems. Although a common species, it is not always present in large enough numbers to result in 10 large thalli at each station site. A further limitation is that the thalli, particularly of old specimens (larger than 30 millimeters), are difficult to measure and that the margins of dry specimens are curled. Individuals larger than 70 millimeters were seldom seen, and the senescence point can be placed at about 50 millimeters.

Umbilicaria hyperborea (Ach.) Hoffm., although abundant, proved to be of limited use in this study. The observers too easily confused it with *Umbilicaria torrefacta* (Lightf.) Schrad. With more careful sampling, these two species could be useful for lichenometry especially when a species — *Umbilicaria proboscidea* for example — is absent or rare.

Xanthoria elegans (Link.) Th. Fr., with its bright orange thallus, is the easiest to recognize but is not suitable for lichenometry in this region because it is not always sufficiently frequent. In a transect study, for example, it may occur only at one station and is thus of little value for sequential studies. The most serious reason for criticizing its use in lichenometry is that it favors bird perches.

Rhizocarpon geographicum (L.) DC. has been divided by Runemark (1956) into 21 species and subspecies. The subunits of this group are very difficult to separate in the field, and in this paper the name *R. geographicum* has been retained in the wider sense. As the most common entity within this group is *R. tinei* (Tornab.) Runem., which grows faster or at the same rate as the other subunits (Beschel, 1961a), there seems to be little danger in this retention. Other species of yellow Rhizocarpons not of the *Geographicum* group—*R. superficiale* (Shaer.) Vain. and *R. inarense* (Vain.) Vain. (*R. chionophilum* Th. Fr.) —were collected in the study area. They are easily confused with *R. geographicum* but are not so frequent. The sampling of maximum diameters would tend to favor the selection of *R. geographicum* s.l. and especially of *R. tinei* and reduce errors in measurement.

R. geographicum is the slowest growing of the common lichens used and remains in the linear phase of the growth curve for a long period. Accordingly, it spans the longest dating range of the lichens used in this study. Although the thallus may have an upper limit of 200 millimeters in this area, it seldom reaches diameters of more than 120 millimeters. This may be attributed to a weathering of the rock surface and should not necessarily be confused with senescence.

Rhizocarpon jemtlandicum Malme is intermediate in growth rate between *A. minuscula* and *R. geographicum*. It has a greyish thallus and was sampled with *Rhizocarpon disporum* (Naeg.) Mull. Arg. and *R. grande* (Flk.) Arn. Allowance must be made for probable confusion of these grey *Rhizocarpons*, which may not have uniform growth rates. *R. rittokense* (Hellb.) Th. Fr., although not a grey *Rhizocarpon*, was sampled with the other lichens in the field and was confused with *R. jemtlandicum*. It is as common but may be easily distinguished by its peltate areoles. In central Baffin Island, *R. jemtlandicum* grows faster than *R. geographicum*, whereas on the coast of West Greenland it grows at two thirds the rate of *R. geographicum* and inland at the same rate (Beschel, 1961a). This variation illustrates that growth rates of different species are not always related in constant proportion from region to region. The growth rates of each species or their ratios must be found again for each new locality.

Other species occasionally measured were: *Umbilicaria virginis* Schaer., *Umbilicaria cylindrica* (L.) Del., *Parmelia alpicola* Th. Fr., *Parmelia incurva* (Pers.) Fr., and *Buellia moriopsis* (Mass.) Th. Fr. Data on these species are very scarce, and the authors have drawn no conclusions about their growth rate and usefulness in lichenometry.

Sampling studies

Quadrats of 8 x 8 meter size were marked out at 25-meter intervals along the proximal side of the outer lateral moraine of the Lewis Glacier. All rock surfaces in these quadrats were searched, and 50 of the largest thalli of each of four species in each quadrat were recorded, as shown in Table II. At a later date, seven observers were sent to take samples at separate stations by the standard procedure. The stations were on the same proximal slope, but the location of the sample sites varied. The results of this study are shown in Table III. A comparison of the two tables reveals a measure of variability, which is very large if only one species — *R. geographicum*, for example — is used. This species, however, is not so common on the moraine as *A. minuscula* and may be less valuable in a recently deglacierized area. Among the causes of variability are slumping and other earth movements. When an ice-cored moraine has slumped, a lichen value must be regarded as a minimal determinant of its age. For age determinations it is therefore important to use more than one species. Although the results of the undertaking described in this paper indicate a tendency toward undersampling, the method used can be regarded as satisfactory in view of the much larger area that would have to be sampled to increase the likelihood of finding the largest lichen.

Basin of Glacial Lake Lewis

Table IV shows the difference in maximum lichen diameters above and below the 510-meter shoreline of Glacial Lake Lewis at various distances from the margin of the ice cap (Figure 1). When the ice front had retreated to within 6 kilometers of the present northwestern margin, the level of this former glacial lake was finally lowered from 510 meters above sea level to a stillstand at 430 meters.

Thus the area between these shorelines was exposed for the simultaneous development of lichens. It represents a standard time base and can be used to evaluate a possible climatic gradient away from the ice cap. In the vicinity of Rimrock Lake (Figure 1) the shoreline of Glacial Lake Lewis is readily visible owing to the difference in lichen development and density of cover, but this visual effect is less pronounced farther down the Isortoq.

TABLE II. QUADRAT STUDIES OF FOUR LICHEN SPECIES ON PROXIMAL SLOPE OF
OUTER MORAINE OF LEWIS GLACIER.

(maximum diameters in millimeters)

Quadrat 8 × 8 m	A. minuscula	U. proboscidea	R. jemtlandicum	R. geographicum
1	42	17	24	10
2	75	35	31	11
3	89	40	25	11
4	57	38	22	11
5	67	27	22	10
6	70	34	28	11
7	83	38	30	13
8	79	35	29	11
Maximum	89	40	31	13

TABLE III. MAXIMUM DIAMETERS OBTAINED FROM REPEATED SAMPLING ON
PROXIMAL SLOPE OF OUTER LATERAL MORAINE OF LEWIS GLACIER.

(in millimeters)

Sample number	1	2	3	4	5	6	7	Maximum diameter
A. minuscula	84	84	79	65	95	79	79.5	95
R. geographicum	6.0	9.0	14.0	9.5	8.0	14.0	11.0	14
R. jemtlandicum	—	—	25	—	25	30	21.5	30

The steady decrease that occurs in maximum diameters above the shoreline as proximity to the ice cap increases (Table IV) is interpreted as a reflection of the progressive retreat of the proto-Barnes Ice Cap (Ives and Andrews, 1963) and in no way as the result of environmental changes occasioned by the degree of proximity. The proof of this lies in the diameters found below the shoreline, which have a maximum variability of only 3 millimeters over the three areas shown in the table. In areas less than a kilometer from the col, which upon exposure led to the drainage of the lake at 510 meters above sea level, the values above and below the shoreline are close (Table 4). These findings are of great importance in an assessment of the validity of lichenometry because they support the basic assumption that lichens on synchronously exposed substrates are about the same in maximum diameter. Thus the conclusion for the area under study is, as Table 4 suggests, that the isophyses are also isochrones. Hence, on the basis of the closeness of maximum lichen diameters, a relative chronology for this area can be sketched. Studies have also shown that what were once snowbank areas came into existence as such after the drainage of Glacial Lake Lewis. In several localities they extend below the shoreline, thus indicating that after the glacial lake was drained, climatic conditions were more severe than at present.

TABLE IV. VARIATION OF MAXIMUM DIAMETERS OF R. GEOGRAPHICUM ABOVE AND BELOW GLACIAL LAKE LEWIS SHORELINE (510 METERS ABOVE SEA LEVEL. (in millimeters)

Approximate distance of stations away from Barnes Ice Cap	30 km NW	12 km WNW	6 km NNW
Location	Rimrock Valley	Middle Isortoq Valley	Striding Valley
Above shoreline	130	71	44
Below shoreline	41	38	38

Calculation of Lichen Growth Rates

One of the main problems was to establish for the more common lichens a preliminary growth rate on which to base the first estimate of the age of the various substrates. On the revisiting of specific lichen stations in five to 10 years, the results would be checked by means of selected true-scale photographs. The growth rates make possible the conversion of relative (Figure 1) into absolute isochrones. Around the Barnes Ice Cap, the technique finds its prime application in the establishment of a relative chronology based on lichens. The calculation of growth rates for comparing the glacial history of the study area with the standard North American, European and Greenland chronologies, although important, is perhaps secondary.

Air photographs taken over the Lewis Glacier in 1948, 1959 and 1961 at scales of 1:40,000 and 1:60,000 make possible the accurate positioning of the ice margin for these years. From this, it is possible to calculate the total vertical and horizontal recession of the glacier for the period covered. Vertical profiles were carried from eight positions along the northern margin of the ice in 1963 and were surveyed by a Wild T 12 theodolite and stadia rod or an Abney level and steel tape. The profiles were taken to the 1948 limit and were then extended to the point where the first *A. minuscula* were noted. The basic data (Table 5) from which the growth rate of this species has been derived indicate that the mean vertical thinning of the glacier has been 1.5 meters a year for the 15-year period extending from 1948 to 1963, with a range of 0.84 meter. The horizontal recession has amounted to 32 meters a year, with a range of 5 meters. The division of these rates into the total distance between the 1963 margin and the point where the first lichens were observed gives, on the assumption of an approximately linear recession prior to 1948, the age of the substrate. Table 5 demonstrates that substrates of 30 to 40 years are associated with *A. minuscula* 10 millimeters or so in diameter. In estimating the growth rate, it was assumed that the lichens would not start to grow immediately upon deglacierization, and a period of 10 years was allowed for the initial colonization. Table 5 illustrates the steps involved in the calculation.

Despite the various assumptions made, the consistency of the results is encouraging. They give an optimum growth rate of 0.52 millimeter a year and an annual mean of about 0.40 millimeter.

TABLE V. CALCULATION OF GROWTH RATE OF A. MINUSCULA.

	1	2	3	4	5	6
Location	Retreat since 1948 (15-yr. period)	Retreat (m per yr.)	Distance from 1948 limit to first Alectoria thallus (m)	Diameter of Alectoria thallus (mm)	Age of substrate (yr.)	Growth rate on assumption of 10 years for initial colonization (mm per yr.)
		$\dfrac{\text{column 1}}{15 \text{ yr.}}$			$\dfrac{\text{column 3}}{\text{column 2}} + 15 \text{ yr.}$	$\dfrac{\text{column 4}}{\text{column 5}-10 \text{ yr.}}$
		vertical	vertical			
Profile 1	21m/v	1.4	41	10	44	.29
Profile 2	22m/v	1.5	15	5	25	.33
" 2	"	1.5	29.5	10	34	.42
Profile 3	24.5m/v	1.6	38	10	39	.34
Profile 4	15m/v	1.0	17	9	32	.41
" 4	"	1.0	61	20	76	.30
Profile 5	24.3m/v	1.62	26	11	31	.52
" 5	"	1.62	58	19	50	.48
Profile 6	27m/v	1.8	27	10	30	.50
		$\dfrac{\text{column 1}}{11 \text{ yr.}}$ horizontal	horizontal			
Lewis snout, profile 1 ('48 to '59)	370m/h	34.0	830	10	36	.38
South edge, profile 2	340m/h	31.0	820	10	37	.38
		$\dfrac{\text{column 1}}{15 \text{ yr.}}$				
Profile 3 ('48 to '63)	425m/h	28.0	450	10	31	.48
Mean		l.5v, 31.0h				.39
Range		.80v, 5.0h				.23
Standard deviation						.028
Number of samples = 12						

v = vertical thinning or recession. h = horizontal recession.

NOTE: The 1948 position of the margins of the Lewis Glacier was traced by means of air photographs.

Three profiles (2, 4 and 5) were extended to include areas beyond the first lichen occurrence so as to make possible a rough check of the expected diameter. The results obtained in two instances suggest a decrease in the rate of lichen growth.

The extent to which extrapolations can be made on the basis of these results depends on the part of the lichen growth curve that is being measured by the method described in this paper. If the values pertain to the "great" period, the estimate of the age of substrates beyond this period will be too small; if they fall in the period prior to the great period, the estimate will be too large. The lack of detailed historical records within the Arctic has so far handicapped research, and the basic form of the growth-rate curve is not known. The growth rate of 0.40 millimeter a year for *A. minuscula* is comparable to the 0.56-millimeter-a-year rate for the Disko area of Greenland (Beschel 1963). In accordance with the general theory underlying the sampling of maximum-diameter lichens, a growth rate of 0.52 millimeter a year should be adopted for the northwestern margin of the ice cap. This, however, has not been used because small-scale local conditions could affect the amount of glacier recession and thus cause error. This study will therefore be based on the mean estimated growth rate. Table VI gives the dates thus obtained for the *A. minuscula* isophyses in the Lewis Glacier vicinity. It should be noted that if the postulated error were excluded from such a date, the result would be the age of the isophyse line as calculated from a growth rate of 0.52 millimeter a year.

TABLE VI. DATES OF ISOPHYSES OF A. MINUSCULA IN FIGURE 2.

Isophyse	Present Age		Date A.D.
10 mm	25	5	1940*
30 mm	75	15	1890
50 mm	125	25	1840
70 mm	175	35	1790
90 mm	225	45	1740

*All values are given to the nearest decade.

These results can be checked by comparing the known ages of willow boughs collected in selected localities with the corresponding isochrones. Diligent searching showed that arctic willow (*Salix arctica* Pall.) colonizes substrates within three years of the disappearance of ice from an area. The lack of a progressive increase in willow age with elevation, however, indicated that a longer period is needed for successful and continual growth. Willow burls were collected and cut into thin sections (50 microns in thickness) in the laboratory on a sliding microtome, stained with safranin and mounted between glass slides. The sections were then photographed and enlarged 10 times so that the number of rings along four radii could be easily counted. A previous study of arctic willow had shown discrepancies in the number of rings along different radii (Beschel and Webb, 1963) because of lenticular growth (Studhälter *et al.*, 1963), but the specimens from central Baffin Island did not differ in any count by more than one ring. The maximum-age willows indicate the minimum age of the isophyses, but as no willow found in the area is more than 40 years old, their usefulness is restricted. Thus, at most, these willows provide a check on the estimated age of the 10-millimeter and possibly on that of the 30-millimeter *A. minuscula* isophyses. A willow that is 20 years old lies near the 10-millimeter line, but no willows approaching 15 years of age could be found immediately above the 1948 ice-margin position. It seems that between five and 10 years must elapse before the growth of a willow is assured. Thus an age of 25 to 30 years can be expected for the 10-millimeter isophyse or the related isochrone. This compares favorably with the age of 25 years given by a growth rate of 0.40 millimeter a year but is perhaps too near the 10-millimeter-line age based on the optimum rate of 0.52 millimeter a year. The indications are that the area bounded by this isophyse has been icefree for 20 years or less.

The willows that lie above the 30-millimeter line have, respectively, 37 and 39 annual rings. These represent minimum ages for the area bounded by the 30-millimeter and 50-millimeter lines. Another willow was collected at a station where *A. minuscula* was measured at 30 millimeters. With allowance for a period of initial colonization, the minimum age for that site is between 42 and 47 years. The mean estimated growth rate dates the 30-millimeter isophyse at 1890. There is thus a 30-year discrepancy between the two estimates, the most likely reason being that the willow sampled is somewhat younger than the substrate. The discrepancy might indicate that the growth rate of *A. minuscula* increases rapidly after a diameter of 10 millimeters has been passed, but the vertical spacing between the isophyses (Figure 2 A and B), which results from the rate of glacier recession, and the variations in the lichen-growth-rate curve show no pattern that can be attributed to growth variations. The geomorphological evidence indicates, in fact, a relatively constant recession, with only a few stillstands or readvances. In short, the results of the attempt to relate isochrones to willow ages are satisfactory for the initial isochrone, which is shown to date from about 1935-40, but the results of the check on the 30-millimeter line are open to reservations.

With the use of the 0.40-millimeter-a-year growth rate in this study, an error allowance of $\pm$ 0.10 millimeter a year is proposed as consistent with the variations in Table V. Though absolute dates are given for specific morainic or other features, the importance of the study lies in its suggestion of the usefulness of lichenometry for relative dating within the 2,800-square-kilometer area.

Because other lichens are rare or totally absent when A. *minuscula* reaches sizes of 5 to 10 millimeters, a method other than the foregoing had to be developed to determine their growth rates. When ratios of one species to another were plotted as frequency diagrams, an apparently normal curve resulted. The fact, however, that a ratio results from two variables differentiates it from the normal curve. When this indirect method of establishing growth rates is used, it has to be assumed that the linear growth phases of both species being examined have a common origin. For four combinations of species (Table VII) the mean ratio and standard deviation of the samples were derived. The median and modal values are all less than the mean value, largely owing to the presence

TABLE VII. DATA USED TO CALCULATE GROWTH RATES OF R. GEOGRAPHICUM, R. JEMTLANDICUM AND UMBILICARIA PRODOSCIDEA.

Species	Mean ratio	Standard deviation	Number of sample ratios	Growth per yr.
R. geographicum: Alectoria[1]	1:7.0	2.8	97	R. geographicum 0.057 ± 0.02
R. jemtlandicum: Alectoria[2]	1:2.98	1.01	105	R. jemtlandicum 0.135 ± 0.04
R. geographicum: R. jemtlandicum	1:2.00	1.06	122	R. geographicum 0.067 ± 0.02
U. proboscidea: Alectoria	1:2.78	1.1	131	U. proboscidea 0.17 ± 0.07

[1]The ratio is based on samples in which the *Alectoria* diameter is greater than 60 millimeters and less than 120 millimeters.
[2]A. *miniscula* is over 30 millimeters.
Note: A growth rate of 0.4 millimeter a year is assumed for A. *miniscula.*

of samples one species of which was poorly represented. The use of the mean ratio makes it possible to estimate the growth rate of an individual species. A check on the relative reliability of the ratio technique can be made by determining the growth of *R. geographicum* directly from its mean ratio with *A. minuscula,* computing that of *R. jemtlandicum* from *A. minuscula* and then deriving the growth rate of *R. geographicum* from the ratio between it and *R. jemtlandicum.* The results can be seen in Table VII. By direct calculation the growth rate of *R. geographicum* is 0.057 millimeter a year whereas the interposition of another species makes the rate 0.067 millimeter. The similarity in the two estimated growth rates indicates the measure of confidence that can be placed in them. For their absolute veracity, of course, they still depend on the determined rate for *A. minuscula,* and thus their accuracy is relative. On the basis of the variations within the sample ratios, confidence limits can be set up; these have been taken from the internal variation within each frequency distribution and are not standard errors of estimate. The suggested ranges of error can be seen in Table VII and Table VIII. The latter lists the estimated age of each of the *R. geographicum* isophyses of Figure 1. Further investigations after an interval of perhaps 10 years will probably result in modifications to Table VIII, but the relative chronology will stand.

TABLE VIII. DATES OF ISOPHYSES OF R. GEOGRAPHICUM IN FIGURE 1.

Isophyse	Present Age	Date
18 mm	315 ± 90 years	A.D. 1645
30 mm	530 ± 150 years	A.D. 1440
70 mm	1,200 ± 360 years	A.D. 750
130 mm	2,300 ± 650 years	350 B.C.

Note: For the period 1400-1960, values are given to the nearest decade. Older values are to the nearest half century.

Recent History of Lewis Glacier

The intensive sampling done in the environs of the Lewis Glacier enables a series of vertical and horizontal profiles of the former glacier to be drawn. Observations were spaced at increments of between 25 and 50 meters. The maximum diameters of *A. minuscula* were plotted on a contour map of the area at an original scale of 1:12,000, and isophyses were drawn by joining points of equal maximum diameters. These lines represent a time picture of the Lewis Glacier during five successive phases represented by the diameters of 90, 70, 50, 30 and 10 millimeters. The regular pattern of the isophyses about the present glacier suggests that they accurately portray the historical Lewis Glacier. Additional confirmation of this is the similarity between the gradients of the marginal and submarginal glacial drainage channels and the isophyses. The picture presented by the 90-millimeter isophyse (A.D. 1740±45 years) is that of the Lewis and Triangle glaciers lying combined immediately behind the outer end moraines and frontal delta, upon which *A. minuscula* reaches a maximum of 114 millimeters (A.D. 1680±60 years). Recession, indicated by the progressively smaller lichen diameters downslope and upvalley, seems to have been fairly constant (Figure 2). Figure 2(C) indicates the rate of horizontal recession. From A.D. 1680 to A.D. 1938 the glacier seems to have retreated at a linear rate of about 10 meters a year, but from 1938 to the present day this figure increased to 43 meters a year. This marked increase, however, is due not so much to climatic factors as to the initiation of the marginal Flitaway Lake drainage, which must have led to considerable undercutting and collapse of the northern margin and snout.

Figure 2(A and B) illustrates the gradient of the isophyses for both the northern and the southern slopes of the Lewis Valley. The isophyses have been constructed by projecting points to a line A-A′ along the centre of the valley and A′-A″ along the crest of the glacier. A most noticeable point is the parallelism between the present glacier surface and margin on the one hand and the various isophyses on the other. Both gradients average 1:15. Figure 2(A), which represents the northern side of the valley, shows very clearly the effects of the small lobe of the Lewis that pushed over Elegans Hill toward the Striding River valley and did not leave completely until after the 50-millimeter phase. Figure 2(A) also suggests that between the 50- and 30-millimeter phases there was a period of rapid vertical recession that ended with a series of small

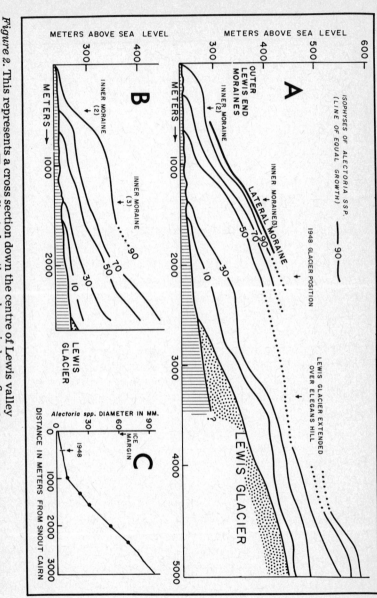

Figure 2. This represents a cross section down the centre of Lewis valley, with isophyses included. The gradient along these, shown in lower figures, is similar to the gradients of such marginal features as lateral moraines and glacial drainage channels. Part on lower right indicates distance from the present snout in meters plotted against the A. *miniscula* maximum diameters.

stillstands, which led, in their turn, to the formation of three end moraines. The period is dated as extending from A.D. 1840 to A.D. 1890, with the stillstand at the latter date. The isophyses from the southern side of the valley (Figure 2(B)) have gradients similar to those of their counterparts to the north but are marked by a noticeable inflexion where the margin of the Triangle Glacier and the ice cap was contained by the hill immediately west of the glacier.

Recent History of Northwestern Margins of Barnes Ice Cap

The margin of the Barnes Ice Cap was studied from Umbilicaria Lake to the Lewis Glacier and then northward to a point 4 kilometers east of the upper Striding River. Figure 1 shows the location and extent of the principal end moraines and former glacial lakes. The drawing of the 130-millimeter isophyse of *R. geographicum* was difficult because of the weathering of rock surfaces. There is thus less probability of finding increasingly larger thalli with increasingly older substrates after the first 80-millimeter thallus has been encountered in a transect. The 130-millimeter line was drawn on lichenometrical and geomorphological evidence. On the plateau surface, west of Umbilicaria Lake and 12.5 kilometers from the present ice margin, there is a massive end moraine upon which *R. geographicum* reaches 133 millimeters. At the lower end of the lake section of the King River there is a series of large end moraines, the outer one being associated with a high outwash terrace, which at present is 16 meters above river level. Lichen stations established on the moraine and on the terrace had maximum readings of 85 millimeters, but the occurrence of maximum values of 89, 98 and 100 millimeters at three sites 10 kilometers to the east indicates that the outer moraine is at least older than the 100-millimeter lichen and is probably contemporaneous with the 130-millimeter isophyse. Similarly, it is possible to map on the air photographs the extension of the outer King River moraines across the interfluve into the Isortoq Valley. Here, a conspicuous end moraine at the junction of the King and Isortoq rivers is associated with an elevated outwash terrace 32 meters above the Isortoq and 16 meters above the King River. The maximum *R. geographicum* measurement obtained on the Isortoq end moraine was 100 millimeters. The correlation of the terrace systems, however, supports the theory that the outer King River moraine and the Isortoq moraine are, in fact, contemporaneous. The latter moraine lies 31 kilometers from the Lewis Glacier. There has been no field investigation in the area between the Isortoq moraines and Rimrock Lake, and the 130-millimeter isophyse is sketched from air-photograph interpretation. In the Rimrock Lake area it seems possible that the 555-meter col was being used as a lake spillway, but there is no evidence to suggest that the Glacial Lake Lewis spillway had been exposed.

The 130-millimeter isophyse corresponds to the age of 2,300 years (Table VIII). This date places the large end moraines associated with this phase within the Sub-Atlantic climatic period.

The 70-millimeter isophyse of *R. geographicum,* estimated as marking the late-eighth-century ice-cap margin, is not significant in the glacial history of the ice cap although a significant still-stand and moraine-construction period seems to have antedated it by about 100 years. Toward the north, the lichen stations indicate that the Glacial Lake Lewis spillway was in use and that ice-dammed lakes were forming between the retreating ice margin and the watershed of the Isortoq and Striding rivers at the head of Arenaria Valley.

The 30-millimeter (A.D. 1440±150 years) line marks the margin of the Barnes Ice Cap as it was immediately after the drainage of Glacial Lake Lewis. By this time the ice-cap margin near the present Umbilicaria Lake must have been very close to the present ice-cored moraines, but near the Lewis Glacier it was 5 kilometers from the present margin. Over the last 500 years the retreat of the ice cap has therefore been progressively more extensive on the north and northwest.

The outer moraine of the Lewis Glacier has a series of associated end moraines indicating an important stillstand that occurred about A.D. 1680 ± 60 years (120-millimeter *A. minuscula*). There are well-defined end moraines at the head of King Valley, and the ice-cored moraines that front the present ice-cap margin to the south appear to have a similar age. The growth of the small ice caps north of the Barnes Ice Cap were also affected by the climatic deterioration. Additional evidence for a deterioration of climate on Baffin Island in the seventeenth century has been noted by G. Falconer, of the Geographical Branch. The age of mosses exposed by the recession of a small ice slab 200 kilometers northwest of the Barnes Ice Cap has been given as 330 ± 75 years (I–1204, GF–63–V–80). The exposure occurred in 1961 (G. Falconer: personal communication, 1964).

The computed ages of the main end-moraine systems are pre-sented in Table IX, where they are compared with readvances or stillstands on Baffin Island (Hale, in Ward, 1952; Thompson, 1954) and Greenland (Weidick, 1959 and 1963; Beschel, 1958). There is close agreement between the records of glacial fluctuations in southwest Greenland and those of the recent moraines of the north-western margins of the Barnes Ice Cap. The date A.D. 1890, ascribed to the third inner Lewis moraine, is of special significance: it seems to have been the date of a world-wide glacier stillstand (Ahlmann, 1953) recognizable in nearly all areas. The date 1920, assigned to the innermost moraine, is also critical, the records indicating a sharp rise in Arctic temperatures after that year (Mitchell, 1961; Ahlmann, 1953).

TABLE IX. SUGGESTED MORAINE DATES FOR BAFFIN ISLAND AND GREENLAND.

	Baffin Island		Greenland	
Barnes Ice Cap (N and NW)	Barnes Ice Cap (SE, Hale, 1952; Sim, 1961)	Pangnirtung Pass (Thompson, 1954)	(SW, Weidick, 1959 and 1963)	(SM, Beschel, 1961a)
A.D. 1920				A.D. 1920–25
A.D. 1905				A.D. 1890–95
				A.D. 1870–80
A.D. 1890		A.D. 1883 ?	A.D. 1890	A.D. 1850
	A.D. 1825	A.D. 1820 ?		A.D. 1820
			A.D. 1800	A.D. 1770–80
A.D. 1740			A.D. 1750	A.D. 1740–50
				A.D. 1680
A.D. 1650				A.D. 1600
A.D. 1300				
A.D. 950				
A.D. 650	A.D. 710			
	A.D. 200			800–500 B.C.
450 B.C.				>2000 B.C.
			Sub-Atlantic	Readvances after thermal optimum

The dates determined by Harrison (1964) for end moraines in the Bruce Mountains of Baffin Island indicate similarities be-tween the date pattern of the Barnes Ice Cap and that historically substantiated for Greenland. Lichen studies made by Hale (Ward, 1952) at the southeastern end of the ice cap and by Sim (personal communication, 1961) indicate that the two ends differ greatly in the history of fluctuations. Sim, in measuring the maximum diameters of 100 *R. geographicum* above and below the prominent former glacial-lake shoreline in the Generator Lake basin near the present glacier margin, found a maximum diameter of 73 millimeters below the shoreline and 100 millimeters above it. This suggests that the lake was drained at 700 A.D. $\pm$ 350 years (if a growth rate of 0.057 mm/year is used), at which time the ice margin could have been only 6 kilometers from its present position. Hale's evidence, since the outer moraine with 50-millimeter *A. minuscula* abuts against a zone of mature lichen colonization, suggests that the southeastern margin had readvanced. Recalculation of his suggested lichen growth rate in keeping with the growth rates established in this paper points to A.D. 1830 as the date of this moraine. In contrast, the western and northwestern margins have been retreating steadily inland since 5,000 years ago, when the Baffin ice may still have reached the inner parts of bays leading into Foxe Basin (Ives, 1964). The margin has responded to climatic fluctuations during which end moraines have been formed, some dating from A.D. 550-850, when glaciers were less extensive in Europe than they are today. There are Alaskan moraines, however, that date from this period (Heusser and Marcus, 1964). Thus the history of the western and northwestern margins is one of linear recession with no major readvances but with stillstands. The southeastern margin, on the other hand, seems to have readvanced extensively at the beginning of the nineteenth century.

Assessment

In the vicinity of the ice cap the topography is subdued, its maximum relief being 300 meters and often much less. The valleys are broad and open and the area has no extreme topographical irregularities, which through insolation or other effects could cause marked differences in lichen growth. Because of these factors and the presence of the former Glacial Lake Lewis drainage basin, which extends north and south from near the ice cap for more than 36 kilometers, the technique of lichenometry could be appraised. The results were very encouraging: three widely separate areas within the Glacial Lake Lewis basin showed no difference in the maximum diameter of *R. geographicum,* and the steady decrease in diameter above the shoreline could be interpreted as a reflection of ice recession. The large number of stations has made it possible to draw a series of isophyses, which represent relative isochrones and permit the correlation of events along the northwestern margin. Absolute isochrones still await firm establishment, but the suggested dates for the principal phases do correlate with

Greenland chronologies. At the very least, it has been established that the 'little ice age' maximum occurred during the seventeenth century or perhaps early in the eighteenth. The fluctuations have been imposed on a general pattern of marginal retreat that has been going on for between 5,000 and 7,000 years, with intermittent stillstands or readvances.

The writers suggest that geomorphologists can have great success in using lichenometry for the relative dating of recent glacial events in areas that lack datable organic material for the time span involved. For future research the stations will be extended right around the margin of the ice cap.

Bibliography

Ahlmann, H. W., son, "Glacier variations and climate fluctuations," *Am. Geog. Soc.* ser. 3, (1953), p. 51.

Beschel, R. E. "Lichenometric in Gletschervorfeld," *Ver Schutz Alpenpfl.*, Jahrb. 22, (1957), pp. 164-85; "Lichenometrical studies in West Greenland," *Arctic*, Vol. XI, no. 4 (1958), p. 254; "Dating rock surfaces by lichen growth and its application to glaciology and physiology (lichenometry)," *Geol. of the Arctic*, Vol. XI, (Proc. First Internat. Symposium on Arctic Geol.), Gilbert O. Raasch (ed.), Univ. Toronto Press (1961), pp. 1044-62; "Botany: and some remarks on the history of vegetation and glacierization," *Jacobsen-McGill Arctic Research Exped. to Axel Heiberg Island, Prelim. Rept. 1959-1960,* F. Müller (ed.), McGill Univ., Montreal (1961), pp. 179-99; "Observations on the time factor in interactions of permafrost and vegetation," *Proc. First Can. Conf. on Permafrost,* Nat. Research Council Can., Assoc. Committee on Soil and Snow Mech., Tech Memo, 76 (1963), pp. 43-56; "Geobotanical studies on Axel Heiberg Island in 1962," F. Müller et al., *Axel Heiberg Prelim. Rept. 1961-1962,* McGill Univ., Montreal (1963), p. 18.

Beschel, R. E., and Webb, Deirdre, "Preliminary growth ring studies," F. Müller et al., *Axel Heiberg Island Prelim. Rept. 1961-1962,* McGill Univ., Montreal, (1963), p. 10.

Dahl, E., "Studies in the macro-lichens of West Greenland", *Medd. Grønland*, Vol. CL, no. 2 (1950), pp. 1-176.

Goldthwait, R. P., "Study of ice cliff in Nunatarssuaq, Greenland," *Tech. Rept. 39,* U.S. Army C.R.R.E.L. (1960) pp. 12-108.

Hale, M. E. "Lichens from Baffin Island", *Am. Midland Naturalist* 51, (1954), pp. 232-64; "Studies in lichen growth rate and succession," *Bull. Torr. Bot.* Club (1959), pp. 126-9.

Harrison, D. A. "A reconnaissance glacier and geomorphological survey of the Duart Lake area, Bruce Mountains, Baffin Island, N.W.T.," *Geog. Bull.*, 22 (1964), pp. 57-71.

Heusser, C. J. and Marcus, M. G., "Historical variations of Lemon Creek Glacier, Alaska, and their relationship to the climatic record," *J. Glaciol.*, 5 (1964), pp. 77-86.

Ives, J. D., "Indications of recent extensive glacierization in north central Baffin Island, N.W.T.," *J. Glaciol.*, Vol. IV, no. 32 (1962), pp. 197-206; "Deglaciation and land emergence in northeastern Foxe Basin, N.W.T.," *Geog. Bull.* 21, (1964), pp. 54-65.

Ives, J. D., and Andrews, J. T., "Studies in the physical geography of north-central Baffin Island, N.W.T.", *Geog. Bull.* 19 (1963), pp. 5-48.

MacKay, D. K., and Cook, F. A., "A preliminary may of continentality for Canada", *Geog. Bull.*, 20 (1963), pp. 76-81.

Mitchell, J. M. J., "Recent secular changes of global temperature", *N.Y. Acad. Sci.*, Vol. 95, no. 1 (1961), pp. 235-50.

Runemark, H., "Studies in *Rhizocarpon*. I—Taxonomy of the yellow species in Europe", *Op. Bot.*, Vol. II, no. 1 (1956), p. 152.

Stork, A., "Plant immigration in front of retreating glaciers with examples from Kebnekajse area, northern Sweden", *Geog. Ann.* Vol. XLV, no. 1 (1963), pp. 1-22.

Studhalter, R. A., Glock, W. S., and Agerter, S. R., "Tree growth: some historical chapters in the study of diameter growth", *Bot. Rev.* 29 (1963), pp. 245-365.

Thompson, H. R., "Pangnirtung Pass, Baffin Island: an exploratory regional geomorphology", Unpub. Ph.D. thesis, McGill Univ., Montreal (1954), p. 227.

Ward, W. H., "The glaciological studies of the Baffin Island Expedition 1950: Part 2—The physics of deglaciation of central Baffin Island, with appendix by M. E. Hale", *J. Glaciol.*, Vol. II, no. 11 (1952), pp. 9-23.

Weidick, A., "Glacial variations in West Greenland in historical time: Part I—southwest Greenland", *Medd. Grønland*, Vol. 158, no. 4 (1959), p. 196; "Ice margins in the Julianehab district, South Greenland", *Medd. Grønland*, Vol. 165, no. 3 (1963), p. 133.

5

On the Vertical Extent of Glaciation in Northeastern Labrador-Ungava

O. Løken

The Labrador coast (Figure 1) north of about 58° N is steep and mountainous; in many places precipitous cliffs rise to altitudes exceeding 600 m. The coastline is cut by numerous fiords extending inland for as much as 70 km between the high mountains, which reach to altitudes of more than 1,600 m just south of Nachvak Fiord, but become lower toward the north, where they reach a maximum height of about 500 m on Killinek Island. The mountains are highest within approximately 45 km of the outer coast, and become lower farther inland as the alpine forms of the coastal area gradually give way to a more rounded and gentler relief. The main fiords continue inland as deep U-shaped valleys until they reach the water divide toward Ungava Bay, a divide which lies well to the west of the highest mountains, except in the area inland from Ramah Bay.

Along the coast a great number of low islands and submerged reefs form the inner part (a strandflat?) of the continental shelf, which extends along the coast. The 200-fathom contour line lies generally about 100 miles offshore, but in the northern section it turns toward the coast, and follows closely the 60½° N parallel to within approximately 20 km off the coast. The continental slope falls off to the great depth of 1,200-1,500 fathoms in the Labrador Sea.

First published in *Canadian Geographer*, Vol. VI, nos. 3-4 (1962), pp. 106-19.

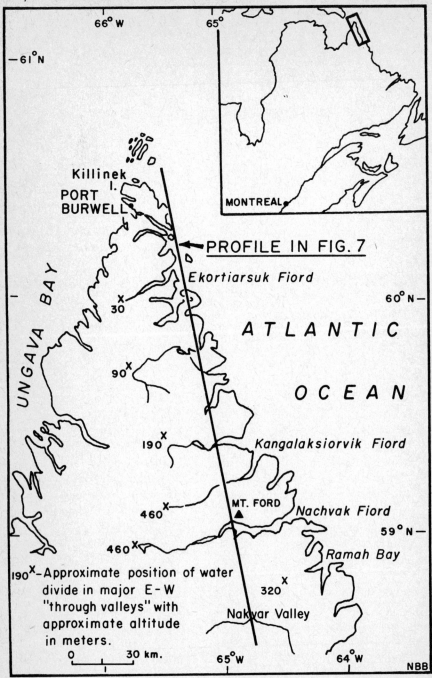

Figure 1. Northeastern Labrador-Ungava showing location of profile in *Figure 7.*

This part of the Labrador coast is very similar to the west coast of southern Norway, and the problem of whether the high coastal mountains situated close to a deep ocean have been glaciated is common to both areas. The question has been, and still is, a matter of considerable controversy. In northern Labrador, the problem has been discussed by several writers, but the number of actual field studies is limited.[1]

The glacial history of the area has been studied mainly by Ives and Løken, and their results show that the last glacial period was dominated by a number of large outlet glaciers in the main valleys.[2] These glaciers flowed eastward from the large ice sheet to the west and toward the Atlantic Ocean. Cirque glaciers occurred in the highest mountains along the coast and formed local centres of ice dispersal, but they were of minor importance compared with the flow from the west.

An abundant amount of glacial-morphological evidence shows that the lower areas were certainly inundated by glaciers, but at higher levels the evidence is considerably sparser. This article is concerned mainly with those areas.

The Area Between Kangalaksiorvik and Ekortiarsuk Fiords

During the summers of 1959 and 1960 the writer worked in this area and several mountains were climbed in order to determine the vertical extent of glaciation. When the distribution of glacial-morphological evidence was considered, it was found that the mountain slopes could be divided into three, mainly horizontal zones.

---

[1] See Coleman, A. P., Extent and thickness of the Labrador ice sheet (*Bull. Geol. Soc, Am.*, 31, pp. 318-28), Northeastern part of Labrador and New Quebec (*Can. Dept. Mines and Geol. Survey*, Mem. no. 124, 1921, p. 68); R. Daly, Geology of the northeast coast of Labrador, (*Bull. Harvard Mus. of Comp. Zool.*, no. 38, 1902, pp. 205-70); J. D. Ives, Glaciation of the Torngat Mountains, Northern Labrador (Arctic, 10 (2), 1957, pp. 67-87), Glacial Geomorphology of the Torngat Mountains, Northern Labrador, (*Geog. Bull.*, no. 12, 1958, pp. 47-75); O. H. Loken, Field work in the Torngat Mountains, Northern Labrador (*McGill Sub-Arctic Res. Papers*, no. 9, 1960, pp. 61-73), Deglaciation and postglacial emergence of northermost Labrador (Unpub. Ph.D. thesis, McGill University, 1961); N. E. Odell, The mountains of Northern Labrador (*Geog. Jour.* 82, 1933, pp. 193-211, 315-26), The geology and physiography of northernmost Labrador in A. Forbes, Northernmost Labrador mapped from the air (Am. Geog. Soc., sp. publ. no. 22, 1938, pp. 187-215).

[2] See above: Ives, 1957, 1958; Loken, 1960, 1961. See also O. H. Loken, Notes on the geomorphology of the Ramah Bay area, northern Labrador, *McGill Sub-Arctic Res. Papers*, no. 9 (1960) pp. 75-82.

The lowest zone is characterized by abundant evidence of glacial action: fine striations, polished rock faces, terminal and lateral moraines, lateral terraces, and a large amount of till and glacio-fluvial deposits which cover the bedrock in most places.

In the middle zone rock exposures are much more frequent, and in general it shows sign of substantial weathering; no striation was observed, but smoothed rock faces, believed to have been ice-scoured, were occasionally found. Figure 2 shows a typical bedrock

Figure 2.

exposure in this area. Small lateral moraines and terraces can be found, but they are much smaller, have subdued forms, and are very different from those in the lowest zone. On the border line between these two zones, a small readvance moraine was found in some places, but it was always poorly developed.

The frequent bedrock exposures of the middle zone give way to mountain-top detritus or block fields, which cover the upper zone. Here again bedrock exposures are few, and those which occur are all restricted to steep slopes. Figure 3 shows a typical locality from this zone with its sharp angular boulders. The size of the boulders and also their "angularity" change from locality to locality, probably reflecting changes in the underlying bedrock.

Figure 3.

The term mountain-top detritus implies that the surface material has been derived from the subjacent bedrock by weathering *in situ*, and a profile should therefore show a continuous transition from the loose detritus on top to the unweathered bedrock below. No profile where this could be studied was dug, nor was any natural profile observed. Nowhere was the boulder field seen in contact with fresh unweathered bedrock, but it is considered that the block field can properly be called mountain-top detritus.

The distribution of the glacial-morphological evidence might be explained by assuming increasing intensity of weathering with increasing height above sea level, but this fails to explain the very rapid transitions which take place at the boundary lines. It is therefore concluded that the boundaries were formed by the valley glaciers during two important phases of the glacial history of the area. The same conclusion has been reached by Ives from areas farther south.[3]

---

[3] J. D. Ives, "Glacial geomorphology of the Torngat Mountains," (1958).

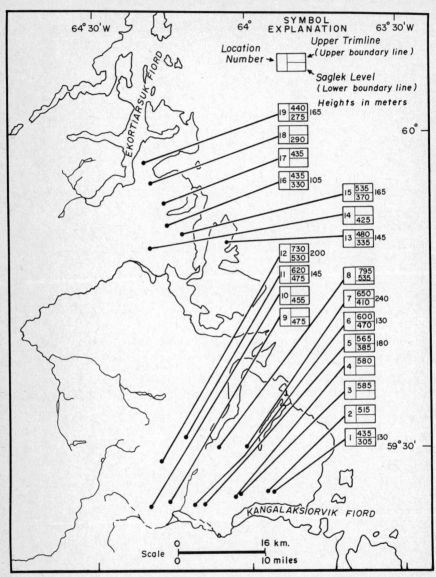

Figure 4. Map showing observations of upper and lower boundary lines.

All places where ascents were made, and one or both of the boundary lines between the zones were found, have been plotted on Figure 4 where the height of the two and also the difference between them are shown. It will be seen that the height of the two boundaries does not change very much. They are almost horizontal, although a slight drop toward the north and east can be recognized.

The height of the intermediate zone varies from 105 m to 240 m, a rather large variation. The maximum difference is, however, to a large extent caused by one unusually high elevation of the upper boundary, and it is worth noting that the area where it occurs is very high and held substantial cirque glaciers during the height of the glaciation.[4] It is believed that these glaciers contributed to the main ice sheet and thus caused the high elevation of the upper boundary. This locality is consequently not considered representative.

No obvious reason explains the spread of the remaining differences, especially not for the low elevation of 105 m immediately adjacent to the elevation of 165 m, but it is believed that the local topography and solifluction have been of importance. The average of the nine differences is 150 m (500 ft).

In the upper zone, no evidence of glacial action such as striation, roche moutonée, moraine, and so on was found. The only evidence which possibly shows that the area was once ice-covered is the occurrence of large boulders, very strange to the general environment, which might be termed erratics. Figure 5 shows several of these. They are gneiss boulders and have possibly been moved by ice.

A similar situation is shown in Figure 6 where the gneiss-granite boulders again are very strange features. In this case, however, the source of at least one of the boulders was evident, as the light rock in the foreground is only weathered loose, not moved from its position in the bedrock where it occurs along a diabase dike. Great care has to be taken in interpreting boulders as erratics, and the last locality made the writer reconsider and change earlier conclusions about erratics.[5] It is very difficult to ascertain whether they are inclusions weathered out of the bedrock below (termed pseudo-erratics by Dahl) or "true" erratics. This difficulty is even more important in an area like this where the geology is only very roughly known, and where the local variation in petrography within

4 O. H. Loken, "Deglaciation and post-glacial emergence of northernmost Labrador," (1961).
5 O. H. Loken, "Field work in the Torngat Mountains" (1960), p. 67.

Figure 5.

Figure 6.

the gneiss-granite complex is very great. Therefore it is almost impossible to determine the source area of a boulder, but this should be known before the boulder is properly termed an erratic.

In sum, it might be stated that the upper zone shows no glacial-morphological feature to prove that it has been glaciated at any time, and, although boulders can be found which might easily be termed erratics, it is not considered that sufficient evidence exists to show beyond doubt that these are erratics.

Dahl has studied the problem of ice-free areas in terms of the clay minerals which occur in the mountain-top detritus.[6] Clay minerals can be formed as secondary minerals owing to a long period of sub-aerial weathering or by hydrothermal activity. In places where the latter way of formation can be excluded, the minerals present in a sample will give an indication of the length of the weathering period.[7]

Four samples were taken from the upper zone of this area. The clay-fraction ($<$ 2 micron) of these has been analysed by X-rays, and the minerals shown in Table I were identified.[8]

TABLE I.

Sample no.	Montmorillonite minerals	Vermiculite	Hydrobiotite	Amphibole	Mica	Chlorite
1525	*		**	*		**
1526	**	**		**	*	**
1527	**	**	**	**	*	*
1528	***	**	**	**	*	**

*Mineral present, but reflection weak.
**Clear reflection.
***Very strong reflection.
The intensity of the reflection is roughly proportional to the quantity of the mineral.

6 E. Dahl, "Weathered gneisses at the island of Runde, Sunmore, Western Norway, Nytt Magasin for Botanikk, 3 (1954), pp. 5-23.
7 M. L. Jackson, S. A. Tyler, A. L. Willis, G. A. Bourbeau and R. P. Pennington, "Weathering sequence of clay-size minerals in soil and sediments. I. Fundamental generalizations", Journ. of Phys. and Coll. Chemistry, 52 (1948), pp. 1237-60.
E. Dahl, "Weathered gneisses at the island of Runde" (1954).
8 O. Gjems, personal communication (1962).
  *: Mineral present, but reflection weak
 **: Clear reflection
***: Very strong reflection
The intensity of the reflection is roughly proportional to the quantity of the mineral.

The samples were taken in four localities all underlain by para-gneisses,[9] and, as no deposit due to hydrothermal activity has been reported from the area, the clay minerals are regarded as weathering products. The presence of the montmorillonite minerals, vermiculite, and hydrobiotite indicates a very long period of weathering.[10] To assess the period involved will necessitate difficult assumptions about past climates and the rate of chemical weathering. This is not yet possible. Dahl, however, on the basis of similar evidence from western Norway states that the weathering must have started prior to the last interglacial period.[11]

The weathering did not necessarily take place in the locality where the sediments are found today; it is possible that they were brought there by a glacier, and for this reason it is important to ascertain whether the block field of the upper zone is mountain-top detritus formed *in situ*. The lack of glacial-morphological evidence in this zone suggests that the last glaciation did not cover this zone, and the clay minerals strongly support this contention.

It was noted above that the boundaries between the three zones were essentially horizontal, but an exception to this has to be made where cirques existed along the main valleys, as several of these have held large cirque glaciers which formed glacial-morphological features well above the upper boundary line.

In several localities it was possible to observe that the main valley glacier had influenced the flow of the cirque glacier. Along the north side of a cirque situated due west of Locality 11 (Figure 2) there extends a well-developed lateral moraine deposited by the former cirque glacier. The trend of this moraine is different from that which it would have had if gravity alone had influenced the flow of the glacier. After having run down part of a slope, the moraine turns toward the north, thus having a convex form toward the cirque rather than a concave. The implication is that something which is no longer in existence blocked the flow of the cirque glacier; this undoubtedly was the former trunk glacier in the main valley.

A similar feature occurs at Locality 4 (Figure 2). On the east side of the opening of a southeast-facing cirque, several moraine

9 M. J. Piloski, "Reconnaissance of northernmost Labrador, Saglek Fiord to Cape Chidley", Unpublished report to British Newfoundland Exploration Co. Ltd., (1954).
10 M. L. Jackson et al., "Weathering sequence" (1948). E. Dahl, "Weathered gneisses at the island of Runde", (1954).
11 *Ibid.*

ridges run in a north-south direction and were evidently formed by a former cirque glacier. West of these moraines a straight ridge of till lies across the small valley leading from the cirque, and this ridge is connected at a sharp angle with still another ridge of till material with which it forms a large L. This last ridge runs southeast and gradually disappears but its trend continues into the boundary between the two upper zones described above.

The following interpretation is given of this locality. The moraines immediately outside the cirque were formed by the cirque glacier, as was the ridge across the valley, but its straight form suggests strongly that another force was involved. It has been pointed out that scattered evidence of glaciation is found throughout the intermediate zone, and it is concluded that the last ridge mentioned above was formed by the valley glacier when it flowed into the re-entrant formed by the small tributary valley. As this ridge is connected with the ridge across the valley it is believed that the straight form of that ridge is accounted for by combined and simultaneous action of the cirque glacier, which flowed toward the southeast, and the re-entrant from the valley glacier, which flowed in the opposite direction. An implication of this is that the cirque glacier and the trunk glacier reached their maximum extent at the same time.

It is of interest to note that one of the moraine ridges mentioned above gradually disappears and continues into the boundary between the two upper zones. This shows that this upper boundary was actually formed by a glacier, as was the moraine.

It was pointed out that a small readvance moraine was occasionally found separating the two lowest zones. Except for the locality just described no such feature was found between the two upper zones. This is rather surprising as glaciers usually form moraines along their outer margin. Dahl suggests that this reflects the position of the firnline at the time.[12] Below the firnline, he points out, the surface of the glacier is convex and the flow vector is directed out toward the periphery of the glacier. This results in a marginal moraine, but above the firnline the surface is concave and the velocity is directed in toward the glacier rather than outward and no marginal moraine will be formed.

---

[12] E. Dahl, "Litt om forholdene under og etter den siste isted i Norge", *Naturen,* no. 7-8 (1948), pp. 232-52.

Comparison with Other Areas

The division of the mountain slopes into three separate zones
with respect to the distribution of glacial-morphological features
has been described earlier from other parts of northern Labrador.
As early as the turn of the century, Daly investigated the area
at the inner part of Nachvak Fiord where he climbed Mt. Ford,
and he states: ". . . below sixteen hundred feet on the west and
south slopes of Mt. Ford, there is a well glaciated zone. . . . At
twenty-one hundred feet and above, the ledges are in sharp contrast
with those lower down. This contrast is such as to enforce the
belief that glacier ice did not reach higher than about twenty-one
hundred feet above the present sea level."[13] He also climbed Mt.
Kaputyat and describes a set of well-developed lateral moraines
found between elevations of 750 ft (230 m) and 1,700 ft (520 m).
The slopes above this level were also glaciated, but between 1,900
and 2,050 ft (580-625 m) "there occurred an abrupt transition
into the region of an interrupted felsenmeer, in no way markedly
different from that on Mt. Ford."[14]

Above the 2,050-ft (625-m) level he found no erratic although
he is well aware of the great variation in rock type, and reports
"fragments of ferruginous gneisses, trap, vein quartz and syenite."[15]

From the area between Kangalaksiorvik Fiord and Komak-
torvik Lakes, Ives[16] reports a distinct border line at 2,000-2,200
ft (610-670 m) a.s.l., which separates a zone with numerous
glacial features from a higher zone where mountain-top detritus
covers all the gentle slopes. He claims that erratics can be found
in the mountain-top detritus.

In Nakvak Valley farther south, Ives[17] found three separate
zones on the mountain slopes. "The upper slopes, except on steep
inclines, are mantled with a deep cover of angular frost-shattered
blocks of bedrock (mountain-top detritus or felsenmeer), which
is so well developed that the underlying bedrock structures are
completely masked."[18] "Between 2,700 ft and 2,300 ft the mountain-top detritus is replaced by glacially-scoured bedrock. Perched
blocks are abundant in this zone and occasional kames, terraces

13 R. Daly, "Geology of the northeast coast of Labrador", (1902), pp. 248-49.
14 *Ibid.*, p. 251.
15 *Ibid.*, p. 247.
16 J. D. Ives, "Glaciation of the Torngat Mountains", (1957).
17 J. D. Ives, "Glacial geomorphology of the Torngat Mountains", (1958).
18 *Ibid.*, p. 53.

and sections of lateral terrace-moraines are found in favoured places; striations have not been preserved."[19] "At an altitude of 2,300 ft., north of Nakvak Lake, the bedrock abruptly gives way to a broad kame terrace . . . and below this, the lower slopes are completely masked with glacial and glacio-fluvial deposits."[20] Also in this area unequivocal erratics were found in the mountain-top detritus. The broad kame terrace which forms the lower boundary was found to slope down toward the east at a rate of 60-90 ft (11-17/1000) per mile, and the lower boundary of the mountain-top detritus was essentially parallel to this.[21]

The similarities between the descriptions of the mountain slopes as they are given by Daly, Ives, and the present writer are striking. The height of the middle zone is also of the same order of magnitude, about 120 m (400 ft) in Nakvak Valley, about 150 m (500 ft) on Mt. Ford, and between 105 m and 200 m in the northern area. The main point of difference lies in the question of erratics in the mountain-top detritus. This discrepancy is not considered to be serious, as the great difficulty involved in identifying erratics has been pointed out. In addition, all observers realize the great variety of boulders which occur in this upper zone, and thus the question of erratics is one of different interpretations of observed facts rather than differing observations.

Another point of difference is the character of the boundary between the two lower zones, which in Nakvak Valley is represented by a broad kame terrace. Small readvance moraines were observed in the corresponding place in the area investigated by the present writer, and the difference seems to be only one of degrees of development.

Due to the similarities in the description of the different zones it is postulated that they are corresponding zones, and that the border lines between them show the position of the ice margin during two distinct phases of the glacial history of this area, which extends from Ekortiarsuk Fiord to Nakvak Valley. Investigations by Andrews in the Okak area farther south gave results similar to those above, and it is thus possible that in the future the two ice margins can be traced over a still larger area.[22]

---

[19] *Ibid.*, p. 54.
[20] *Ibid.*
[21] J. D. Ives, personal communication, (1962).
[22] J. T. Andrews in J. T. Andrews and E. M. Matthew, *Geomorphological Studies in Northeastern Labrador-Ungava*, Ottawa, Dept. of Mines and Tech. Surveys, Geographical Paper no. 29, (1961), pp. 5-16.

Ives called the upper boundary the "upper trimline,"[23] and this term will be used in the following remarks. The lower boundary was named the main kame terrace level, but as this does not fit the area farther north the name Saglek level is used, after the area where it was first described.[24]

In order to show the regional trend of the two levels, the observations available have been plotted on a profile (see Figures 1 and 7). The single observations from the southern areas have been projected into the profile, but in the north several observations have been grouped together, and the average values, considered representative for the geographical centre of each group, have been plotted.

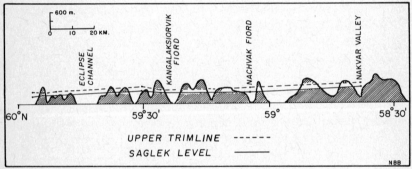

Figure 7. Profile showing regional trend of the upper trimline and the Saglek level.

It will be seen that both levels slope toward the north rather regularly except for the area just north of Kangalaksiorvik Fiord where special conditions existed (see above). The average slope is about 1:350 in the direction of the profile, i.e., N-10-W. This slope is steeper than the one which existed when the two levels actually were formed, because it has been shown that a considerable tilting of the land has taken place in northernmost Labrador in post-glacial times.[25] A maximum upward tilt of 1:1000 in the direction S-25-W has been observed, and when adjustment for this is made the slope of the two lines in the profile should have been about 1:500, but even this is steeper than the original slope, as the tilt of 1:1000 almost certainly does not comprise the total late-glacial and post-glacial tilting of the area.

[23] J. D. Ives, "Glacial geomorphology of the Torngat Mountains", (1958).
[24] O. H. Loken, "Deglaciation and postglacial emergence of northernmost Labrador", (1961).
[25] O. H. Loken, "The late-glacial emergence, and the deglaciation of northernmost Labrador", *Geog. Bull.*, no. 17 (1962), pp. 23-56.

Although the profile in Figure 7 shows the conditions on the east side of the mountains, it is considered that some information about the slope of the ice sheet farther west can be obtained, as the main ice flow came across the Ungava Bay-Atlantic Ocean divide. The outlet glaciers of the southern part of the area occupied troughs with a much steeper slope toward the east than those farther north, and it is believed that these glaciers also had steeper slopes. This means that a profile parallel to the one made, but taken in the water-divide area (see Figure 1) approximately 30 km farther west, would show a steeper sloping ice surface, which implies that the ice surface sloped down toward the Hudson Strait during the periods represented by the upper trimline and the Saglek level. This conclusion is of interest when it is compared with the results of Ives, who has shown that a considerable ice-body remained over the present-day Ungava Bay until a very late phase of the deglaciation.[26]

The profile of the two levels cannot be used to determine the height of the former ice surfaces at any point, because it should be remembered that all the observations so far available are located close to one of the major through-valleys, which lead across the water-divide from west to east. If areas lying between these valleys had been investigated, the indications are that both levels would drop in altitude as the distance away from any of the major valleys increased, that is, as one came into areas which were in the shadow of large mountain complexes.

As shown above, local cirque glaciers existed during the period when the ice reached the upper trimline, and they caused the ice surface to rise locally in limited areas. Only in a few cases, however, did these cirque glaciers overspill the sides of their basins and it is not considered that these glaciers contributed very significantly to the main ice sheet.

The Age of the Upper Trimline and the Saglek Level

Ives has suggested that the ice sheet during the last glaciation reached the upper trimline, and during an earlier part of this period, or in pre-Sangamon times, another more extensive ice sheet deposited the erratics on the highest mountains.[27] Another interpretation of the observations is given below.

---

[26] J. D. Ives, "Former ice-dammed lakes and the deglaciation of the middle reaches of the George River, Labrador-Ungava", *Geog. Bull.*, no. 14 (1960), pp. 44-70.

[27] J. D. Ives, "Glacial geomorphology of the Torngat Mountains", (1958).

Because of the uniformity of the lowest zone with respect to the appearance and distribution of the glacial-morphological features, it is believed that the whole of this zone was glaciated during the Wisconsin period, as at least the Kangalaksiorvik outlet glacier reached tide water along the Atlantic coast less than 9000 years B.P.,[28] and thus part of this zone was ice covered during "classical" Wisconsin.

There is a very noticeable difference in the surface characteristics of the middle and the lowest zones, indicating that the area above the Saglek level was last glaciated at a considerably earlier date than the area below. Dating this glaciation is extremely difficult, but if a parallel to the "pre-classical"-Wisconsin[29] glaciation of the central U.S.A. ever occurred in northern Labrador this would be a reasonable suggestion, as any later date is considered very improbable.

The interpretation of the upper zone is even more difficult, as has been pointed out already. It is contended that sufficient evidence to show that this zone was ever glaciated is not yet available, and, if it were, it must have been in pre-Sangamon time or earlier.

It has been implied above that, because of the regular and consistent relationship between the different zones, the distribution of mountain-top detritus, on which the outlining of the upper zone is mainly based, is a useful tool in determining areas which were ice-free during the last glacial period. This has been suggested earlier by Dahl[30] in Scandinavia and by Ives[31] in Labrador-Ungava, although strong opposition has been expressed by Holtedahl[32] and E. Bergstrøm.[33] The evidence presented in this article demonstrates strongly, in the view of the writer, that the study of mountain-top detritus is an important and useful tool, at least in the area covered by this study. The changes in the appearance of the mountain-top detritus from locality to locality indicate, however,

28 O. H. Loken, "The late-glacial and postglacial emergence of northernmost Labrador", (1962).
29 R. F. Flint, "Glacial and Pleistocene Geology", New York, John Wiley & Sons, Inc., (1957).
30 E. Dahl, "Refugieproblemet og de kvartaergeologiske metodene", Svensk Naturvetenskap, (1961), pp. 81-96. He states that areas with mountain-top detritus were ice-free during at least the two last glaciations, and possibly during the whole Pleistocene.
31 J. D. Ives, "Mountain-top detritus and the extent of the last glaciation in northeastern Labrador-Ungava", Can. Geog., no. 12 (1958), pp. 25-31.
32 O. Holtedahl, "Norges Geologi", Norges Geol. Undersokelse, no. 164 (1953), p. 1118.
33 E. Bergstrom, personal communication (1962).

that care should be taken, and where possible other evidence also be considered. To what extent the results from this area are valid in general is still an open question.

Discussion

J. D. Ives

I would like to take this opportunity of congratulating Dr. Løken on his excellent work in the Torngat Mountains and to wish him success in his forthcoming field season in the same general area.

Dr. Løken's title refers to "northeastern Labrador-Ungava" yet his text is restricted to a part of the Torngat Mountains. In making this restriction he has omitted reference to the vital Kaumajet and Kiglapait mountain areas. As these areas provide definitive information, which lends support to a criticism of his main conclusion, this discussant feels that there is a good case for broadening the commentary to take in the wider area, as the title would imply.

With one exception, I wholeheartedly agree with Løken's general conclusions. As this point forms the pivot upon which the entire paper is balanced, however, I will restrict myself to building up a discussion around it. Have or have not the high coastal mountains ever been overridden by the inland-ice during the Pleistocene? Daly, Coleman, Dahl, and Løken have said "no," while Odell, Wheeler, Tomlinson, Johnson, and I have said "yes"! (See references and also Løken's text.)

As Løken suggests, the controversy hinges upon the interpretation of certain boulders to be found but rarely on some of the higher Torngat summits. I have already agreed that the Torngat erratics do not stand as absolutely conclusive evidence, once Dahl's strict criteria are adopted (Dahl, 1961). I therefore turn to the Kaumajet and Kiglapait mountains where Tomlinson, Johnson, and Wheeler have provided irrefutable evidence, granitic gneissic blocks resting on Proterozoic basic volcanics on the former, and garnetiferous gneiss resting upon anorthosite above 1,000 m on the latter.

Of course, it might be argued that, as these mountain groups are distant from the Torngat, they may well have been glaciated completely while the Torngat summits were not. The Kaumajet Mountains, however, are less distant from the southern Torngat, to which point Løken is prepared to extend his extrapolations by utilizing my field data (Ives, 1958), than the southern Torngat are from Løken's field work area. To further this case I shall utilize and extend Løken's own reasoning concerning the slope of the upper trimline, or Koroksoak level. Figure A shows that, by extending Løken's line southward to the Kiglapait, assuming that it does roughly represent the slope of the inland-ice surface at the Koroksoak phase, then the Kaumajet summits would have risen well above it. It is stressed that the series of points on Figure 1 do not lie along a straight line. Also, a less steep slope has been added to the diagram to tie in with the lower limit of mountain-top detritus in the Kiglapait. Moreover, Løken has taken a single observation from my 1957 work in Nakvak Valley. In fact, the trimline in this valley slopes eastward at approximately 20 metre/km. Thus, if a point only 5 km farther down-valley were selected, a value lower by 100 m would be obtained, and the resulting point would provide a better "fit" onto the north Torngat-Kiglapait line. This would place even more of the Kaumajet above the upper limit of the Koroksoak Glaciation; in addition, the mountain-top detritus and glacial erratic on Man o' War Peak would also be above this limit.

I must stress that these lines are drawn in a highly subjective way and that we probably do not have sufficiently wide-spread data to draw them in a convincing manner. Precision is not claimed, however, and the motive is rather to use Løken's line of argument to present an alternative interpretation. Whatever the actual slope of the Koroksoak level between the critical points shown on the diagram, it is apparent that the definitive glacial erratics on the high Kaumajet summits lie well above the upper limit of this glaciation. Thus Løken's claim that the Koroksoak level represents the absolute upper limit of glaciation is not well founded. By extending this reasoning there is no further need to discount the claim that the high-level Torngat boulders are glacial erratics. Even Dahl has accepted some as unequivocal glacial erratics (Dahl, personal communication). An important addendum is that Løken's discussion was rather restricted to the actual mountain tops. I can quote many examples of occurrence of large erratics above the Koroksoak level, which further weakens the contention that this level represents the absolute upper limit of glaciation.

A final point is a piece of new evidence which has an important bearing upon our existing knowledge of late-glacial conditions in northeastern Labrador-Ungava. A radiocarbon age of 10,450 ± 250 years before present has been obtained for marine molluscs collected from Deception Bay at 86 m above sea level by Mr. Barry Matthews. (Matthews, 1962. Marine molluscs collected by B.

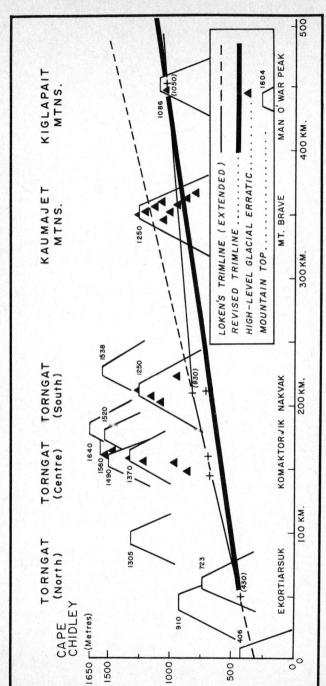

Figure A. The extended "trimline" profile, Cape Chidley to the Kiglapait Mountains.

Matthews, graduate student of the Geography Dept., McGill University, during independent field work in the Sugluk-Wolstenholme vicinity, 1961. Radiocarbon dating arranged through Geographical Branch; Lab. no. J.D.I-61-6S; Isotopes Inc., no. I-488.) This implies that at the time of the Valders Readvance (10,500 B.P.) this sector of Hudson Strait was not only open to late-glacial marine incursion, but that it had been inundated for a sufficient interval to allow a maximum of some 60 metres of relative vertical isostatic recovery prior to 10,450 years ago. This point is discussed at greater length elsewhere (Ives, 1962). In brief, it is argued that the Saglek lateral moraines, originally interpreted as representing the maximum of the last glaciation by Andrews (1961), are definitely older than the Valders moraines in the south, that Hudson Strait may have been partially or completely ice-free at the classical Wisconsin maximum, and that the Koroksoak Glaciation probably pre-dates it. Thus the Torngat Glaciation, which virtually resulted in total submergence of the high Torngat summits, is much older and may have pre-dated the Sangamon Interglacial.

References

Andrews, J. T., "The glacial geomorphology of the northern Nain-Okak section of Labrador", Unpublished M.Sc. thesis, McGill University, Montreal (May, 1961), 280 pp.
Dahl, E., "Refugieproblemet og de Kvartaergeologiske Metodene", Sartryck ur Svensk Naturv. Fjortonde argangen, pp. 81-96.
Ives, J. D., "Glacial geomorphology of the Torngat Mountains, Northern Labrador", Geog. Bull. no. 12, pp. 47-75, Ottawa, Geographical Branch.
Ives, J. D., "The deglaciation of Labrador-Ungava: an outline", Cah. de Geog. de Que. IV (8), pp. 323-43
Ives, J. D. "Field problems in determining the maximum extent of Pleistocene glaciation along the eastern Canadian seaboard: a geographer's point of view", Acta Natur. Islendica, in press. Paper presented to the symposium on the history of the North Atlantic Biota, Reykjavik, Iceland, July 1962.
Johnson, J. P.: personal communication.
Matthews, B., "Glacial and Post-Glacial Geomorphology of the Sugluk-Wolstenholme Area, Northern Ungava", McGill Sub-Arctic Research Papers, no. 13, in press.
Tomlinson, R. F., "Glacial geomorphology in the Kaumajet Mountain and Okak Bay areas of northeastern Labrador". Unpublished M.Sc. thesis McGill University, Montreal, August 1961, 193 pp.
Wheeler, E. P. "Pleistocene glaciation in northern Labrador", Geol. Soc. Am Bull., Vol. 69, pp. 343-44.

Comment
O. Løken
Although my paper deals with the part of northeastern Labrador-Ungava which is to the north of Saglek Fiord, I fully recognize Ives's right to extend the discussion to the Kaumajet and Kiglapait mountains, but in doing so each step should be carefully considered.

Ives's Figure 1 is definitely of interest, but it is not correct when he claims that the extrapolation from the southern Torngat to the Kaumajet Mountains is comparable to the extrapolations which I made in my Figure 3. The latter figure shows clearly that there is nowhere as much as 50 km between two adjacent observations, while the distance between the southern Torngat and the Kaumajet Mountains as shown on Ives's Figure 1 is c. 125 km — 2½ x 50!

This difference is large in itself, but becomes even more significant when the topography to the south of the Torngat is considered, because here are found the Saglek and Hebron fiords, two of the largest along the Labrador coast, and no relevant observation is made along any of them. The Nachvak Fiord farther north is of comparable size, but here Daly's observation shows the position of the ice margin.

These points are made to show that the extension of my profile is not as simple as it might appear. It is conceded that future observations might prove Ives to be correct, and in that case I shall have to revise my interpretation of the highest zone, for I have not questioned the interpretation of the high-level erratics in the Kaumajet-Kiglapait area. It is worth noting, though, that along the outer part of Nachvak Fiord the presence of the sedimentary Ramah Series (Proterozoic) provides conditions under which it should be easy to identify erratics from the Archaean formations farther west. Coleman, however, who has worked in the area states that such erratics are not present above 2,100 ft (630 m).

Future observations might provide conclusive evidence whether or not the highest zone ever was glaciated, and this made me conclude in my paper that ". . . sufficient evidence to show that the highest zone was ever glaciated is *not yet* available."

6

Cross-Valley Moraines of the Rimrock and Isortoq River Valleys, Baffin Island, N.W.T., A Descriptive Analysis

J. T. Andrews

Introduction

The program of air photograph interpretation and field research in north-central Baffin Island, N.W.T., initiated by the Geographical Branch in 1961, has already led to a considerable increase in knowledge of this hitherto geographically unexplored area. The interpretation and plotting of glacial erosional and depositional forms has led to a realization that this region, and in particular the area around the present margin of the Barnes Ice Cap, is a natural laboratory for the study of glacial-morphological processes. The importance of studies originating from these observations cannot be sufficiently stressed. Glacial geomorphology as a branch of geography has been primarily concerned with erosion and deposition under temperate glacial conditions, with the ice mass at the pressure melting point. Glacial forms occurring in arctic areas have been attributed to the same set of processes as those occurring under the vastly different glaciological conditions of more southerly latitudes, and rarely has any attempt been made to examine the structure and composition of these forms. It is

First published in *Geographical Bulletin*, no. 19, May (1963), pp. 49-77. Reprinted by permission of the Department of Energy, Mines and Resources.

possible, even probable, that some glacial features have a like origin regardless of glaciological conditions, but it is also possible that forms, superficially similar in external characteristics, differ in the mechanics of their formation.

This study is a detailed description of the cross-valley moraines of north-central Baffin Island, glacial depositional features that have not been adequately described or examined previously. They were initially observed on air photographs as a series of cross-valley ridges located around the margins of the Barnes Ice Cap. During the summer of 1961 a three-man party from the Geographical Branch examined in detail the glacial forms in a small area some 25 miles north of the ice cap (Figure 1). From the outset it was realized that an adequate knowledge of the de-glacierization of the area could be gained from a. a study of certain lateral, marginal, and frontal forms, and b. an examination of the cross-valley moraines. It should be noted that this latter term is used only in a descriptive sense, and it is not intended that any mechanism of formation be presupposed. Ives (1964) investigated the terrace forms, and the writer dealt primarily with the cross-valley moraines. The chief aim of the writer's field survey was to furnish material for a detailed description of the cross-valley moraines, and to provide data for laboratory and statistical manipulation.

The information thus gained during the summer can be broken down naturally into two separate component parts, both distinct entities and able to stand as separate studies, but at the same time both contributing toward an understanding of the origin of the cross-valley moraines. This paper is concerned essentially with describing the cross-valley moraines and associated deposits; the description is mainly detailed and is intended to serve as a guide in the eventuality that similar moraines are found elsewhere.

Fraser (1962, personal communication) has observed cross-valley moraines on Boothia Peninsula below the marine limit. Lee (1960) has observed what appears to be, at least superficially, similar forms in the Great Whale River area above the marine limit, and there appears to be some resemblance in form between the cross-valley moraines of Baffin Island and the 'Rogen' type moraine in Sweden (Lundqvist, 1937; Hoppe, 1952). It is again emphasized that although all these glacial forms appear super-ficially alike, a conclusion that they are similar would be warranted only after a careful examination of the internal structure of the moraines.

In this paper, five working hypotheses of origin are advanced. A second paper (Andrews, 1963), dealing with the quantitative analysis of certain aspects of cross-valley moraines, examines the hypotheses in detail and reaches certain tentative conclusions on their origin; the results of till-fabric analysis on the moraines are discussed, and the sedimentary characteristics of cross-valley moraines are analyzed.

Location of the Cross-Valley Moraines

The initial air photograph interpretation phase of the Cockburn Land 1:500,000 map sheet revealed that in addition to the normal glacial forms, such as end moraines and glacial drainage channels, some unusual forms existed. It became apparent that a discussion of the deglacierization of the area would involve the introduction of new concepts, and a reappraisal of the now classical concept of the deglacierization of an area by the downwasting of the ice mass, at least as applied to this single arctic area.

In many of the valleys around the margin of the Barnes Ice Cap transverse ridges literally cover the valley floor and sides. Goldthwait (1951a) was the first to describe these ridges, calling them 'sub-lacustrine moraines'. No detailed observations, however, were included, nor was any argument advanced on their origin. The detailed air photograph interpretation (Ives and Andrews, 1963) of the Cockburn Land sheet, and a rather more cursory examination of the areas to the north and south, has extended the known occurrence of the cross-valley moraines, and it has also led to a more exact definition of the glaciological conditions that must, it seems, exist before cross-valley moraines are formed.

No cross-valley moraines have been seen on the western parts of the map area, though on the west coast of Steensby Inlet there is a belt of minor moraines similar to those described by Lee (1960) and Craig (1961). In the north, cross-valley moraines have been traced on the floor of the Pilik valley, and some of the north bank tributaries of this river (Figure 1). The moraines are virtually absent from the Rowley River system, except for a small area in the upper south bank tributary. The cross-valley moraines are best developed in the Isortoq valley, upstream from the Lewis Glacier. In the upper section of the valley, a transition occurs between end and lateral moraines and the cross-valley moraines as the regional watershed is crossed. The upper Striding valley shows a very good degree of development of the moraines. Other examples occur as partially submerged ridges in tributary bays of Conn and Bieler lakes. The main shores of these lakes are devoid of moraines.

The virtual absence of the cross-valley moraines in the Pilik and Rowley valleys, two major west-flowing rivers, is the more noteworthy because of their considerable development in the Isortoq drainage system; this absence possibly reflects a change in conditions during deglacierization.

The moraines can be traced southward along the eastern margin of the Barnes Ice Cap to the Clyde River. They are especially prominent on the sides of Generator Lake, in a south bank tributary of the Sam Ford valley, and in tributaries of the Clyde River.

South of these localities the air photograph interpretation has been of a reconnaissance nature, and was undertaken with the aim of delimiting the areal extent of the Cockburn end moraines and the cross-valley moraines. The whole of Baffin Island was examined during this survey. Cross-valley moraines were noted in nearly all the upper valleys of westward-flowing streams (Figure 1), and they were eventually traced to within 30 miles of the northwestern margin of the Penny Ice Cap. A hiatus of 200 miles in moraine occurrence was observed between this latter point and the west-flowing tributaries of the McKeand River, which flows into Cumberland Sound.

Air photograph interpretation indicates that a strong correlation existed between the location of the cross-valley moraines and the areas characterized by abandoned high-level shorelines. From these shorelines it is inferred that former glacial lakes had been ponded between the main watershed and the westward-retreating late-Pleistocene ice cap. In fact, in all cases where cross-valley moraines were located, elevated shorelines were also observed. The location of the cross-valley moraines thus forms a discontinuous band, running from the Pilik River in the north, to the McKeand River in the south, a distance of 650 miles. The moraines are situated east of a former ice-divide and west of the major watershed. In a few instances cross-valley moraines lie to the west of the ice-divide, and, in the lower Isortoq valley, they have been formed in tributary valleys where the evidence suggests that a lake was ponded against the main trunk glacier.

The cross-valley moraines are thus intimately associated with former glacial lakes, and the evidence of the shorelines, plus the slope of glacial drainage channels and other indicators of glacial directional movement, suggests that only when a glacier or ice cap is retreating down a regional or local slope and into water are the necessary conditions for the formation of the moraines realized. The moraines are restricted to the interior upland physiographic

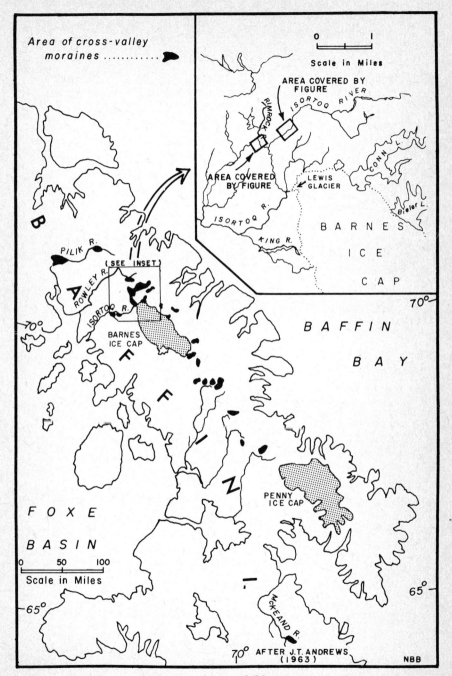

Figure 1. Location map. Inset map shows field areas.

region, a region characterized by broadly convex hills rising from a gently undulating upland surface. Relief is 1,000 feet or less. The relief conditions precluded the formation of long tongues of ice in the main valleys; instead, it would appear that the margin of the late-Pleistocene ice cap was analogous to the eastern edge of the present Barnes Ice Cap. The retreat of the ice cap into Conn and Bieler lakes epitomizes the conditions under which the glacial lakes were formed, and if one can judge from Generator Lake, the way in which the cross-valley moraines were created. The existence today of glacial lakes around the Barnes Ice Cap is especially interesting because of current work in Scandinavia (Holdar, 1952) and Britain (Sissons, 1958); in both areas, early hypotheses of deglacierization demanded the existence of glacial lakes, but present workers are refuting these ideas.

The seeming probability that the cross-valley moraines are sub-lacustrine forms suggests that there is a genetic relationship between these features and the sub-marine De Geer moraines (Hoppe, 1959), the estuarine ridges (Bergdahl, 1953, 1959, 1961), and the minor moraines (Lee, 1960; Craig, 1961). This suggestion requires that precise quantitative studies be performed on the internal structure of these various forms; moreover, these studies should be presented in such a way that the data is accessible to other workers. A recognition of similar and dissimilar characteristics would play a vital part in the understanding of certain glacial depositional processes. In regions where ice is calving into a lake, local conditions of ground temperature and ice basal temperature might be similar to those in more temperate regions.

The relationship between the cross-valley moraines and the former lake shorelines is not always clear. In many instances there are a number of abandoned shorelines, and without detailed field investigations it is impossible to say whether the moraines actually crossed a shoreline while they were being formed, or whether those instances where shorelines obviously post-date the moraines are simply due to wave action in a glacial lake, both younger and lower than the shoreline existing at the time of moraine formation. In many instances, the latter hypothesis is an adequate explanation, but in other cases, as on the south side of Bieler Lake, the cross-valley moraines cross the single, high-level shoreline.

Because of the obvious relationship between glacial lakes (it should be noted that not all former glacial lakes are associated with cross-valley moraines) and the cross-valley moraines, their distribution pattern should shed considerable light on the problem of locating late-glacial ice-divides. A change in glacial features, such as that occurring at the junction of the Isortoq and Lewis rivers where cross-valley moraines and glacial lake shorelines are superseded by glacial drainage channels and lateral moraines, denotes a radical alteration in the relationship between the local ice-divide and natural drainage direction.

The air photograph interpretation has revealed an immediate and obviously crucial relationship between glacial lakes and cross-valley moraines, which, although it does not answer the question of how the moraines were formed, does provide a basic point of departure from which to work.

Cross-Valley Moraines of the Rimrock River

This section provides a detailed descriptive analysis of the cross-valley moraines in the Rimrock and Isortoq valleys (Figure 1); it is based on the observations of the 1961 field survey.

Physical Setting

The broad physiographic regions of north-central Baffin Island have already been outlined (Ives and Andrews, this Bulletin). The area lying north of the Barnes Ice Cap presents a landscape of moderate relief, with wide, open valleys, and gently convex hills rising above a rolling upland surface approximately 2,000 feet above sea level in the field area. Maximum relief is about 1,000 feet, but this is generally spread over a considerable linear distance. The Rimrock valley is broad and open, and drains east and then south to the Isortoq River. Downstream from the lake outlet is a series of poorly developed cross-valley moraines, seemingly formed entirely of large angular boulders. Farther downstream, for a distance of approximately 4 miles, a series of small flat-topped, conical kames (Holmes, 1947) occupies the centre of the valley. On either side, the ground moraine, consisting of angular blocks 2 to 5 feet in diameter, forms a network of intersecting ridges and enclosed hollows. Two miles south of the upper bend these deposits are replaced by linear, transverse ridges that superficially appear as rather simple forms, but upon closer inspection are revealed as being extremely complex (Figure 2). Deposits of sands and gravels, generally rising above the crest of the moraines, in places straddle them and in other places appear as discrete deposits. Hereafter these stratified features are referred to as the *central kames*. Two miles farther down valley the cross-valley moraines become less pronounced, and the most striking feature is a belt of coalescing central kames which traverse the valley at right angles to the moraines (Figure 2). Two massive moraines, noticeably convex toward the south, extend across the valley. There is a dearth of glacial features between this point and the junction with the Isortoq River, though very faint ridges 1 to 2 feet high can be discerned in a west-bank tributary, both on the ground and on the air photographs. They are associated with a series of small kames.

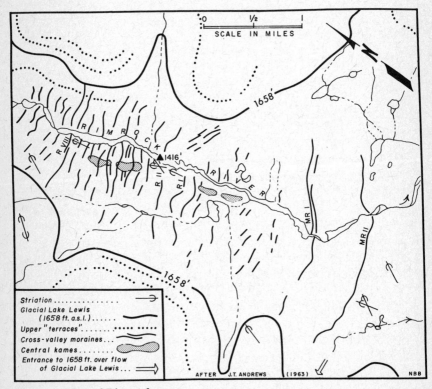

Figure 2. Map of Rimrock area.

Direction of Ice Movement

Roches moutonnées, glacial lineations, and the disposition of glacial grooves and striations on ice-molded bedrock indicate that at one time the regional ice movement was directed toward the northeast, with the orientation of striations varying from N 25° E to N 50° E. During this period the ice moved obliquely across the Rimrock River valley from a centre of dispersal somewhere to the southwest of the area. Striations near river level show that even the basal layers of the ice were affected and that topography played little part in determining the direction of flow. Three exposures suggest, however, that these conditions did not prevail during the final stages of the deglacierization, for striations trending N 10° W were observed to truncate northeasterly trending striations, implying that the former were the younger of the two sets. There is thus evidence for a late-glacial ice movement directed north along the valley. This movement probably occurred after the thinning of the ice, with the concomitant result that topography increased in importance as a controlling agent on the direction of ice movement. However, it further suggests that the centre of dispersion of the ice had shifted from a position southwest of the area to a more southerly location.

Description of the Cross-Valley Moraines

A description, however detailed, has certain inherent disadvantages because of the complex pattern of the moraines and the difficulty of generalization. Most of the aspects dealt with in this paper, such as slope, angle, height, width, angle of junction with other ridges etc., are susceptible to greater quantitative treatment than is accorded to them in this paper. It is recognized nevertheless, that information of this nature is needed, and it will be collected in the future surveys. The main reasons for the collection of these data would be the setting up of confidence limits, and the analysis of recurring trends and patterns. To date only the internal composition and structure of the moraines has been dealt with quantitatively.

In the middle and lower Rimrock River valley the moraines are generally orientated at right angles to the last recorded direction of ice movement (Figure 2). Certain moraines, noticeably MR I and MR II* are noticeably lobate and, being convex toward the former direction of ice movement, are thus radically different in this respect from normal end moraines. In detail, however, the moraines show no rigid organization; they are gently sinuous, or else consist of small sections slightly offset from each other. Many of the smaller intervening ridges are not orientated at right angles to the last ice movement, and may even be parallel to it.

The height of the moraines varies within wide limits, both along any single moraine, and from ridge to ridge. There is, however, a certain discernible pattern; the cross-valley moraines commence on the hillside as small ridges approximately 2 to 3 feet high. They increase in height toward the valley floor and reach a maximum height of between 15 to 45 feet. A point of maximum height is reached on either side of the central axis of the valley; beyond that point the moraines decline progressively toward the river. By the time the bank of the river is reached they are usually less than 10 feet high. There are exceptions: RV III is 30 feet high where it is breached by the river, and MR I and MR II are at least 40 feet high in a similar situation. In most instances they are insignificant near the river, but the location of rapids, leading directly across the river from a moraine suggests that at one time they were continuous cross-valley features. In some localities the moraines are only noticeable during the strong evening light, presenting a pattern similar to a well-ploughed field.

No attempt was made to measure the distance between the separate moraines over the entire field area, though selected measurements were made. These indicated an interval of approximately 150 to 200 ft., although in areas where the moraines have a small amplitude this figure is reduced to between 20 to 40 feet. In the middle section of the valley the moraine ridges appear to fall into four main groups, with a series of high, distinct features being succeeded by a number of smaller ones, which in turn are followed by another group of the more massive ridges. Whether any pattern exists in the spacing of the ridges will only be seen after a detailed, quantitative sudy.

One of the most striking features of the moraines is the asymmetry of their cross profile, numerous measurements revealing that the distal slope, in this case the north-facing slope, is inclined at 30° to 40° while the proximal slope varies between 17° and 25°. The asymmetry was even observed on the very small moraines. It might well be argued that the steeper distal slope is not a primary characteristic, but is a result of secondary modifications due to periglacial activity. Late-lying snowbanks occupy many of the north-facing slopes but disappear early on the southern slopes, and thus this argument must be examined carefully. It is submitted, however, that the asymmetry is a result of the original depositional process. This thesis is advanced after a careful consideration of the evidence on both slopes. Both the south- and north-facing slopes illustrate certain aspects of periglacial activity, with minor flow features located mainly on the northern slopes. These slopes, however, meet the ground moraine with a very sharp angle and there is very little evidence of any slump material at their base. The till-fabric analysis adds further support for the proposal that the asymmetry is a primary feature though this is discussed in a later paper (Andrews, 1963). The asymmetry of the moraines, therefore, is considered an inherent characteristic.

Numerous exposures were examined, and many pits excavated to provide a true estimate of the composition of both the cross-valley moraines and the central kames. This aspect will be dealt with in detail in the later study, but results show that the cross-valley moraines are formed of a sandy till, and that the central kames are formed of bedded sands and gravels.

The surficial cover of the moraines varies both as regards the percentage area effectively masked and the nature of the surficial deposits. In their upper sections the inner core of the moraines is entirely masked by a mantle of large angular boulders with diameters of 2 to 4 feet. The boulder cover is not restricted to the moraines; the intervening areas are similarly covered. In places the moraines might be composed entirely of this material. Many of the boulders rest in unstable, and even perched positions, and some boulders appear to have fallen into the moraine crest. The angularity of the material is very marked, and it appears that this material has been little affected by transportation. In places it is possible to map the underlying bedrock geology by the occurrence of a particular rock type in the ground moraine.

The area between the moraines covered by large boulders decreases toward the centre of the valley. This might be due in part to burial of boulders under a layer of clay. The moraines themselves, however, still have a marked boulder cover, involving 30 to 70 per cent of the surface area. The morphometry of the boulders is still rather angular, despite an increase in the roundness factor, although the measurements on the roundness and area covered need to be defined more rigorously. During the numerous excavations into the moraines, large boulders were encountered at depth, in some cases making further excavation very laborious. In most cases, however, boulder cover is predominantly surficial, and is not caused by the removal of fines from the moraine crest and sides; deposition of most of the boulders therefore post-dates that of the moraine formation.

Experience in the field indicated that it is very difficult to generalize on the distinctive traits of the cross-valley moraines. Nevertheless, certain facets of development are sufficiently repeated to justify some comment, and this allows them to be divided into four main units from their commencement on the hillside to the centre of the valley (Figure 3). They begin as small ridges and gradually emerge from the maze of boulders which are found at the higher elevations. Several ridges, with a fan-shaped gathering ground

(Figure 3A) will then lead to one main ridge, noticeably larger than any of its component ridges, which can be followed down the hillside, sometimes as a single continuous ridge, but most often presenting a series of sharply terminated ridges, continuing in a slightly offset position to each other (Figure 3A 2). In the third section the moraine changes its form from a steep, rather narrow ridge with a surficial cover of angular boulders, to a distinctly higher central kame composed of bedded sands and gravels with angular and rounded boulders littered over the surface. In many instances there is no apparent morphological break between the moraine and the central kame except for a distinct increase in height, though in other places the central kames stand as discrete units, obviously unrelated to any one particular cross-valley moraine. The central kame section is followed by a moraine development comparable to the second unit (Figure 3A 2).

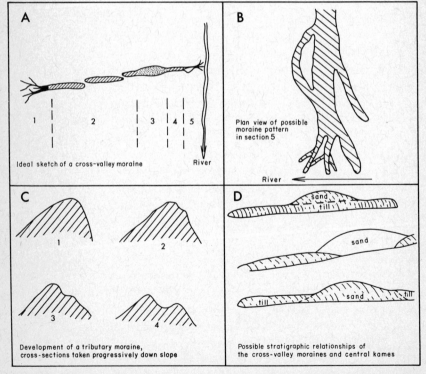

Figure 3. Sketches of certain aspects of cross-valley moraines.

As the moraines approach the river, they often lose their individual form, with distributory ridges branching from the main ridge and occasionally rejoining it (Figure 3B). The moraines are generally quite small near the river, and often terminate in a maze of ridges running in all directions. However, as previously mentioned, several of the moraines are massive in their lower portions, and are dissected by the river. There is no doubt that at the time of formation these features extended across the valley without a break. Where they become very low near the river, the presence of rapids also suggests that these were at one time continuous features, although the reduction in height appears to be a primary feature and little evidence was noted of secondary modifications by running water. Sometimes one of the main moraines is paralleled by a smaller ridge; an excellent example of this form of development is seen with MR I (Figure 3C). In places the branching of a distributory ridge from the main moraine can eventually be traced to a faint terrace on the main ridge which eventually widens and gradually forms a distinct ridge-like form (Figure 3C).

The crests of the moraines are gently undulating, with an occasional col or notch incised into the ridge. Sim (personal communication, 1962) has noted similar features on the cross-valley moraines of Generator Lake.

The stratigraphic relationship between the moraines and the central kames is obviously of major importance in ascertaining the depositional history of both forms. Two possibilities exist; either the till of the moraine, and indeed the moraine itself, continues beneath the central kames (Figure 3D), or else the moraine is entirely absent, and the till and bedded sands are in juxtaposition, with one deposit entirely displacing the other. As has been previously noted, the relative position of the central kames to the cross-valley moraines does vary. In places the two forms merge without any evidence of a morphological break, whereas in other areas the high knolls of sands and gravels are disassociated from the moraines. This is especially true of the Isortoq valley. If the two forms were irrevocably linked spatially in all instances, then the origin of the two main elements must be a product of two different processes operating concurrently along the same deposition line, or else a change in process along the length of the moraine. However, there is a sufficient number of observations where there is no connection between the two forms to suggest that they are the result of two different processes (this is necessary to explain the difference between the till and the bedded deposits) acting at slightly different times, and in places, probably, concurrently.

No answer to the problem of the age relationship of the two forms was gained after a study of exposures in the Rimrock valley, although the results of a detailed study of air photographs, including enlargements at a scale of approximately 1:16,000, implies that the central kames are smeared onto the cross-valley moraines and are thus younger in age. A somewhat similar association of forms was observed by Hoppe (1952) in northern Sweden, where the Rögen type of moraine is found in association with eskers that are younger than the moraines.

Description of R II

In order to draw together certain of the previously noted generalizations of the cross-valley moraines, and to clarify further some of their more important characteristics, a short description is given of R II. This moraine is in many ways typical of the Rimrock valley moraines and since four till-fabric studies were undertaken on this ridge a detailed description is warranted.

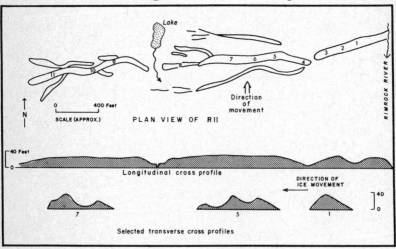

Figure 4. Detailed profiles and plan of R11 moraine, Rimrock Valley.

Figure 4 has been drawn both from field sketches and measurements, and from photographs. The moraine is 11 feet high where it is cut by the river, and it rises to 35 or 40 feet above the adjacent swales *1*. The distal face has a slope of 38°; the proximal face has a slope of 24°. The boulder cover is moderate, and mantles 20 to 50 per cent of the total moraine surface. The crest of the moraine falls to a low col *2* and then rises again, so that at

3 it is about 25 to 30 feet wide with a very steep distal slope. The crest then descends to another col *4*, and at this point the main ridge is joined by a subsidiary ridge on its proximal side. The col is floored with sands and gravels, heavily impregnated with silts, and is 15 feet above general ground level. The moraine increases in height to *5* and the subsidiary proximal ridge is orientated approximately parallel to the main ridge. The very steep distal slope is a continuous feature along the entire length of R II. A minor, distal, sub-parallel moraine disintegrates into a maze of rectilinear ridges. At *6* the main proximal subsidiary ridge almost joins R II, while at *7* a secondary moraine joins it, and a series of small ponds have been dammed up between the two ridges. R II then swings south and ends abruptly, but is continued by a smaller moraine which joins R II between points *7* and *8*. The subsidiary ridge on the proximal side becomes progressively fragmented, and eventually disappears in a confused network of boulder-covered ridges and hollows. As R II ascends the hillside the surficial boulder cover increases, and eventually blankets the moraines entirely. At *9, 10,* and *11* a series of small moraine ridges join and leave R II, the main ridge eventually fading out into an area of ridges and slight hollows, completely covered with angular shattered rock.

Cross-Valley Moraines of the Middle Isortoq River Valley

The investigation of the cross-valley moraines in the middle Isortoq River valley (Figure 1) was rather more cursory, and neither the descriptive analysis nor quantitative measurements were as detailed as those undertaken in the Rimrock valley. Two areas were studied: the first area lies 8 miles upstream from the junction of the Rimrock and Isortoq rivers (Figure 5) whereas the second area lies to the south of this junction.

In the first study area glacial striations and *roches moutonnées* on the valley floor show that the last recorded ice movement was directed along the axis of the valley, the orientation of the striations varying as the axis of the valley shifted. Orientations thus varied from N 50° E to N. No crossing striations were found. Above the valley the indications were that the regional ice movement had been in a general northeasterly direction.

Figure 5 illustrates the broad pattern of the cross-valley moraines. One of the most striking aspects of this pattern is their orientation in tributary valleys. In many cases they do not run transverse to the valley, but parallel to its axis, although in other cases they are transverse. However, if they are examined in re-

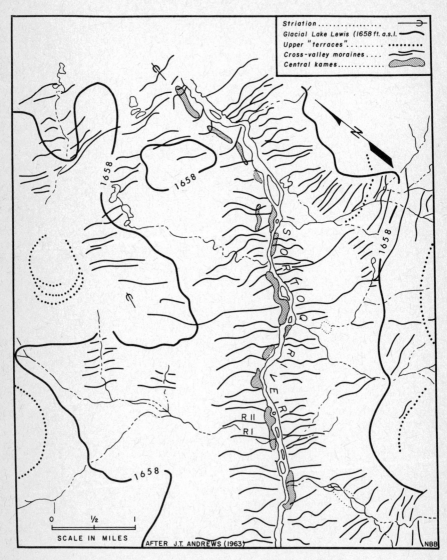

Legend:
Striation ....
Glacial Lake Lewis (1658 ft. a.s.l.) ....
Upper "terraces" ....
Cross-valley moraines ....
Central kames ....

1658

ISORTOQ RIVER

R II
R I

0   ½   1
SCALE IN MILES
AFTER J.T. ANDREWS (1963)
NBB

Figure 5. Map of Isortoq area showing cross-valley moraines and major terraces.

lation to the Isortoq moraines, then a pattern of broadly curving morainic arcs is discernible (Figure 5). The curvature is directed inward to the former direction of ice movement, and a similarity cannot help but be drawn between this pattern and the broad, curving, calving bay of Conn Lake. The moraines are not so massive as in the Rimrock valley, and they vary from 2 to 15 feet in height. They have a more linear form, and exhibit only a slight tendency to bifurcate, though one ridge crosses directly over another and both continue their course without any noticeable alteration. The distal slopes, that is the north- or north-easterly-facing slopes, are appreciably steeper than the proximal slopes. Certain moraines reach the river, though in most cases the areas adjacent to the river are characterized by a series of central kames, of which the more massive members are 100 feet high. Several of the central kames are intimately associated with individual cross-valley moraines, but on the other hand, many are isolated knolls of sand. Several exposures through the moraines indicate that they are formed of identical material to the ground moraine.

The stratigraphic relationship of the central kames and cross-valley moraines was clearly seen in one exposure where bedded sands overlie and overlap a central core of till that is being exposed. The moraines are thus the older of the two deposits and have been blanketed by younger deposits of sands and gravels.

In the second study area the situation is different: the central kames are non-existent, but river-bank exposures show that the moraines were formed of sands and gravels, poorly bedded, with a litter of well-rounded boulders and cobbles. The river-bank exposures also showed that the composition of these moraines did not differ from the material flooring the intervening swales. On the west side of the valley the moraines are rather simple linear forms, with an average height of 10 feet and less. However, on the bend of the Isortoq River opposite the Rimrock River valley, they are visually the most complex that have been seen. Ground observations are lacking, due to inability to ford the river.

General Discussion
The Relationship of the Cross-Valley Moraines
to Former Glacial Lake Shorelines

In the Rimrock valley the cross-valley moraines are all located below the level of Glacial Lake Lewis, 1,658 feet above sea level though the inter-relationship of ice movement, ice gradient, and location of the main overflow of this lake proves that the moraines must have been formed in an earlier and higher glacial lake. A deeply cut gorge to the south of Rimrock Lake has an elevation of 1,817, ± 10 feet above sea level, and appears to be the first main spillway above the 1,658-foot level. Once the ice retreated from the Isortoq-Rowley river watershed, lakes were dammed up between the ice front and the height of land; as the ice withdrew to the south, successively lower cols were uncovered, thus promoting successive falls in lake level. It is postulated that the majority of the cross-valley moraines were formed in a frontal or sub-glacial position under approximately 400 feet of water, though MR I and MR II, because of their position opposite the 1,658-foot overflow were possibly formed under approximately 300 feet of water.

The continued expansion of an ice-dammed lake in the Rimrock drainage basin is postulated on theoretical grounds: studies in the immediate field area (Ives, 1964) suggest that the upper terraces were formed principally by lateral meltwater or wave action or both in confined lateral glacial lakes. No ice-pushed ridges were noted. In some areas, small sections of dissected kame terraces are associated with the terraces, proving their marginal origin. The shoreline of Glacial Lake Lewis marks two well differentiated zones: the upper zone has well-developed crustaceous lichens growing on the rocks, with *Rhizacarpon geographicum* having maximum diameters of 100 mm; whereas below the shoreline this lichen species has a maximum diameter of 30 mm. The shoreline of Glacial Lake Lewis is also marked by well-developed ice-pushed ridges. The lake is probably entirely frontal, but it is related to a col which must have been used by earlier frontal and lateral lakes. Thus, it is impossible to ascertain at which phase the cross-valley moraines were notched by lake water.

The situation is much more complicated in the Isortoq River valley, and the restricted area of the field survey makes a complete evaluation impossible at this stage. Several of the moraines are notched by the shoreline of Glacial Lake Lewis, and several have been traced onto the summits of the hills bordering the Isortoq River, which rise to about 1,800 feet above sea level (by aneroid measurements). The series of terraces above this level has been studied by Ives (1964), and their complicated history is all too evident. Field work in the upper Isortoq drainage basin will help in the elucidation of this problem. However, the withdrawal of the ice cap from the eastern Isortoq watershed would have resulted in the creation of a glacial lake ponded to the height of the eastern col. As the ice cap receded toward the west and south, the glacial lake would expand in size, and probably in depth, though this depends on the availability of lower cols. All that can be stated now is that the moraines in the first Isortoq study area were formed prior to the opening of the 1,658-foot col on the west side of the Isortoq. This is evident as the moraines are cut by the shoreline. The moraines in the second area were probably formed as the ice retreated down the valley, and the col of Glacial Lake Lewis must have been in operation. The moraines of these two areas were thus formed at different times and under glacial lakes of different elevation.

The Pattern of the Cross-Valley Moraines in Other Areas

It would be a mistake to regard all references to cross-valley moraines as related only to the previously mentioned forms, although it is true that there are certain common characteristics. However, these do not hide the differences, and any thesis on the origin of the moraines must at present be limited to the area of study. There is evidence to suggest that these similar forms resulted from different glaciological environments.

In the upper Isortoq and its main upper tributary (Figure 1) a simple pattern of cross-valley moraines was traced on the air photographs. The moraines are broadly curving, linear ridges, trending transverse to the valley axis. The pattern is similar to a broadly curving, calving bay (cf. Hoppe, 1952). In the main tributary valley there is a regular sequence to the pattern, with a broad, massive ridge followed by a series of 10 to 15 smaller ridges. Interspersed with the moraines is a series of conical kames, with one main system occupying the centre of the valley and broken sub-parallel series in other areas, notably the valley sides. They form a sequence of features trending perpendicular to the moraines. They appear to have some of the characteristics of the so-called 'beaded eskers' of De Geer (Bergdahl, 1953). Air photograph interpretation of the area suggests that here, even more than in the field areas, the kames are not related to the cross-valley moraines, and very few are associated with an individual moraine. In the upper portion of the Isortoq valley it is noteworthy that the main terrace, below which the cross-valley moraines terminate, is dissected by glacial drainage channels. This fact implies that the ice in the valley was retreating as a calving bay, with ice on the hillside thus being somewhat in advance of the ice cliff fronting the lake.

South of Bieler Lake the receding ice margin slopes gently into the rather shallow lake containing a whole series of linear, cross-valley moraines. They appear to have been formed by the concentration of dirt from the marginal shear planes, and they are similar in formation to the Baffin type moraines described by Goldthwait (1951b). Here, however, the lake is shallow, whereas in the middle Isortoq and Rimrock river valleys the glacial lakes were very deep. It is thus suggested that the origin of the two types of cross-valley moraines differs fundamentally.

Suggested Mechanisms of Formation of the Cross-Valley Moraines

The analysis of the moraines and associated glacial forms has been descriptive, and little attempt has been made to interpret the significance of certain characteristics. This approach has been intentional. The distribution of the moraines, limited to areas of former glacially dammed lakes, indicates that any hypothesis of origin must take into account the undoubted fact that they are sub-lacustrine glacial deposits. The development of cross-valley moraines south of Bieler Lake shows one possible mechanism of formation, though it is considered that this cannot be applied to all examples.

Five working hypotheses for the formation of the cross-valley moraines are proposed. No attempt is made at this point to fully justify the selection of these hypotheses, although the basic reasoning is discussed briefly. It is further suggested that each explanation is feasible and that any conclusion as to the precise origin of the moraines must be considered within this broad framework of multiple-working hypotheses. They are intended to provide a basis for a quantitative analysis (Andrews, 1963) of the moraines to be published separately, as well as to rationalize the mass of descriptive data that has been included in this paper.

The first is based on the sinuosity of the ridges and their tendency to form tributary and distributary ridges. These facts taken in isolation, are suggestive of an origin similar to that attributed to sub-glacially engorged eskers (Mannerfelt, 1945). However, this hypothesis, although it adequately describes the external characteristics of the ridges, must be precluded as it does not explain their internal structure nor the lack of stratified material in the actual ridge.

The second proposes that the cross-valley moraines are frontal features emplaced under deep-water conditions. This hypothesis should be considered carefully as it would appear that Goldthwait

(1951b) also considered the Generator Lake cross-valley moraines as a linear series of frontal deposits. In the field areas, the material in the moraines is undoubtedly till and it exhibits none of the sorting that might be expected under deep-water sedimentation. The use of the term outwash deposits by Goldthwait is not considered justified; nearly all the tills in the area are rather sandy with a low silt/clay content, and the falling of material from the steep ice cliffs would produce some degree of sorting. Further, if meltwater streams were finding their way down through the terminal crevasse zone near an ice cliff, the resulting deposits would be aligned at right angles to the ice margin and not parallel to it.

The third hypothesis is similar in some ways to the second in that it is suggested that the cross-valley moraines are frontal glacial features, but it proposes that they are formed by the movement of material along shear planes, with deposition in shallow water. This hypothesis has at least a possible present-day example in the area south of Bieler Lake. However, these examples are in places obviously ice-cored and none of the features of the moraines in the field area suggested that they contain an ice core. Furthermore, it is certain that the moraines in the Rimrock valley were formed under deep-water conditions, so this hypothesis cannot apply to all areas. It is possible that the very broad, cross-valley moraines in the upper headwaters of the Isortoq and its tributary were formed in this manner as their height above sea level relative to the lowest col indicates that the former glacial lakes were shallow.

The fourth hypothesis takes into account the asymmetry of the cross-valley moraines which could be explained by assuming that they are 'satzendmoranen', or push moraines. This proposal does help to explain the asymmetry of the moraine profiles, but the number of the moraines and their branching pattern are basic factors in the rejection of this hypothesis, though it is suggested that this mechanism might operate in conjunction with another process.

The final hypothesis is that the cross-valley moraines are formed by the squeezing of basal till into a crevasse system; behind the ice front in a calving bay the crevasse pattern is a product of increased flow into a glacial lake and tension produced by the tendency to buoyancy of the ice front. The broadly arcuate pattern of the cross-valley moraines, which resembles a calving bay, is a very strong point in favor of the proposal, as is the series of offset moraines, which closely resemble the crevasse pattern on a calving glacial tongue. However, the general complexity of the moraine pattern is rather difficult to explain by this thesis, especially the pattern of the distributary ridges.

There is of course a sixth possibility: that the cross-valley moraines are a product of any of a series of combinations of the above mechanisms.

The setting up of a series of working hypotheses, which have been examined, albeit cursorily, indicates the problems that must be faced in seeking a valid explanation for the formation of the cross-valley moraines. It is realized that the advantages and disadvantages of each hypothesis have been only outlined, and a more critical appraisal is left to the second paper which deals with the quantitative aspects. Even then, caution must be exercised, for without more detailed and precise field work, solutions to the problem of the cross-valley moraines must be limited to the immediate field area. Factors which seem striking in one field area must be shown to be significant in other areas before a general thesis of formation can be proposed or accepted. The cross-valley moraines of north-central Baffin Island are a distinct glacial form that developed under distinct glaciological conditions, and study of their various facets will illustrate the depositional role of a retreating arctic glacier discharging into, or fronted by, a glacial lake.

Bibliography

Andrews, J. T., "The cross-valley moraines of north-central Baffin Island: a quantitative analysis", *Geog. Bull.*, no. 20 (in press), Geographical Branch, Ottawa (1963).

Bergdahl, A., "Marginal deposits in south-eastern Sweden, with special reference to the oses", *Lund Studies in Geography*, Royal University of Lund, Sweden, Series A, no. 4 (1953), p. 24; "Glaciofluvial estuaries on the Narke Plain", *Svensk Geogr. Arsbok*, Arg. 35 (1959), pp. 47-71; "Glaciofluvial estuaries on the Narke Plain, II, the lower Kvismar valley", *Svensk Geogr. Arsbok*, Arg. 37 (1961), pp. 71-91.

Craig, B., "Surficial Geology of Northern District of Keewatin, Northwest Territories", *Geol. Surv. Can.*, Paper 61-5 (1961), 8 pp.

Goldthwait, R. P., "Deglaciation of north-central Baffin Island (abstract only)", *Geol. Soc. Am. Bull.*, Vol. LXII (1951), pp. 1443-44; "Development of end moraines in east-central Baffin Island", *Jour. Geol.*, Vol. LIX (1951), pp. 567-77.

Holdar, G. G., "Problemet Torne-Issjon", *Geogr. Annal*, Vol. XXXIV (1952), pp. 73-88.

Holmes, C. D., "Kames", *Am. Jour. Sci.*, Vol. 245 (1947), pp. 240-49.

Hoppe, G., "Hummocky moraine regions with special reference to the interior of Norbotten", *Geogr. Annal.*, Vol. XXXIV (1952), pp. 1-72; "Glacial morphology and the inland ice recession in northern Sweden", *Geogr. Annal.*, Vol. XLI (1959), pp. 193-212.

Ives, J. D., "The terrace forms of the upper Isortoq River, north-central Baffin Island — a study of the deglacierization process", *Geog. Bull.* (in press), *Geographical Branch*, Ottawa, (1964).

Ives, J. D. and Andrews, J. T., "Studies in the physical geography of north-central Baffin Island, N.W.T.", *Geog. Bull.* No. 19, Geographical Branch, Ottawa (1963), pp. 5-48.

Lee H. A., "Sakami Lake surficial geology", Map 52-1959, *Geol. Surv. Can.*, Ottawa (1960).

Lundqvist, G., "Sjösediment fran Rögenomradet i Hajedalen", *S.G.U.*, Series C, no. 408 (1937).

Mannerfelt, C. M., son, "Nagra glacialmorfologiska formelement", *Geogr. Annal.*, Vol. XXXIII (1945), pp. 166-209.

Sissons, J. B., "Supposed ice-dammed lakes in Britain with particular reference to the Eddleston valley, southern Scotland", *Geogr. Annal.*, Vol. XL (1958), pp. 160-87.

7

Crevasse Fillings and Ablation Slide Moraines, Stopover Lake Area, N.W.T.

J. R. Mackay

A particularly intricate pattern of unusual ridges and terraces covers 40 square miles of the Stopover Lake area, 80 miles north of Smith Arm, Great Bear Lake (Figure 1). The complex network of ridges, enclosed depressions, pitted flat-topped terraces, scalloped ice-contact slopes, overflow melt-water channels, and associated features have probably resulted mainly from deposition of sediment against stagnant ice, seamed with fissures and penetrated by tunnels. Most of the larger ridges are believed to be crevasse fillings (Flint, 1928; 1929; 1930; cf. Kupsch, 1956) although a few seem to be eskers. Some of the small sub-parallel ridges occupying the floors of kettle depressions may be ablation slide moraines left by seasonal chuting of ablation moraine down the slopes of wasting ice-blocks onto dry land. The entire assemblage of topographical forms has many points of similarity with the dead-ice landscapes described by Hoppe (1952), Mannerfelt (1945), Tanner (1944) and others. The writer carried out field work in the area during the first week of July 1951, when a stop was made on a flight from Great Bear Lake to the Arctic coast. In the preparation of this report field observations were supplemented with air-photo interpretations. The writer wishes to thank Dr. W. H. Mathews for helpful comments and discussions.

First published in *Geographical Bulletin*, no. 14 (1960), pp. 89-99. Reprinted by permission of the Department of Energy, Mines and Resources.

Figure 1. At centre left, most of the north-south trending ridges are part of an extensive corrugated morainic belt. The most prominent area of crevasse fillings extends from northeast to southwest as a band, one mile wide, marked by numerous looped and branching ridges. The remaining linear elements are eskers, ice-contact slopes and ablation slide moraines. The large abandoned channel crosses the area in the southeast.

The crevasse fillings, ablation slide moraines, and associated features lie at an altitude of from 1,150 to 1,250 feet above sea level. The sand and gravel crevasse fillings and terraces are well drained and relatively dry. Permafrost is doubtless present although frozen ground was not encountered in a pit dug 6 feet deep on a south-facing gravel slope on July 3. However, frozen ground lay within a foot of the surface in wet peaty areas. The crevasse fillings and terraces support an open woodland cover of scattered white spruce, aspen and balsam poplar copses, bush alder and willows, heaths, mosses, lichens, grasses, etc. The bottoms of swales and basins tend to be sedgy to peaty.

The area of ridges and terraces grades northward into a higher tundra-covered rolling landscape rising to over 1,300 feet above sea level. To the south and west of the ridged area, there is a higher corrugated morainic belt from 1 to 2 miles wide and 1,250 to 1,450 feet in altitude. Thus, with higher land to north, west, and south, a blockage of drainage to the east apparently caused a large lake to form in the Stopover Lake depression.

To the east, there is a large north-south trending, abandoned drainage channel, whose flow entered the Horton River about 10 miles northeast of Stopover Lake. The abandoned channel formerly carried a substantial flow, as shown both by the length of the abandoned channel and by its size. Although the total length of the channel is unknown, it can be traced upstream from Stopover Lake for over 80 miles with virtually no diminution in size. The abandoned meander scars and oxbows have radii of 1 to 2 miles, a size fully equal to the meander loops and scrolls of the lower course of the modern Horton River which has a length exceeding 330 miles.

In the section east of Stopover Lake, flow in the abandoned channel started when the river was at the present 1,100 to 1,150 foot level, as shown by the highest terraces that bear flow marks at that altitude. The absence of higher river terraces is in keeping with the view that flow from the channel was initially into a water body, 1,150 to 1,250 feet in altitude, occupying the Stopover Lake depression. Only after the lowering of the lake(s) to about 1,150 feet did river flow to the east, past Stopover Lake, commence. As the area became partially drained and the river began downcutting, incised meanders developed to a depth of 50 to 75 feet below the upper river terrace levels. Eventually, the channel was abandoned.

Crevasse Fillings

The crevasse fillings, which are believed to be sand and gravel casts of former fissures in stagnant ice, resemble undulating to flat-topped eskers. They have a preferred north-south orientation, nearly at right angles to the last regional direction of ice movement which was west to northwest. Some of the ridges are relatively straight and occur singly, but most of them branch in wishbone fashion, loop or hook around kettle depressions, intersect in an irregular network, or merge into pitted outwash plains at accordant levels. Many of the looped ridges have a flat terrace-rim on the inner side, encircling the kettle depressions or dead-ice hollows. The hollows are often lake-filled, or wet and sedgy in the centres.

The network ridges lie in a mile-wide, north-south band to the west of Stopover Lake, with a small concentration on the east side of the lake. The ridge junctions are not distributed at random but are grouped into clusters with a range of from two to eight junctions. A significant tendency to clustering or grouping is observed in statistical tests (Clark, 1956). The tendency towards clustering of ridge junctions shows that local conditions which favored one junction also favored others. As the ridges are interpreted as crevasse fillings, the ridge junctions appear to locate the intersection of two crevasse systems, the north-south ridges reflecting transverse crevasses, and east-west ridges longitudinal crevasses.

So far as could be ascertained from surface exposures, natural sections, and excavations, the ridges and terraces are of sand and gravel without a veneer of glacial till, although scattered erratics, bearing striations and facets, occur on their surfaces. Most of the ridges range from 20 to 50 feet in altitude. The long profiles of the smaller and narrower ridges are usually irregular, but many of the larger ridges have even crests. Some of them broaden to merge, with an accordant level, into flat-topped terraces. A single massive ridge may show, in cross-profile, several parallel ridge crests, of which the centre-most one is typically the highest.

The crevasse fillings are believed to have formed during the last stage of deglaciation when a mass of fissured, crevassed, and tunneled stagnant ice lay immobile in the Stopover Lake depression with the interconnecting passageways between the ice-blocks filled with water up to the level of closure of the depression. As there is higher land to the north, west, and south of the depression, ponding of waters would probably have occurred if there were an ice dam to the east, towards Horton Lake. Under such conditions, debris from the wasting stagnant ice-blocks would have been washed into the fissures, crevasses, tunnels, ponds, etc. to form ice-contact features such as crevasse fillings, kame terraces, outwash deposits, and so forth. If there were many interconnecting passageways, as the pattern of ridges indicates, there would have been a controlling water level over large areas of the Stopover Lake depression. Thus, at any one time, deposition of sands and gravels would have had an effective upper limit determined by this water level. This would explain the flat crests of many ridges and the accordant levels at which they merge into flat-topped terraces.

About 25 to 30 per cent of the Stopover Lake area is covered with sand and gravel terraces and ridges with an estimated thickness of 20 to 45 feet. If the material were spread uniformly over the entire depression, the thickness would range from roughly 5 to 15 feet. Although little is known about the thickness of till in adjacent areas, there is some doubt as to whether it would average as much as 5 to 15 feet. This suggests, therefore, that some of the sand and gravel comprising the ridges and terraces may have been transported into the Stopover Lake depression by inflowing streams. This suggestion is supported by the fact that the most extensive and continuous systems of ridges and terraces occur in the south, bordering the large abandoned drainage channel, previously mentioned, which could have provided a major source of transported sediments. Significantly, the ridges and terraces become smaller and more isolated with increasing distance away from the abandoned channel.

Inasmuch as the upper limit of terrace and crevasse deposition was determined by the prevailing water level, any lowering of the water level would have left the depositional features above it high and dry. At Stopover Lake, there is a series of terraces and crevasse fillings whose tops lie between 1,150 and 1,250 feet in altitude—the range is probably smaller, but it cannot be determined more accurately with existing maps. Once the level fell below 1,150 feet, most of the area would have acquired its present over-all lake pattern because of drainage by the river that formerly flowed in the abandoned channel to the east. The lowering of the water level seems to have been rapid, because there are no successive flights of terraces and crevasse fillings as would have been expected to develop in a lake with a gradually lowered water level. Rapid lowering may have been caused, for instance, by the failure of an ice dam to the east or the uncovering of a lower outlet. As the water level was lowered in the Stopover Lake depression the margins of ice-blocks melted back from the sand and gravel banked against them to give ice-contact terraces and crevasse fillings. Narrow crevasse fillings slumped and lost their altitude upon removal of the ice support, the broader ones remaining flat-topped. Melt-water from some of the ice-blocks escaped in overflow channels across terraces and crevasse fillings. Melt-water courses are now preserved as high level, abandoned, channels extending across terraces and linking one kettle hole with another. The channels are typically less than 10 feet deep and 200 feet wide.

Ablation Slide Moraines

Groups of sub-parallel ridges, here referred to as ablation slide moraines, occur on the sloping to flat bottoms of some depressions formerly occupied by stagnant ice. The ridges average only a few feet in height and about 50 feet in width. Their steepest slopes, when discernable, tend to face towards the position of the former ice-block from which the material is believed to have slid. The ridges curve in response to the relief of the ground and, in general, follow the contour. Ridge spacings range from about 70 to 110 feet. For example, at the southwest side of Stopover Lake there are 12 ridges in a horizontal distance of 1,050 feet, the average spacing being 88 feet.

An ablation slide moraine is belived to have formed by the chuting, slumping, and sliding of ablation material down the slope of a stagnant ice-block onto dry land, or, at least, onto the ground with a minimum of reworking by water. With annual wastage of ice-blocks, a series of sub-parallel ablation slide moraine ridges would have been formed. The ridges would have reflected the general ground plan of the ice-block from which their material came. Large ice-blocks that wasted away to form two or more smaller ice-blocks would be so shown by the eventual separation of ridges around two or more centres.

The formation of the ablation slide moraines is believed to have been, in general, as follows: A rapid lowering of water level left many large and small ice-blocks high and dry. For awhile, there was flow from some ice-block-occupied areas to others, as is shown by abandoned melt-water channels crossing terraces and linking kettle depressions. In time, the ice-blocks melted back from the ice-contact faces and the chuting, sliding, and slumping of ablation material built a series of sub-parallel ridges which, initially, closely followed the trend of the ice-contact faces. As the spacing of the ridges are some 70 to 110 feet apart, this suggests a retreat of the ice-block edge by that amount each year. This is in reasonable agreement with rates of melting observed elsewhere (Charlesworth, 1957). As recession would be expected to occur on all sides of the ice-block, the net horizontal reduction in size would be roughly 140 to 220 feet a year. On this basis, most of the ice-blocks in the Stopover Lake basin would have lasted no more than 10 to 20 years.

Conclusion

Most of the sand and gravel ridges in the Stopover Lake area are crevasse fillings built by deposition of sediment in open fissures and crevasses of stagnant ice with a controlling water level of 1,150 to 1,250 feet above sea level. Much of the sand and gravel comprising the crevasse fillings and the associated terraces was transported into the area by a large river whose abandoned channel lies to the east of Stopover Lake. Following a rapid lowering of the water level to below 1,150 feet, ablation slide moraines formed by the chuting, sliding, and slumping of ablation material down the sides of wasting ice-blocks, to build up a series of sub-parallel small ridges. The ridges, whose spacings range from about 70 to 110 feet, may be annual features.

Bibliography

Charlesworth, J. K. *"The Quaternary Era"*, London, Edward Arnold (Publishers) Ltd., Vol. II (1957), p. 1151.
Clark, P. J., "Grouping in spatial distributions", *Science*, Vol. CXXIII (1956), pp. 373-74.
Flint, R. F., "Eskers and crevasse fillings", *Am. J. Sci.*, Ser. 5, Vol. XV, (1928), pp. 410-16; "The stagnation and dissipation of the last ice sheet", *Geog. Rev.* Vol. XIX, (1929), pp. 256-89; "The origin of the Irish 'eskers' " *Geog. Rev.*, Vol. XX (1930), pp. 615-30.
Hoppe, G., "Hummocky moraine regions, with special reference to the interior of Norrbotten", *Geog. Annaler*, Vol. XXXIV (1952), pp. 1-72.
Kupsch, W. O., "Crevasse fillings in southwestern Saskatchewan, Canada", *K. Nederlandsch geologisch mijnbouwkundig genootschap, Verhandelingen geologische series*, Deel 16 (1956), pp. 236-40.
Mannerfelt, C. M., "Nogra Glacialmorfologiska Formelement", *Geog. Annaler*, Vol. XXVII (1945), p. 239.
Tanner, V., "Outlines of the geography, life and customs of Newfoundland-Labrador (The eastern part of the Labrador peninsula)", *Acta Geog. Fenn.*, Vol. VIII (1944), p. 909.

8

The Meaning of Till Fabrics

S. A. Harris

Introduction

Since the orientation of pebbles in tills was first measured quantitatively by Richter,[1] fabric analysis has become one of the more commonly used tools of anyone wishing to reconstruct past depositional environments. Wadell[2] applied it to eskers and deltaic sediments and since then it has been applied to the study of almost all the sedimentary deposits that contain pebbles.

Pebble orientation is often mentioned as one of the diagnostic properties of tills. When carrying out a study of the till sheets near Waterloo, Ontario (see Table I), the writer started collecting data on the till fabrics using the standard methods. The more data he collected, the more he thought about the interpretation of the results and the more doubtful he became about both the methods and

TABLE I. STRATIGRAPHY OF THE ERIE-SIMCOE LOBE TILLS NEAR WATERLOO.[3]

Lobe	Till	Suggested Correlation	Classification by Karrow[4]
Simcoe	Parkhill	Ashtabula or Halton Till?	Wentworth
Erie	Port Stanley	Port Stanley	Port Stanley
Simcoe	Bamberg	....	Wentworth
Erie	Catfish Creek	Catfish Creek	....
Erie	Flamborough	U. Bradtville	Wentworth and Catfish Creek tills
Erie	Canning	L. Bradtville	Canning

First published in *Canadian Geographer*, Vol. XIII, no. 1 (1969), in press.

[1] K. Richter, "Die Bewegungsricjtung des Inlandeises rekonstruiert aus den Kirtzen und Langaachesen der Geschiebe", *Zeits. f. Geschiebeforschung*, 8 (1932), pp. 62-66.
[2] H. Wadell, "Volume, shape and shape position of rock fragments in open work gravel", *Geog. Annaler*, 1 (1936), pp. 76-92.
[3] S. A. Harris, "Origin of part of the Guelph drumlin field and the Galt and Paris Moraines, Ontario", *Canadian Geog.*, XI (1967), pp. 16-34.
[4] P. F. Karrow, "Pleistocene Geology of the Hamilton-Galt map area", Ontario Dept. Mines Geol. Ref. No. 16 (1963), p. 68.

the conclusions reached in such studies. This paper sets forth these doubts, which are just as applicable to his own data, used here as examples, as to the data obtained by anyone else. Nevertheless, the study of till fabrics appears to be a very valuable tool provided that adequate care is taken in the interpretation of deformation fabrics.

Field Work

In determining till fabrics, the finer material in a deposit is carefully scraped away from a smooth surface with a knife until a bladed or rod-like pebble is found. Removal of part of the martix enables either the direction of the long axis of the pebble or the direction of the most elongate part of the pebble in a given plane (horizontal or vertical) to be measured. The direction of the most elongate part of the pebble in a horizontal plane was used by the writer because the fabric on the tills generally lies with two modes in this plane.[5] The horizontal plane is also a standard plane of reference from which possible rotations may be determined. Differentiation of the pebbles into the two modes is presumably achieved by differential rates of flow of the ice on either side of the pebble. The vertical differences in ice velocity could also produce a rolling action. There appears to be only one record where a single mode was obtained from a till.[6]

A third mode was detected normal to the horizontal plane by Holmes when he averaged many determinations, but he was unable to explain satisfactorily the presence of this mode by ice movement. An additional problem, first noted by Harrison,[7] is that the pebbles may be deposited with a distinct dip of their longest axis relative to the ice movement. This gives a monoclinic symmetry to the distribution of pebbles when plotted on a polar projection, instead of an orthorhombic symmetry. In the cases described by Harrison, the dip was upstream relative to the presumed direction of ice movement, but recently Harris[8] has found evidence that there is a tendency for the dip of the long axis of the pebbles in the primary mode to occur in the direction towards the ice front near at least one present day ice margin. The difference may be due to deformation by overriding ice in the case studied by Harrison.

5 D. C. Holmes, "Till Fabric", *Bull. Geol. Soc. Am.*, 52 (1941), p. 1314.
6 W. C. Krumbein, "Preferred orientation of pebbles in sedimentary deposits", *J. Geol.*, 47 (1939), pp. 673-706.
7 J. T. Andrews and K. Shimizu, "Three-dimensional vector technique for analyzing till fabrics: Discussion and Fortran program", *Geog. Bull.*, 8 (1966), pp. 157-65.
8 S. A. Harris, "Till fabrics and speed of movement of the Arapahoe Glacier", *Prof. Geog.*, XX (1968), pp. 195-98.

Representation of the Results

The easiest form of representing the results is to plot the data in the form of a two dimensional rose-diagram. The accuracy of measurement of the angular orientation of the pebble has been regarded by other workers (e.g. Andrews and Shimizu[9]) as being ±5°. It is the practical difficulties of the method that limit the accuracy. This raises problems with the choice of class intervals for two dimensional rose-diagrams. 10° or 20° classes are commonly used though the latter has the disadvantage of providing only nine classes. Of the observations, 50-90 per cent usually fall in the primary mode and 5-45 per cent in the secondary mode. The degree of dispersion about these modal classes varies but is generally very small.

The most usual alternative method of representing the results is to plot the data on a three dimensional polar stereographic projection.[10] This overcomes the problem of representing a three dimensional pattern on a two dimensional surface, but it also introduces additional problems in applying statistics to the data in this form. The rose diagram is quite suitable for plotting on maps, which is one of the ultimate uses of the data. This is difficult when using a three dimensional projection. Accordingly, rose diagrams will be used in this account.

Statistical Representation

In the statistical treatment of the results of the field work, there are three basic questions that must be answered. (1) Is there a preferred orientation, (2) What is the degree of orientation (dominance in the modal class) in this direction, and (3) What is the probability that the preferred orientation is real and not imaginary? The treatment may involve the data obtained from one site, or it may involve the study of the variations of the fabric over a given area occupied by a particular till.

There are at least four ways of answering the three questions. Krumbein used an arithmetic method to study the reproducibility and significance of data from a given source.[11] This study suggested that either the arithmetic mean or the mode were good

[9] P. W. Harrison, "New technique for three-dimensional fabric analysis of till and englacial debris containing particles from 3 to 4.00 mm. in size", *J. Geol.*, 65 (1957), pp. 98-105.

[10] W. C. Krumbein, "Preferred orientation of pebbles in sedimentary deposits", *J. Geol.*, 47 (1939), p. 681.

[11] W. C. Krumbein, "Preferred orientation of pebbles in sedimentary deposits", *J. Geol.*, 47 (1939), p. 686.

indicators of the direction of ice movement. Curray[12] used vector analysis as a measure of significance of the primary mode, while Harrison[13] used a Tukey chi-square test. The method of Krumbein has the disadvantage of involving the choice of a suitable source on which to base the statistical work. All three methods produce low values of the primary peak due to the fact that between 5 per cent and 45 per cent of the pebbles may fall in the secondary mode at right angles to it.

This last disadvantage may be overcome by using the conventional chi-square method.[14] Here the numbers of pebbles which would be in each class if all the classes were equally filled are compared with the number of pebbles actually occurring in a given mode. Andrews, and Andrews and Smithson[15] have used it for comparing paired samples, one from each of two environments.

The conventional chi-square method can also be adapted to ensure that the degree of significance reaches at least a 95 per cent level of probability for the primary mode in all the cases examined. This can most easily be done by means of a graph (Figure 1). First, calculations were made of the number of pebbles needed in the primary mode to give a 95 per cent level of probability at a number of different total pebble counts. Then these were plotted on the graph and a smooth line was drawn through them. The number of pebbles requiring to be counted to achieve the 95 per cent level of probability for a count at any site can be determined from the graph, this value being called the *minimum significant orientation count* or *strength* (henceforth called the MSOC). The latter measures the degree of dispersion of the pebbles outside the primary mode, and it varied from 10 to 88 in 118 cases from Southern Ontario (Table 2). Low values of MSOC predominated indicating a high degree of orientation of the pebbles in the tills. A count of 100

12 J. R. Curray, "The analysis of two-dimensional orientation data", *J. Geol.*, 64 (1956), pp. 117-31.

13 W. Harrison, "A clay till fabric; its character and origin", *J. Geol.*, 65 (1957), pp. 275-308.

14 L. K. Kauranne, "A statistical study of stone orientation in glacial till", Finland, *Comm. Geol. Bull.* No. 1881 (1960), pp. 87-97; J. T. Andrews, "The cross-valley moraines of north-central Baffin Island: a quantative analysis," *Geog. Bull.* no. 20 (1963), pp. 82-129; J. T. Andrews and B. B. Smithson, "Till fabrics, of the cross valley moraines of north-central Baffin Island, Northwest Territories, Canada", *Geol. Soc. Amer. Bull.*, 77 (1965), pp. 271-90.

15 J. T. Andrews, "The cross-valley moraines of north-central Baffin Island: a quantitative analysis", *Geology Bull.*, no. 20 (1963), pp. 82-129; J. T. Andrews and B. B. Smithson, "Till fabrics ofthe cross-valley moraines of north central Baffin Island, Northwest Territories, Canada", *Geol. Soc. Amer. Bull.*, 77 (1965), pp. 271-90.

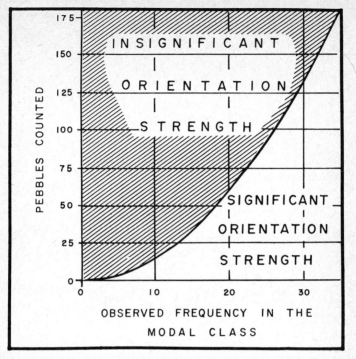

Figure 1. Limit of significance at the 95% level of pebble orientation strengths as a function of the number of pebbles counted for 20° classes.

TABLE II. DISTRIBUTION OF MINIMUM SIGNIFICANT ORIENTATION COUNT AT THE 95% LEVEL OF PROBABILITY FOR 118 SAMPLES OF TILL FABRICS FROM TILLS NEAR WATERLOO, ONTARIO.

MSOC	Number of Samples
0-14	10
15-29	56
30-44	24
45-59	24
60-74	3
75-99	1
Total	118

pebbles was never needed to arrive at a total which was significant at the 95 per cent level of probability for the primary mode, in contrast to the advice of Krumbein and Pettijohn[16]. In fact, this

[16] W. C. Krumbein and F. J. Pettijohn, *"Manual of Sedimentary Petrography"*, Appleton-Century-Crofts, Inc., New York, (1938), p. 218.

level of significance was achieved by counting only 30 pebbles in over half the cases studied. If we are merely concerned with the primary mode, then the question arises as to how few pebbles we may count in order to determine the MSOC. The time and energy that would be spent in digging out, say, another 50 pebbles could be spent in making at least another complete determination at an extra site. As a general rule for the tills of Southern Ontario, a check should be made after 30 pebbles have been examined to see if the 95 per cent level of probability has been reached. If so, no further measurements need to be made at that site.

Results of Analyses

1. Natural variability.

As indicated in Table 2, this may be quite small for a given site. The best method of confirming this is to plot the development of the primary mode on a pebble by pebble basis, as was done in a recent study.[17] Again the determinations at a given site may be collected in batches of 25 and plotted in the form of a rose-diagram. Figure 2 shows an example from a single site in the Parkhill till at G.R. 352, 127, just west of Waterloo, Ontario. Apparently good consistent results are a characteristic of the tills and outwash gravels examined by the writer near Waterloo, Ontario.

TABLE III. REPLICATE DETERMINATIONS OF PEBBLE ORIENTATION IN THE KANSAS TILL AT THE PAPER FACTORY SITE, G.R. 3040,3191, LAWRENCE, KANSAS.

| Sample No. | Soil Horizon | Based on 50 Pebbles | | |
		Pebbles in Primary Mode	MSOC	Direction of Primary Mode
LR22	Cg	17	60	N153°
LR24	B₂	19	49	N152°
LR34	CG	19	49	N153°
LR41	Cg	22	38	N155°
LR42	A₁/A₂	20	44	N155°
LR43	B₂	18	53	N156°
Mean		19	49	N154°
Standard Deviation		1.34		1.56

N.B. The minimum significant orientation count (MSOC) increases the apparent differences from one sample to the next by about 4X in this particular range of values.

[17] S. A. Harris, "Till fabrics and speed of movement of the Arapahoe Glacier", *Prof. Geog.*, XX (1968), pp. 195-98.

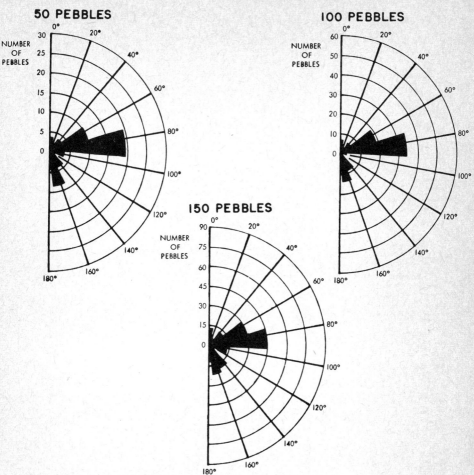

Figure 2. Test of variability of fabric from the same site in Parkhill till.

If the results are consistent for a given block of the deposit, how about local variations laterally and vertically? Six replicate determinations were made along a three metre-high face of Kansas till that extended 50 metres at the Paper Factory site, G.R. 3040, 3191, Lawrence, Kansas. The results are shown in Table 3. The standard deviation of the azimuth of the primary mode was only 1.56°. Soil formation of Yarmouth age has not affected the result, as far as one can tell. However, a similar study in Kansan outwash gravels believed to be the product of ice stagnation[18] produced different results at every location, including some completely random fabrics!

[18] H. G. O'Connor, "Geology and Groundwater Resources of Douglas County, Kansas", *State Geol. Surv. Kansas, Bull.*, 148 (1960), p. 200.

Clearly then, the natural variability of the fabric is an essential part of the properties of a deposit. It needs to be examined as well as the general fabric direction.

2. Deformation fabrics.

Deformation fabrics can be formed in various ways by post-depositional stresses. One such stress appears to be the readvance of an ice sheet over a soft, unconsolidated till, which results, in the formation of tight folds.[19] This produces a rotation of the pebbles in the till in the limbs of the fold so that they come to lie with their long axes at right angles to the fold axis. In the axial region of the fold, the pebbles became aligned parallel to the fold axis. Clearly this must be watched for, since it can cause cross-fabrics, viz., cases where the dominant mode lies at right angles to the probable direction of ice movement of the present ice sheet. Presumably cases where the dominant mode lies obliquely to the direction of probable ice movement may be produced where two ice sheets moved in directions not parallel or at right angles to one another, although no such cases have been described.

The cases described by Banham[19] are from East Anglia, under a present-day climate where the ground rarely freezes. Due to the influence of the Gulf Stream Drift, it is doubtful whether the ground would have frozen to any great depth at any time in the past except during the coldest part of a glaciation. However, Southern Ontario lies near the centre of a continent. Every winter the ground freezes to a depth of between 35 and 120 cms. Thus the likelihood of an ice sheet advancing across previous tills and producing folds in them is appreciably less, even under the present climatic regime. In fact no signs of folding in the tills were observed during the field work. A notable feature of deposits is the lack of displacement of thin underlying tills by later ice sheets. However, ice push structures have been described from Kansas and Iowa.[20]

Instead of folds in the tills of Southern Ontario, there is good evidence for upturning of pebbles into a vertical position in the upper layers of some till sheets. Clearly this could present at least as great a problem in tracing the fabrics for a given till sheet as the folding discussed above. An example is found in the surface

---

[19] P. H. Banham, "The significance of till pebble lineations and their relation to folds in two Pleistocene tills at Mundesley, Norfolk", *P.G.A.*, London, 77 (1966), pp. 469-74.

[20] L. F. Dellwig and A. D. Baldwin, "Ice-push deformation in north-eastern Kansas", *State Geol. Surv. Kansas Bull.*, 175, pt. 2 (1965), pp. 1-16; P. R. Lamerson and L. F. Dellwig, "Deformation by ice push of lithifield sediments in south-central Iowa", *J. Geol.*, 65 (1957), pp. 546-50.

layers of the Flamborough till (probably equivalent to the Upper Bradtville till) at G.R. 457, 113, (Universal Transverse Mercator Grid, Zone 17) east of Kitchener. Studies of the fabrics at different levels within till sheets in this area suggest that there is little variation in orientation strength, direction of orientation, or MSOC with position vertically in the till at a given site except where frost rotation has occurred. Assuming this to be true prior to frost action in the Flamborough till at this site east of Kitchener, fabrics were obtained for 45 cms. and 180 cms. below the present ground surface in this one till. In addition to measuring the orientation of the longest part of the pebble in a horizontal plane, a check was also made on the inclination of the longest axis of the pebble in a vertical plane.

The results of this study are shown in Figure 3. The lower parts of the figures show the marked rotation or upturning of the pebbles in the upper part of the till. However the plots of the azimuths of the pebbles as measured in a horizontal plane show similar fabrics to one another with only a minor reduction of the size of the primary mode in the fabric from 45 cms., i.e., where the pebbles were vertical. However the secondary mode has disappeared from the fabric exhibiting rotation of pebbles.

Could pebble rotation be the real explanation of the tertiary mode in a vertical direction recognized by Holmes? If so, the tertiary mode may be induced by including deformation fabrics in the average fabric, rather than being part of the original apposition fabric of the till. As an induced feature it would not demand special explanation as regards ice movement. In a similar fashion, the third component of the orthorhombic or monoclinic fabrics described by Andrews and Shimizu[21] may perhaps be due to frost action. Certainly the results suggest this.

Turning to the reliability of the data, the primary modes in the fabrics from both 45 and 180 cms. are significant at the 95 per cent level of probability yet the difference in angle of dip in the direction of the long axes of the pebbles is extreme. Clearly pebble reorientation by frost action is unimportant in modifying the preferred azimuth of the till fabric in this example, and the direction of movement of the ice determined from the modified till fabric is not changed. Frost action is also relatively unimportant in its effect on the MSOC of the fabric.

---

[21] J. T. Andrews and K. Shimizu, "Three-dimensional vector technique for analyzing till fabrics: Discussion and Fortran program", *Geog. Bull.*, 8 (1966), p. 153.

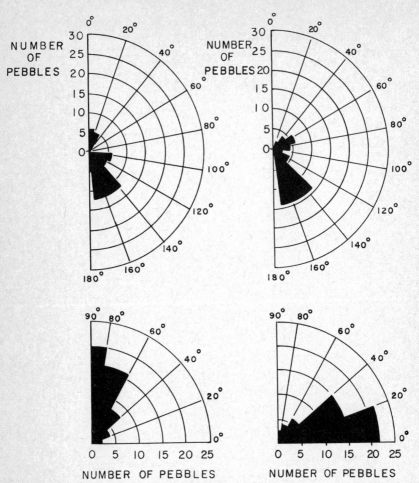

Figure 3. Local relief based on a square kilometre unit for the area around Waterloo, Ontario. The upper parts of the diagram show the azimuth of the pebbles while the lower parts show the angle of declination of the long axes of the pebbles. Note the lack of serious modification in the dip of the pebbles in the surface layers of the till, but the lack of effect on the mode in the upper diagrams.

Upward movement of pebbles and stones within the zone of freezing and thawing continues at many places in Quebec and Ontario under present-day climatic conditions. Similar pebble movement undoubtedly occurred in the upper parts of the tills in early postglacial times in Southern Ontario. The only way to measure its effect is to compare a till fabric with the fabric of pebbles which have moved upwards into younger sediments that originally lacked pebbles.

It is realized that these results obtained in two tests of the effect of frost action represent too small a number of tests to be statistically significant. These results may either represent a fortunate coincidence or may be typical of what would be obtained in a more extensive study. Time and the availability of suitable exposures precluded a more extensive examination of the problem. However, taken with the consistency of the results of plotting large numbers of till fabrics on a map, the present interpretation appears to be reasonable.

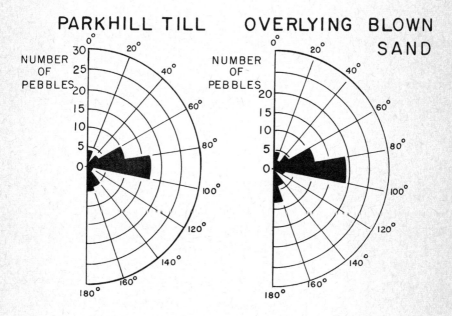

Figure 4.

Figure 4 shows paired fabrics for the pebbles in the Parkhill till and for the overlying windblown sands at G.R. 352, 127, 1 mile west of Waterloo, Ontario. The fabrics are remarkably alike, the primary mode being slightly more marked in the overlying sands. A second study was made of paired fabrics of pebbles in clayey Port Stanley till and in the overlying lake silts at G.R. 249, 062,

Baden, Ontario (see Figure 5). Once again there is little difference between the fabrics, though in this case, the secondary mode has strengthened somewhat at the expense of the primary mode in the overlying silts. Since the difference between the fabrics in the different materials is so small, it seems reasonable to assume that the movement of pebbles within the till has had little effect on the fabric as measured here.

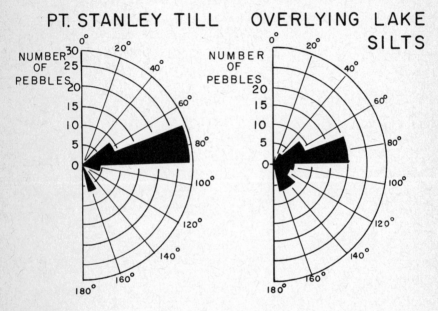

Figure 5.

The presence of a till-like fabric in a sediment is clearly a poor indicator of origin of the sediment in this environment and must never be used on its own. Identification of depositional environment must involve other methods such as grain size analysis, and presence or absence of structures such as bedding, etc. By using the fabric technique, it should be easy to distinguish between a deposit where pebbles have moved upward as a result of frost action and an ice-rafted deposit containing pebbles. The ice-rafted deposits are most unlikely to have the same fabric as the till. The only exception to this might be where ice stagnation has produced a till with a randomly orientated fabric pattern in space. However this has yet to be demonstrated.

Frost action may also cause polymodal fabrics when the pebbles from one till have moved up into a thin overlying younger till with a different fabric. This would have a bimodal fabric in the lower till and a polymodal fabric in the upper till. The polymodal fabrics found so far by the writer could all be explained in this way. Modification of the surface of the underlying till by ice probably explains the infrequent occurrence of zones of upturned pebbles, though this is also a function of climate in some cases.

While frost action does not appear to affect markedly the orientation of the axes, mass wasting is quite another matter. Two studies were carried out on the effect of colluviation on till fabrics. The first involved paired samples of Canning till and overlying colluvium at G.R. 246, 171, just west of Bamberg, Ontario. The results are shown in Figure 6. Figure 7 shows the results of a similar study of

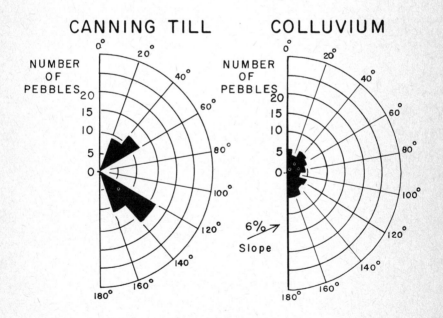

Figure 6.

loamy Bamberg till and its overlying colluvial layer at G.R. 323, 123, one mile south of Erbsville, Ontario. Complete destruction of the fabric has taken place in both cases. The products of mass wasting should not produce inaccuracies in mapping till fabrics since the two sediments are so obviously different. Likewise, a check on the relationship between local slopes and the fabric should establish whether any realignment due to earth flowage may have occurred. Once again, the possibility of ice stagnation must also be considered.

Lundquist[22] has shown that in the case of earth flowage the pebbles will be reoriented parallel to the new direction of movement, the modes of the original depositional fabric having been obliterated. In the cases of the tills of Southern Ontario, only on the margins of the older kettle holes was this type of reorientation observed.

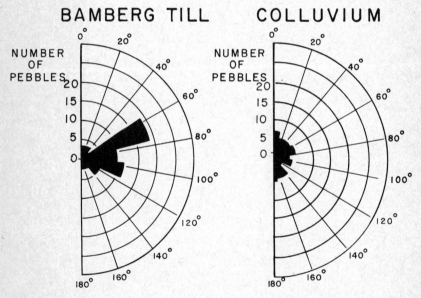

Figure 7.

From this discussion, it would appear that deformation fabrics can be readily differentiated from primary depositional (apposition) fabrics in tills. Even though frost action may cause rotation of pebbles into an upright position, the primary mode generally seems to remain the same.

22 A. Lundquist, "The orientation of block material in certain species of earth flow", *Geog. Annaler*, 31 (1949), pp. 335-67.

3. Relation of the fabric to ice movement.

Taking the preceding discussion into account, it is possible to make maps of the depositional fabrics for a given till.[23] The exact density of observations needed to produce an adequate map will be a function of the complexity of the situation so that minimum sampling density is a variable factor. Near Waterloo, the data needed to be obtained from sites which were within 2-5 kilometres of one another, but given at least this density of observations, the results produce a characteristic pattern for a given till sheet. Andrews and Shimizu[24] have inferred that there may be more than one pattern for a given till sheet, i.e., the ice changed its direction of movement across an area without retreating, but this remains to be thoroughly demonstrated.

The pattern produced by the fabrics for a given till permits the drawing of flow lines parallel to the bulk of the long axes of the primary modes. In many cases, some of the fabrics appear to be transverse to the rest, i.e., the primary mode lies in a direction at right angles to the bulk of the fabrics. Similar evidence comes from ice margins; the normal arrangement of the majority of the pebbles is with their long axes parallel to the direction of the movement of the ice, though cases of transverse fabrics have also been described.[25] Holmes ascribed the occurrence of long axes of pebbles parallel to the ice movement to lodgement of pebbles in the till as they were being slid along beneath the ice. The pebbles lying across the direction of movement were regarded as being rolled along. Glen, Donner, and West[26] concluded that if negligible reorientation occurred during deposition, material that travelled a great distance will show a transverse primary mode whereas pebbles that have travelled a short distance will show a dominant mode parallel to the direction of movement. So far the available data on the percentage of pebbles from the rocks of the Shield and the percentage of trans-

[23] R. G. West and J. J. Donner, "The glaciations of East Anglia and the East Midlands; a differentiation based on a stone orientation measurements of the tills", *Q. J. Geol. Soc. Lond.*, 112 (1956), pp. 69-91; S. A. Harris, "Origin of part of the Guelph drumlin in field and the Galt and Paris moraines, Ontario", *Canadian Geog.*, Vol. XI, (1967), pp. 27-29.
[24] J. T. Andrews and K. Shimizu, "Three-dimensional vector technique for analyzing till fabrics: Discussion and Fortran program", *Geog. Bull.*, 8 (1966), pp. 151-65.
[25] S. A. Harris, "Till fabrics and speed of movement of the Arapahoe Glacier", *Prof. Geog.*, Vol. XX (1968), pp. 195-8; J. J. Donner and R. G. West, "The Quaternary Geology of Brageneset, Nordaustlandet, Spitzbergen", *Norsk. Polarinst. Skr.*, 109 (1956), pp. 1-37.
[26] J. W. Glen, J. J. Donner and R. G. West, "On the mechanism by which stones in till became oriented", *Amer. J. Sci.*, Vol. 255 (1967), pp. 194-205.

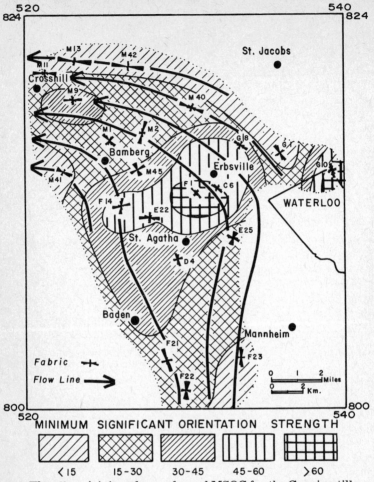

Figure 8. Flow lines joining places of equal MSOC for the Canning till near Waterloo. Note the low values where the flow lines are close together, further suggesting a correlation with greater relative speed of movement of the ice.

verse modes fails to indicate such a relationship (Table IV). Likewise plots of MSOC versus percentage of pebbles from the Shield fails to yield anything but a random scatter. There is, however, a noticeable relationship between the appearance of transverse modes and local relief (Figures 8 and 9). There is little relationship with height of the land above sea level (Figure 10). The swing in the direction of ice movement in the case of the Canning till is presumably due to collision with Huron lobe ice sheets.

27 A Luttig "Eine, neue, einfache geröllmorphemetrische Methode", *Eiszeitalter und Gegenwart*, 7 (1956), pp. 13-20; and "The shape of pebbles in the Continental, Fluviatile and Marine Facies", Int. Assoc. Sci. Hydrol., Publ. 59 (1962), pp. 253-58.

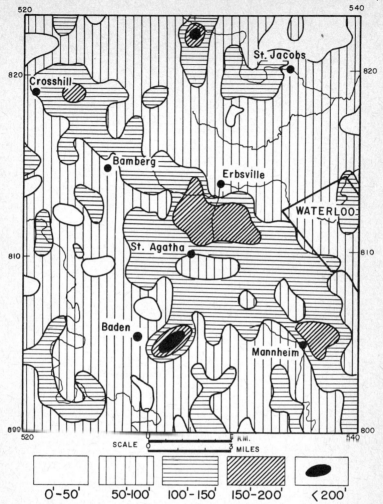

Figure 9. Local relief based on a square kilometre unit for the area around Waterloo, Ontario. Note the good correlation between roughness of terrain and high values for minimum significant orientation strength for the Canning till.

TABLE IV. COMPARISON OF PERCENTAGE OF TRANSVERSE MODES MEASURED WITH MEAN CONTENT OF CRYSTALLINE PEBBLES DERIVED FROM THE CANADIAN SHIELD.

Till	% Transverse Modes	Average % Crystalline Pebbles
Parkhill	0	30
Port Stanley	5	18
Bamberg	4	20
Catfish Creek	10	33
Flamborough	0	15
Canning	10	22

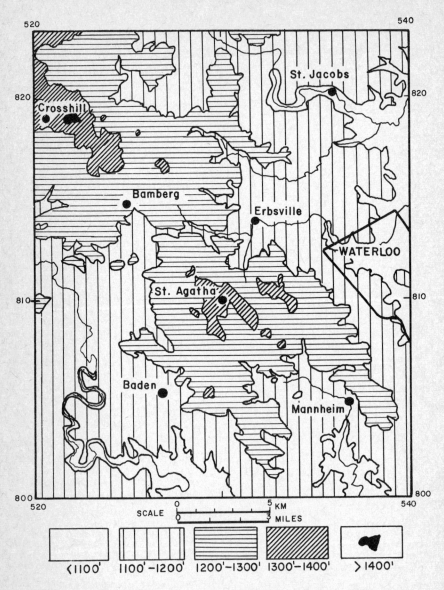

Figure 10. Actual relief of the area around Waterloo, Ontario (after the 1:50,000 series map of the Department of National Defence). Note the lack of correlation with the minimum significant orientation strength.

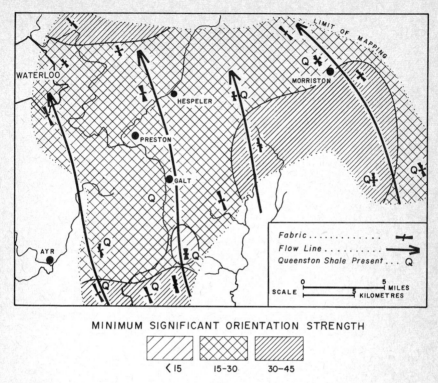

MINIMUM SIGNIFICANT ORIENTATION STRENGTH

⟨15   15-30   30-45

Figure 11. Variations in the MSOC of fabrics from the Flamborough till across the Galt map sheet.

It is also possible to map the values of MSOC. Figure 11 shows an example where this has been done for the Flamborough till east of Waterloo. The results show a moderate range of values in this case, but for the Canning till west of Waterloo, wide variation may be seen (Figure 8). It has recently been shown that there is a close relationship between MSOC and speed of ice movement at the terminus of the Arapahoe Glacier in Colorado.[28] If this is typical of other glaciers, then we should expect that the ice sheet that deposited the Canning till slowed down at the sites where the MSOC is greatest. Comparison of Figures 8, 9, and 10 indicates that this occurs where there is the greatest amount of local relief, not where the land is highest and presumably where the ice was thinnest. This would therefore appear to agree satisfactorily with the published data. Similar results are found in the cases of the later tills in the same area.

---

[28] S. A. Harris, "Till fabrics and speed of movement of the Arapahoe Glacier", *Prof. Geog,,* Vol. XX (1968), pp. 195-98.

Holmes and others[29] have ascribed variations of development of the primary mode to the physical nature of the pebbles rather than to the nature of the ice sheet that produced the till. Table V shows the main physical properties and the MSOC for ten randomly chosen samples from the tills near Waterloo. No significant relationship between the properties and the MSOC is apparent. Local variations within the till sheets are much larger than any differences related to MSOC. Thus it seems likely that the variations must be connected with some factor in the movement of the ice.

Provided that the shape of the pebbles in a series of tills is about the same, it should be in order to compare values for the MSOC in each till. This is certainly the case for the tills near Waterloo and those sampled at the Arapahoe Glacier.

Comparison of the values of MSOC for the Ontario tills with those for the tills in front of the Arapahoe Glacier suggests that if the Arapahoe results are typical of glacier in general, the maximum speed of movement of the basal zones of the continental ice sheets ranged from one to six feet per year. These figures are interesting if only to point out a possible major discrepancy between the rate of ice advance and the rate of movement of the basal zone of the ice sheets. The currently assumed rate of advance of the Port Stanley ice sheet requires a minimum coverage of 300 kilometres in 400 years, i.e., a rate of 2600 feet per year. This gives us a discrepancy of nearly 500 times! It would seem to suggest very fast rates of flow in the upper layers of the ice sheet with an overturning of these layers in front of the terminus during an ice advance. A similar motion is seen in avalanches and in fast-moving lava flows. This overturning motion would, of course, be absent from the terminii of ice sheets when they were at their maximum extents. If this type of movement did not occur, then this implies that the data from Arapahoe are not transferable, in other words, different values of MSOC will be found by studying different glaciers whose rate of movement of the basal zone of ice is the same. Obviously more information is necessary.

---

29 D. C. Holmes, "Till Fabric", *Bull. Geol. Soc. Am.*, 52 (1941), p. 1340; J. W. Glen, R. G. West and J. J. Donner, "On the mechanism by which stones in till become oriented", *Amer J. Sci.*, Vol. 255 (1967), pp. 194-205; J. T. Andrews and C. A. M. King, "Comparative till fabrics and till variability in a till sheet and a drumlin: a small scale study", *Proc. Yorks. Ged. Soc.*, Vol. 36 (1968), pp. 435-661.

TABLE V. MINIMUM SIGNIFICANT ORIENTATION STRENGTH, MEAN SIZE, MEAN SHAPE, MEAN ROUNDNESS, AND PEBBLE COMPOSITION FOR RANDOMLY CHOSEN SAMPLES OF PEBBLES FROM DIFFERENT TILLS NEAR WATERLOO, ONTARIO. SHAPE AND ROUNDNESS ARE DETERMINED USING THE LÜTTIG METHOD.26

Till	Sample No.	Pebbles Counted	Mean Length of axes m.m.			Ratios		p%	Breakage %	p-8%	MSOC	Pebble Composition (%) of > 2 mm. fraction.		
			a	b	c	c/a	c/b					Evaporites#	Clastics	Cryst.
Parkhill	L1	147	14.5	11.0	6.8	.496	.671	41.9	n.d.	10	23	84	6	10
Bamberg	N11	171	15.7	12.0	8.1	.524	.686	32.4	50	10	28	61	18	21
	N24	100	12.1	9.3	7.0	.575	.743	36.9	39	12	34	87	8	5
	N41	73	14.2	10.5	7.5	.545	.740	27.3	93	19	13	37	40	23
	M49	159	14.9	11.4	7.0	.455	.602	27.8	82	20	40	26	49	25
	Mean 4 samples	—	14.2	10.8	7.4	.524	.693	31.1	66	15	—	—	—	—
Catfish	G2	178	17.2	11.3	7.6	.437	.690	26.9	54	24	16	36	31	33
Canning	G17	158	14.7	11.0	6.9	.486	.624	34.0	46	13	52	49	28	23
	N3	108	17.4	13.4	9.2	.530	.686	34.4	44	13	16	89	4	7
	E22	173	12.4	9.0	5.9	.470	.663	30.6	72	15	45	40	38	22
	F14	119	14.1	10.7	7.5	.519	.689	32.8	81	13	50	28	44	28
	Mean 4 samples	—	14.6	11.0	7.4	.501	.666	32.9	61	13	—	—	—	—

*"Evaporites" includes limestone, dolomite and chart pebbles.

Conclusions

1.   Considerable care is needed in the interpretation of till fabrics. There are major gaps in our knowledge, e.g., in the total variation of fabrics at a given site, and in the nature of the average orientation of all three axes of individual pebbles.

2.   Till fabrics can be mapped to a constant degree of accuracy by using the chi-square method of Kauranne (1960). This appears to be the most satisfactory statistical method devised so far.

3.   Using this method it is possible to determine the minimum significant orientation strength or MSOC for each sample at a chosen degree of significance (95 per cent level in this case). This gives us a measure of dispersion of the pebbles about the primary mode.

4.   Frost action involving rotation of the pebbles into a vertical plane and actual upward movement of the pebbles into overlying younger sediments failed to modify the direction of the primary mode in the cases studied here (Figures 3-5). The minimum significant orientation strength changes only slightly in the process.

5.   Partial alignment of the pebbles in a vertical direction by frost action may account for the third vertical mode noted by Holmes.

6.   Upward movement of pebbles from one till into another is a possible explanation of polymodal fabrics.

7.   Colluviation under post-glacial conditions results in complete destruction of the original fabric.

8.   Plotting of the fabrics for a given till produces consistent patterns which vary considerably locally from one till to the next, but reflect directions of the ice flow. Flow direction within one till may vary appreciably over a distance of 20 kilometres. Richter was obviously fortunate to obtain such consistent results.

9.   There is appreciable variation in minimum significant orientation strength in a horizontal direction from one part of a till to another (Figure 8). This correlates well with degree of local relief within the study area (Figure 9) but only poorly with actual relief or with probable ice thickness (Figure 10).

10.   The variations in MSOC fail to correlate with mean pebble size, mean pebble shape, and mean roundness or lithology.

9

Lithology of the Erratics Train in the Calgary Area

A. V. Morgan

Introduction

The city of Calgary is situated on a roughly north-south trending line of quartzite erratics known as the Foothills Erratics Train. The quartzite blocks are of considerable size and have thus attracted the attention of many geologists since the time of the first observations by Hector (Palliser *et al.*, 1863) in mid-nineteenth century. Stalker (1956) suggested that the Foothills Erratics Train demarcates the approximate limit of Cordilleran and Laurentide ice.

This paper discusses the lithology of tills in the Calgary area with the intent of relating the lithology of glacial deposits to the position and other characteristics of the Foothills Erratics Train. It was hoped that this would aid in the interpretation of the Train. The lithology of the till was also envisioned as an aid in interpreting the extent and origin of glaciers formerly covering the study area and the sequence of glacial events in the map region.

Previous Studies

The limits of the different lithological indicators in the Calgary area were described by Hector (Palliser *et al.*, 1863) and Dawson and McConnell (1885, 1895). Nichols (1931) was the first writer to report Laurentide pebbles west of Calgary. In 1955 Gravenor and Bayrock used three indicators; Athabasca sandstone, McMurray tar sand, and the Presq'ile dolomite, to illustrate ice movement in central and northern Alberta. The existence of large blocks of quartzite forming the Foothills Erratics Train has been noted by many authors since Hector. Rutherford (1941), Allan (1943) and Horberg (1954) are some of the more recent authors who have contributed to the description of the Train, its relationship to other

Part of an unpublished M.Sc. Thesis presented to the Department of Geography, University of Calgary, 1966.

deposits, and the origins of the large blocks. By far the most comprehensive account is by Stalker (1956) who described details of the Train, and gave a good historical summary of the theoretical origins of the quartzite erratics. Tharin (1960) has also described the position of some of the quartzite erratics in the Calgary area.

Similar lithology studies have been conducted elsewhere using indicator erratics; Mackintosh (1879), Knechtel (1942) and Gillberg (1955). Flint (1957) gives a brief summary of some of the early work in this field. Papers by Gillberg (1955, 1965) outline some of the problems in this type of study.

Sampling Methods

Sample sites were chosen by picking land sections from a random number chart (Wolfe, 1962) and although 171 sites were examined in an area of 1,634 square miles, only 155 were regarded as being suitable for analysis.

When the site localities had been selected on the township and range grid each land section was visited and a pebble sample taken from the first roadcut bank available. If the section did not produce a suitable exposure, the nearest exposure adjoining this section was taken; often this was just across the road. A sample site was passed over if a suitable exposure was not found within three-quarters of a mile of the appropriate land section. For example, sample sites which were placed by the random number chart in the centre of lacustrine deposits were ignored if a suitable exposure was not found within three-quarters of a mile of the section which should have been sampled. This sampling method was subjective, in that the sample sites chosen were in till and not, for example, in lacustrine silts or pre-glacial gravels.

Each sample usually consisted of 50 or 100 erratics randomly selected from the face of the exposure. Most of the erratics collected were between 4mm. and 256mm. Samples were brought back to the laboratory, washed and placed into one of eight lithologic categories. Pebble size was measured in millimetres and roundness was classified using a chart by Krumbein (1941) for visual roundness. When this was complete the percentage of lithological types for each locality was calculated.

During the mapping of the study area 1,612 erratics estimated at over two cubic feet (herein called mega-erratics) were plotted on 1:50,000 topographic maps. Erratics smaller than two cubic feet are usually moved by farmers from the surface of the field to the edge of the nearest road section. Erratics in excess of this size can be seen in the centre of fields. Three categories of mega-erratics were used in field mapping: igneous and metamorphic; dolomites and limestones; and the quartzitic conglomerates of the Foothills Erratics Train. As with the study of pebble lithology, the study of mega-erratics was undertaken with the hope that their distribution would cast light on the inter-relationship between Cordilleran and Laurentide ice sheets.

Distribution of Selected Lithological Groups

When all the representative erratic lithologies at each site had been expressed as percentages certain lithological groups were used as areal indicators. For example, igneous and metamorphic pebbles were chosen as representing erratics brought into the study area from the Canadian Shield by Laurentide ice. The merits of the various indicators will be discussed later.

The percentages for any one of the lithological groups were plotted on a map of the study area, and the distribution of that rock type indicated by isolithological lines. Isolithology maps were prepared for igneous and metamorphic erratics; limestone and dolomite erratics, which were combined and expressed as total percentage carbonate; and finally for brown and purple quartzites.

Isolithological lines were drawn at 5 per cent intervals to show the percentage distribution of the various lithological groups in the map area. The resultant maps were complex and showed unnecessarily high peaks. They were simplified by using 10 per cent line intervals and grouping readings over 50 per cent into one category (e.g. Figure 1).

Observations and Values of the Lithological Indicators

Igneous and Metamorphic Erratics

Figure 1 shows the percentage distribution of igneous and metamorphic pebbles within the map region. The granites, pegmatites, gneisses and schists typical of the igneous and metamorphic rock groups are good general markers. The question arises as to whether these igneous and metamorphic erratics could have been brought into the map area from the west. Only one possible source area is known immediately to the west of the map region. Allan and Carr (1947, p. 30) report an igneous pebble conglomerate, 1,200 feet above the basal Blairmore conglomerates in section 17, Tp.17, R.6, W.5. It reaches a thickness of 85 feet at the forementioned locality but thins to 10 feet at Loomis Creek, one and a half miles to the northwest. The pebbles consist of pink, purple and green syenite, pink to purple feldspar porphyry and rare pink granite. The present writer feels that if any igneous pebbles were brought into the area from

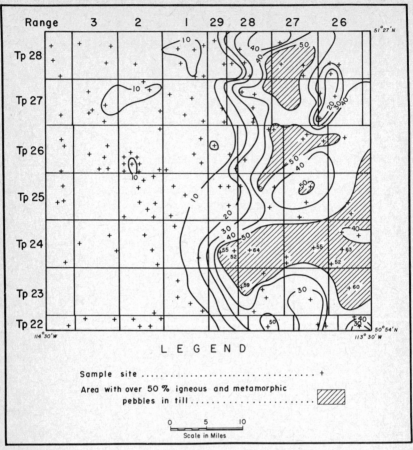

Figure 1. Percentage of igneous and metamorphic pebbles in till.

the west, from deposits like those mentioned above, their numbers would be so small as to make them practically unnoticeable.

Rutter (1965) working in the Banff area of the Bow Valley has not found any igneous or metamorphic pebbles, and presumes that ice did not cross from the western side of the Continental Divide. There is a possibility that metamorphic pebbles may have been brought across the Divide by glaciers in the Athabasca area (Roed et al., 1967) but their numbers should be limited by the time they reached the Calgary region.

Since the percentage of igneous and metamorphic pebbles is negligible in the west, but rises appreciably to the east, all igneous and metamorphic erratics in the study area are considered to have been derived from the Canadian Shield over four hundred miles to the northeast. In this regard the source of transportation must have been Laurentide ice which scoured the pebbles from the Shield area

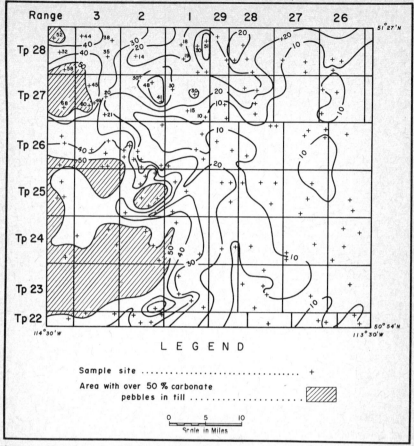

Figure 2. Percentage of carbonate pebbles in till.

and carried them to the Calgary region. It follows from this conclusion that the last Laurentide ice sheet must have covered the eastern half of the map area, at least as far as the fifth meridian (114°W.). The presence of Shield-derived pebbles west of the fifth meridian is discussed later in this paper.

Carbonate Erratics

The percentage distribution of carbonate pebbles is shown in Figure 2. Some carbonates in the eastern half of the map area could have been derived from the sedimentary areas fringing the Shield. However, the concentration of carbonates in the west, and the continuity of distribution towards the mountains, leads to the conclusion that the carbonates of the Western Cordillera provided the main contribution of erratics of this lithology in the map area. Further details of the carbonates and their distribution east of the fifth meridian are given later.

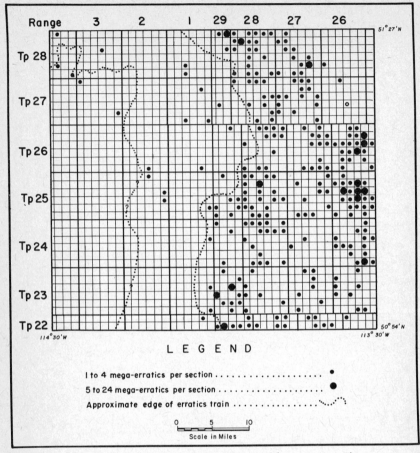

Figure 3. Distribution of igneous and metamorphic mega-erratics.

Brown Quartzite Erratics

The term "brown" in brown quartzite refers generally to the colour of the brownish-yellow weathering patina, although the rocks may be yellow or even white on unweathered surfaces. It seems likely that these erratics were derived from the Western Cordillera since they occur west of the study area, and are similar in lithology to the Lower Cambrian quartzites. Comments on their distribution and origin will be made later in this paper.

Purple Quartzite Erratics

Although these are believed to have been derived from valleys west of the study region their distribution appears to be concentrated east of the fifth meridian. Their distribution will also be discussed later.

Observations of the Mega-erratics

Igneous and Metamorphic Mega-erratics

The edge of the Laurentide drift has been arbitrarily delimited by the superimposition of Figures 3 and 5. The junction of the Laurentide drift and the eastern edge of the Foothills Erratics Train was defined by drawing a line between the easternmost occurrence of the quartzite erratics and the westernmost occurrence of Laurentide mega-erratics adjacent to it. The western edge of the Erratics Train is delimited by the 4,100 foot contour, believed to be the surface level of the glacier in the Calgary area for reasons given later. The area between the western and eastern limits of the Train is herein referred to as the zone of Erratics Train ice.

Shield-derived mega-erratics in the map area are located in Figure 3. The areas east of the Foothills Erratics Train, which appear to be devoid of mega-erratics, are probably lacustrine regions where the erratics have probably been covered by lake deposits. The mega-erratics west of the zone of the Erratics Train have been suggested to be remnants of a pre-Wisconsin or early Wisconsin Laurentide advance, (Nichols, 1931; Stalker, 1956; Tharin, 1960).

The Shield erratics consist of igneous and metamorphic rocks, primarily red granite and feldspar pegmatites, as well as gneisses and schists. Several occurrences of garnet gneiss were noted with large garnet inclusions. The Shield erratics are generally rounded and a few are striated.

Carbonate Mega-erratics

The distribution of mega-erratics which are believed to have been derived from the Cordillera west of the map area, is shown in Figure 4. The erratics consist of a variety of limestones and dolomites and, although a few have yielded fossils, none of the mega-erratics in this category are sufficiently characteristic to be traced to a definite geographical locality. These mega-erratics have been transported at least 50 miles from the front ranges of the Western Cordillera.

The carbonate mega-erratics east of the Erratics Train are problematical as they occur well into the zone of Laurentide till. These carbonate mega-erratics will be mentioned later.

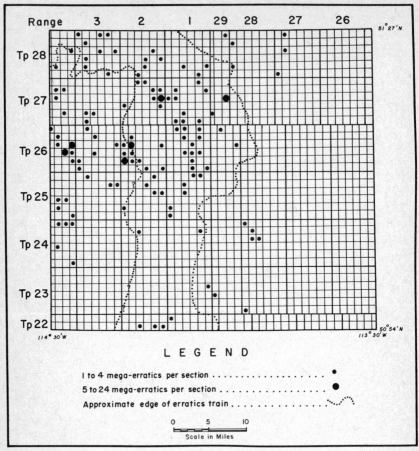

Figure 4. Distribution of carbonate mega-erratics.

Quartzite Mega-erratics

The quartzites of the Foothills Erratics Train were the third lithological group mapped, and the distribution of these erratics is shown in Figure 5. The largest known erratic in the Train occurs 13 miles south of the map area in the southeast quarter of section 21, Tp.20, R.1, W.5. Stalker (1956) refers to the rocks in the Train as quartzites, and this description is adhered to in this paper. One of the most useful characteristics for field identification was the presence of rose quartz grains and pebbles in the blocks.

The source of the erratics has been a major problem since the middle of the last century (Hector, in Palliser *et al.*, 1863). Stalker (1956, p. 14) summarizes the earlier literature in which Hector and Dawson venture the opinion that the source is north-

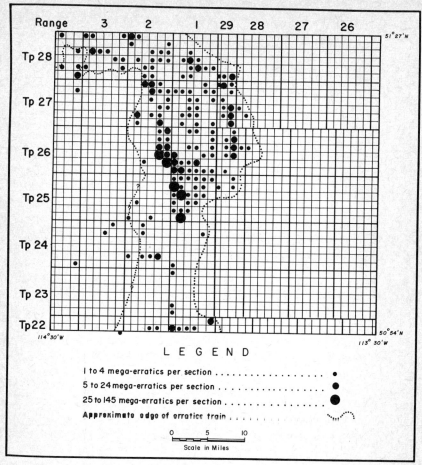

Figure 5. Distribution of Erratics Train quartzite (mega-erratics).

eastward, whilst writers since 1900 have stated that the source is from the west. E. W. Mountjoy (1958, pp. 218-226) suggested, on the basis of a study of the lithological details, that the erratics could have been derived from the Cavell Formation near Jasper.

The transportation of the erratics has been a second problem. One of the characteristics of the erratics is that the quartzite blocks are angular, bedding plane fragments. As Stalker (1956, p. 10) has reported:

. . . no glacial striae were observed during careful examination of many of the blocks and it seems probable that none were produced during the transportation of the blocks to their present positions.

The blocks are nearly always situated on the surface of the ground, usually on ground moraine, but there are localities where the erratics are found within the ground moraine. Two localities, section 7, Tp.28, R.3, W.5, and section 6, Tp.23, R.1, W.5, were found where pebbles of the same lithology as the large quartzites were noted in the till. These rounded cobbles, both less than one foot in diameter, probably fell into crevasses and were incorporated in basal till. In order to satisfy conditions above, it is now generally believed that the blocks were transported on or near the surface of an ice sheet (Stalker, 1956, p. 15). This would help to explain the absence of striations, in that subglacial material could not erode the blocks and it would also prevent the blocks from grounding in low country. Stalker (1956, p. 16) states:

> The difficulty of raising such blocks to the surface of a glacier, if they were quarried near its base, without rounding and striation, indicates that the blocks were originally received by the transporting glacier upon or near its surface.

Earlier writers believed that blocks of rock which were as large as those of the Train could not have been carried far without being broken up, rounded and striated. Consequently the earliest workers looked for methods other than glacial transport, and overthrusting and ice rafting in glacial lakes were both suggested, but later discarded, (Hume, 1931; Rutherford, 1941; Horberg, 1954). Stalker (1956, p. 17) noted that quartzitic erratics occurred up to 4,100 feet and, with the exception of two erratics mentioned below, this altitude agrees with the present writer's observations. Since the cut-off altitude of the quartzitic erratics in the Calgary area is approximately 4,100 feet, and assuming that the erratics were carried to this position on the surface of the glacier, the margin of the Erratics Train ice during the time of emplacement must have been near 4,100 feet.

The igneous and metamorphic mega-erratics that occur in the map area west of the Foothills Erratics Train are higher than the 4,100 foot limit of the quartzites. Dawson and McConnell (1885, p. 148), Nichols (1931, p. 53), and Stalker (1956, p. 17) are a few of the authors who have reported Laurentide erratics above 5,000 feet in the foothills west of the Erratics Train. This suggests that at least one earlier Laurentide glaciation advanced into the foothills and that the ice was at least 1,000 feet thicker than the glacier which emplaced the Erratics Train. The highest igneous and metamorphic erratics recorded west of the Erratics Train, within the map area, occur at 4,125 feet.

Other Methods of Analysis of Drift Lithology

Heavy Mineral Analysis

From inferences made by the examination of the distribution of mega-erratics and isolithological lines, the writer suspected that a similar demarcation might be shown in heavy mineral suites within the Cordilleran and Laurentide drift areas. This research could not be undertaken because of time restriction, however, the heavy minerals of the Cordilleran till have been described by Rutter (1965), whilst Bayrock (1962) has described heavy minerals in the till of central Alberta.

X-Ray Diffraction

This procedure was not used by the writer but the clay mineral contents of the tills present in the western half of the map area have been studied using this method by Tharin (1960, pp. 86-102).

Carbonate Analysis

The Chittick gasometric determination of percentage carbonate in a till has been used by Dreimanis and others (1953, 1962). Tharin (1960, pp. 93, 94, 105-108) found that the percentage carbonate in the western tills was higher than in eastern tills, the percentage carbonate also decreasing from the west within

TABLE I. CARBONATE CONTENT OF TILLS IN THE CALGARY AREA.

Author	Locality	% Calcite	% Dolomite	Total % Carbonate
N.60	Banff	41.5	37.9	79.4
N.87	2.5 miles N.W. of Canmore	35.0	26.6	61.6
N.94	West of Bow at Canmore	24.5	42.6	67.1
T.	S.34, Tp.25, R.7, W.5.	—	—	54.0
M.27	S.11, Tp.27, R.4, W.5.	19.5	8.9	28.4
M.89	S.15, Tp.25, R.4, W.5.	12.4	11.3	23.7
T.	S.14, Tp.28, R.1, W.5.	—	—	9.0
T.	S.1, Tp.26, R.1, W.5.	—	—	21.0
M.16	S.29, Tp.28, R.28, W.4.	5.8	3.9	9.7

N. Rutter (1965)

T. Thorin (1960)

M. Morgan (1966)

the same till sheet. In the western Bow Valley the carbonate percentage of the tills was as high as 54 per cent, whilst in the Lochend Hummocky Moraine area the percentage dropped to 8 per cent carbonate. Rutter (1965) also sampled exposures for carbonate content in the Bow Valley tills. Selected samples from Tharin and Rutter's work, as well as from the present study are shown in Table 1.

The percentage carbonate in the tills steadily decreases eastward from 79.4 per cent at Banff (Rutter: N-69) to 9.7 per cent (M-16) east of the fifth meridian approximately five miles into Laurentide drift. Whilst preparing sample M-27 fragments of feldspar were found below the two millimetre grain size, although no signs of igneous or metamorphic pebbles had been found in the till at this site.

Glacial History of the Study Area

Evidence has been presented showing a division of till areas in the study region based upon the distribution of lithological indicators. The inter-relations of the various tills and associated deposits makes possible an interpretation of the glacial history of the map region and the relations and extent of the glaciers in the study area, particularly during the last glaciation.

The glacial history of the region is believed to have been predated by the deposition of Tertiary gravels. These are represented by brown and purple quartzites, which are assumed to have been derived from similar rock types located to the west of the map area. A Tertiary gravel deposit would help to explain why Cordilleran derived material is found well into the zone of Laurentide drift. Assuming that the original mixture of pebbles nearer the flanks of the Western Cordillera would be composed of both carbonates and quartzites, it seems likely that the quartzites, because of superior resistance, would travel further to the east. The quartzites would then be reincorporated in the westerly Laurentide advance and mixed into the basal till. This might explain the concentration of quartzites along the eastern flank of the Foothills Erratics Train.

A continuation of gravel deposition at various erosion levels, found east of the study area, is believed to have occurred throughout the Tertiary. The last gravel sequence deposited in the Calgary area was the pre-glacial gravels believed to be of early Pleistocene age.

Figures 1 and 3 illustrate that igneous and metamorphic pebbles occur west of the Foothills Erratics Train. Earlier writers have commented on the distribution of Laurentide erratics at altitudes of almost 1,000 feet above and to the west of the highest quartzite erratics of the Train. It has been shown that these erratics vary in size from mega-erratics to fragments of less than two millimetres. Tharin (1960) has also shown that in the Lochend Hummocky Moraine there are anomalies in both X-ray diffraction and carbonate gasometric readings. These facts have led to the conclusion that a Laurentide advance occurred earlier than the emplacement of the Foothills Erratics Train, probably in early Wisconsin times. Observations made whilst mapping west of Calgary agree with the observations and conclusions of earlier writers, such as Nichols (1931), and Tharin (1960). The quartzite blocks of the Foothills Erratics Train were emplaced at the end of the last glaciation.

Ice Sheet Relationships in Southwestern Alberta and the Calgary Area

Although most of the earlier writers suggest that two ice sheets existed over southwestern Alberta during parts of the Pleistocene period there is a certain amount of confusion as to whether the Cordilleran and Laurentide ice sheets were contemporaneous, or whether they reached their acme at different times. Stalker (1956) suggests that the Foothills Erratics Train demarcates the junction of the two major glaciers, at least as far south as Calgary. Further south he suggests that Cordilleran ice had withdrawn by the time the erratics were emplaced, and then appears to leave the distribution of the eastward trend of the Train in the extreme south of the province to Laurentide ice. The Cordilleran glacier is believed to have carried the blocks, with the Laurentide ice controlling their distribution, an observation which shows the two ice sheets to be contemporaneous at least as far south as Calgary, in late Wisconsin times.

Three glaciers appear to have been present in the study area during the last glaciation; these are from west to east, the Cordilleran ice sheet, the Erratics Train glacier, and the Laurentide ice sheet.

There is a possibility that two further zones of mixed ice lie between the Erratics Train and Laurentide glaciers. Fourteen land sections contain carbonate mega-erratics east of the Erratics Train and these appear to be grouped into two areas. The first group, containing seven erratics, lies one to five miles east of the Train, whilst the second group, also containing seven erratics, occurs seven to eleven miles into the zone of Laurentide drift. Christiansen (personal communication) has stated that large erratics of dolomitic limestone are rare in western Saskatchewan. As the map area is two hundred miles west of the Saskatchewan border it is unlikely that these mega-erratics are derived from the sedimentary fringe of the Shield. The writer suggests that these blocks, containing brachiopods and stromatoporoids, are Devonian and derived from the valleys north of the Athabasca, the probable source valley of the Foothills Erratics Train. This would then place the dolomitic mega-erratics to the east of the Erratics Train. They could then be mixed with Laurentide drift by the interdigitation of Cordilleran and Laurentide ice north of the map area.

The carbonate pebbles in the till east of the fifth meridian are believed to have been derived from the edge of the Shield, carried westwards by the ice, and deposited in the area of Laurentide drift together with the igneous and metamorphic pebbles.

The zone of the Erratics Train is believed to represent the main zone of mixing between Cordilleran and Laurentide ice. Bow Valley ice appears to have been retreating during the emplacement of the quartzite erratics in the Calgary area, and probably retreated west of the study area by the time of the 3,875 foot level of glacial Lake Calgary (Morgan, 1966, p. 60).

Altitudinal Relationships of Erratics Train Ice in the Study Area

Stalker (1956) and Tharin (1960), have both mentioned that the surface of the Erratics Train ice in the Calgary area was about 4,100 feet. Stalker (1956, p. 18) allows an extra thickness of ice "necessary to carry these erratics to their present position." During this study the position of 936 quartzites of Erratics Train lithology were plotted. Only two erratics were found above 4,100 feet, however, the present writer believes that the Erratics Train ice was at least one hundred feet thicker than the estimated 4,100 feet surface. If the ice surface in the Calgary region stood at 4,100 feet, the topographic high north of Calgary would have deflected the ice flow toward the east. Because of the direction of glacial flutings and the position of igneous and metamorphic erratics (Morgan, 1966, pp. 36, 61 and 62) ice is believed to have passed over Nose Hill from the northeast. Quartzite blocks of Erratics Train lithology and quartzite grits found associated with the Foothills Erratics Train occur on the top of Nose Hill at elevations of 4,075 feet. A minimum thickness of two hundred feet of ice is suggested for flowage.

The location of two small quartzite erratics in section 24, Tp.27, R.4 and section 9, Tp.28, R.4 at 4,195 feet and 4,135 feet respectively, indicates that the surface of the Erratics Train ice rose above 4,100 feet in local areas. As these are small mega-erratics (2-3 cubic feet) it is probable that they could have been moved during farming or grading operations. However, the nearest point below 4,100 feet that they could have been brought from is at least half a mile in the first example and nearly two miles in the second example. It seems unlikely that they would have been carried uphill if they had been moved.

From the above evidence it appears that the Erratics Train ice was at least a hundred, and probably two hundred, feet thicker than the 4,100 foot level given by the majority of quartzitic erratics in the Calgary area. If this was the case the question then arises as to why the erratics appear to terminate abruptly, with the exception of the two quoted above, at 4,100 feet.

It is probable that the ice level represented by the emplacement of the erratics does not show the maximum thickness of the Erratics Train ice but merely a retreat level. Stalker (1956, p. 18) notes that the altitude of the lower erratics falls 150 feet between Calgary and Claresholm. This height change could either be due to Erratics Train ice moving eastward into areas of lower elevation or to continued ablation of the glacier as the erratics moved southward. Moreover, if the two erratics above 4,100 feet are *in situ*, why are there not more of them at this level?

Origin of the Foothills Erratics Train

Before answering this question an attempted explanation of the derivation of the erratics should be given. A number of large rockslides have been described in India (Holland, 1894), Switzerland (Heim, 1932), Iran (Harrison and Falcon, 1936, 1937, 1938), and Norway (Holtedahl, 1960). The Western Cordillera has produced several rockslides, notably in Montana (Hadley, 1959; Mudge, 1965), Alberta (Bell, 1904; McConnell and Brock, 1903; Brock, 1910, 1911), and British Columbia (Hope Slide, 1965).

The rockslides are usually caused by agencies which disturb the stability of the mountain side, for example, earthquake activity as in the "Good Friday" earthquake in Alaska (Tuthill *et al.*, 1964; Post, 1965; Garrett, 1965). Various writers such as Conway (1900), Varnes (1958), Mudge (1965) and Kent (1966) have suggested that air trapped between masses of rock in the slides could facilitate rapid transport which typifies all recorded rockslides. Another notable feature of the rockslides is that the blocks involved in the debris all appear to be quite angular even after flowing considerable distances; over nine miles in Iran and nearly two miles at Frank, Alberta. It seems likely that the Foothills Erratics Train originated as a rockslide onto the surface of a glacier.

Returning now to the question of the two erratics located above 4,100 feet, the writer suggests that before the major slide or slides occurred, quartzites were being deposited on the surface of the ice by glacier erosion. If these were carried from the Jasper area at the maximum extent of the Wisconsin, they would have arrived in the Calgary region when the ice was several hundred feet thicker than the level indicated by the main mass of the erratics at 4,100 feet. By the time the majority of erratics arrived in the map area the ice level probably had fallen to approximately 4,100 feet. Erratics carried south of the Calgary area should be found at progressively lower levels as ablation continued. There is, of course, a probability of more than one slide occurring in the source area in the same formation, if not in the same locality. Mountjoy (1958) suggests that the Mount Edith Cavell region is a likely source. In a later paper by Roed *et al.* (1967) the authors state that:

> . . . it appears that the Jasper area was the prime source for the Erratics Train, but with some contributions of quartzite erratics from glaciers that occupied the North Saskatchewan, Brazeau, Snake Indian Valleys, and Smoky River Valleys . . .

A small quartzitic erratic of the same lithology as the Foothills Erratics Train was observed in a road embankment four miles from the Angel Glacier terminus on the Mount Edith Cavell road. Other blocks of quartzite found in association with the Foothills Erratics Train in the Calgary region also occur in the Mount Edith Cavell area. If the actual slide scar or scars of the Train are ever found, it would be of interest to know if the quartzitic beds occur above or below the maximum extent of Wisconsin ice in the source area. If the latter is shown to be the case then the erratics could not have been derived at maximum Wisconsin. Finally, there is also a possibility of a second source area in the Waterton Lake region contributing erratics to the southern section of the Train, as suggested by Williams and Dyer (1930) and Rutherford (1941).

Bibliography

Allan, J. A., "General geology of Alberta", *Res. Coun. Alberta*, Rept. 34, (1943).

Allan, J. A., and Carr, J. L., "Geology of the Highwood-Elbow area, Alberta", *Res. Coun. Alberta*, Rept. 49 (1947).

Bayrock, L. A., "Heavy minerals in till of Central Alberta", *Alberta Soc. Petrol. Geol.*, Vol. X, no. 4 (1962), pp. 171-84.

Bell, R. A., "Notes on the rock slide at Frank, Alberta", *Geol. Surv. Can. Summ. Rept 1903*, pt. A (1904), p. 8.

Brock, R. W., "Notes on Turtle Mountain, Frank, Alberta", *Geol. Surv. Can. Summ. 1909*, (1910), pp. 29-30.

Conway, W. M., "The Alps from End to End", Westminster Press, Philadelphia, (1900), p. 200.

Dawson, G. M. and McConnell, R. G., "Report on the country in the vicinity of the Bow and Belly Rivers, N. W. Territory, Canada", *Geol. Surv. Rept. of Prog.* 1882-4, pt. C., (1885), p. 168; "Glacial deposits of southwestern Alberta in the vicinity of the Rocky Mountain", *Geol. Soc. America Bull.*, Vol. VII (1895), pp. 31-66.

Dreimanis, A. and Reavely, G. H., "Differentiation of the lower and upper till along the north shore of Lake Erie", *Jour. Sed. Petrol.*, Vol. XXIII (1953), pp. 283-59.

Dreimanis, A., "Quantitative gasometric determination of calcite and dolomite by using Chittick apparatus", *Jour. Sed. Petrol*, Vol. XXXII (1962), pp. 520-29.

Flint, R. F., *"Glacial and Pleistocene Geology"*, John Wiley and Sons Incorp., New York, (1957), 553 pp.

Garrett, W. E., "Alaska's Marine Highway", *Nat. Geog. Mag.*, Vol. CXXVII (June 1965), pp. 776-819.

Gillberg, G., "Den glaciala utvecklingen inom Sydsvenska höglandets vastra randzon", *Geologiska Föreningens I Stockholm Förhandlingar*, Vol. LXXVII (1955), pp. 481-524; "Till distribution and ice movements on the northern slopes of the south Swedish Highlands", *Geologiska Föreningens I Stockholm Förhandlingar*, Vol. LXXXVI (1964), pp. 483-84.

Gravenor, C. P. and Bayrock, L. A., "Use of indicators on the determination of ice movements in Alberta", *Geol. Soc. America Bull.*, Vol. LXVI, no. 10 (1955), pp. 1325-28.

Hadley, J. B., "The Madison Canyon landslide", *Geotimes*, Vol. IV (1959), pp. 14-17.

Harrison, J. V. and Falcon, N. L., "Gravity collapse structures and mountain ranges as exemplified in southwestern Iran", *Geol. Soc. London Quart. Jour.*, Vol. XCII (1936), pp. 91-102; "The Saidmarreh landslip, southern Iran", *Jour. Geography*, Vol. LXXXIX (1937), pp. 42-47; "An ancient landslip at Saidmarreh in southwestern Iran", *Jour. Geology*, Vol. XLVI (1938), pp. 296-309.

Heim, A., "Bergstrug and Menschenleben", Fretz and Wastmuth, Zurich, Switzerland (1932), p. 218.

Holland, T. H., "Report on the Gohna landslip, Garwahl, India", *Geol. Surv. India Records*, Vol. XXVII, pt. 2 (1894), pp. 55-64.

Holtedahl, O., "Geology of Norway", Norges Geologiske Undersøkelse NR. 208, I Kommisjon. Hos. H. Aschehoug and Co., Oslo (1960).

Horberg, L., "Rocky Mountain and continental Pleistocene deposits in the Waterton region, Alberta, Canada", *Geol. Soc. America Bull.*, Vol. LXV (1954), pp. 1093-1150.

Hume, G. S., "Overthrust faulting and oil prospects of the eastern foothills of Alberta between the Bow and Highwood Rivers", *Econ. Geol.*, Vol. XXVI (1931), pp. 258-73.

Johnson, C. E., "The position of the Erratics Train as the zone separating Cordilleran and Continental drift", 407 Term paper, Geog. Dept., The University of Calgary, Alberta (1964).

Kent, P. E., "The transport mechanism in catastrophic rockfalls", *Jour. Geology*, Vol. LXXIV, no. 1, (1966), pp. 79-83.

Knechtel, M. M., "Snake Butte boulder train and related phenomena, north-central Montana", *Geol. Soc. America Bull.*, Vol. LIII (1942), pp. 917-36.

Krumbein, W. C., "Measurement and geological significance of shape and roundness of sedimentary particles", *Jour. Sed. Petrol*, Vol. XI, (1941), pp. 64-72.

Mackintosh, D., "Results of a systematic survey, in 1878, of the directions and limits of dispersion, mode of occurrence, and relation to drift-deposits of the erratic blocks or boulders of the west of England and the east of Wales, including a revision of many year's previous observations", *Geol. Soc. London Quart. Jour.*, Vol. XXXV (1879), pp. 425-55.

McConnell, R. G. and Brock, R. W., "Report on the great landslide at Frank, Alberta territory (Canada")", *Canada Dept. of the Interior, Ann. Rept. 1903*, pt. 8 (1904), pp. 1-17.

Morgan, A. V., "Lithological and glacial geomorphological studies near the Erratics Train, Calgary Area, Alberta", Unpub. M.Sc. thesis, The University of Calgary, (Sept. 1966).

Mountjoy, E. W., "Jasper area, Alberta, a source of the Foothills Erratics Train", *Jour. Alberta Soc. Petrol. Geol.* Vol. VI, no. 9 (1958), pp. 218-25.

Mudge, M. R., "Rockfall avalanche and rock slide avalanche deposits at Sawtooth Ridge, Montana", *Geol. Soc. America Bull.*, Vol. LXXVI, no. 9 (1965), pp. 1003-14.

Nichols, D. A., "Terminal moraines of the Pleistocene ice sheets in the Jumpingpound-Wildcat Hills area, Alberta, Canada", *Roy. Soc. Trans.*, Ser. 3, Vol. XXV, sect. 4 (1931), pp. 49-59.

Palliser, J., "The journals, detailed reports and observations relative to the exploration by Captain Palliser during the years 1857, 1958, 1859, and 1860", G. E. Eyre and W. Spottiswoode London (1863).

Post, A. S., "Alaskan glaciers: recent observations in respect to earthquake advance theory", Science, Vol. CXLVIII (1965), pp. 366-68.

Roed, M. A., and Mountjoy, E. W., and Rutter, N. W., "The Athabasca Valley Erratics Train, Alberta, and Pleistocene ice movements across the Continental Divide", *Canadian Jour. Earth Sci.*, Vol. IV (1967), pp. 625-32.

Rutherford, R. L., "Some aspects of glaciation in central and southwestern Alberta", *Roy. Soc. Canada Trans.*, Vol. XXXV, sec. 4 (1941), pp. 115-24.

Rutter, N. W., "Surficial geology of the Banff area, Alberta", Unpub. Ph.D. thesis, Univ. of Alta, Edmonton (1965), p. 105.

Stalker, A. MacS.., "The Erratics Train Foothills of Alberta", *Geol. Surv. Can., Bull.* 37 (1956).

Tharin, J. C., "Glacial geology of the Calgary, Alberta area", Unpub. Ph.D. thesis, University of Illinois, Urbana (1960).

Tuthill, S. J., Laird, W. M., and Freers, T. F., "Geomorphic effects of the Good Friday, March 27, 1964 Earthkquake in the Martin River and Bering River area, south-central Alaska", *Geol. Soc. America Bull.*, Vol. LXXV, Soc. Proc. (1964), p. 182.

Varnes, D. J., "Landslide types and processes" *Landslides and Engineering Practice*, Chapter 3, Eckel, E.B., (ed)., Nat. Res. Coun. (U.S.A.) Spec. Rept. 29 (1958), pp. 20-47.

Williams, M.Y., and Dyer, W.S., "Geology of southern Alberta and southwestern Saskatchewan", *Geol. Surv. Can. Mem. 163*, (1930).

Wolfe, F. L., *Elements of probability and statistics*, McGraw-Hill, New York (1962).

10

Periglacial Mass-movement on the Niagara Escarpment Near Meaford, Grey County.

A. Straw

Introduction

About 800 metres south of Griersville, at 43°32′N, 80°34′W, the somewhat acute convergence of the southern and western margins respectively of the Bighead and Beaver river valleys has produced on the Amabel scarp of the Niagara Escarpment the prominent salient referred to by Chapman and Putnam (1951, p. 137) as the "Griersville Rock" (Figure 1). At this point the Amabel scarp faces a little west of north but 800 metres both to the east and the west it swings south to overlook, on the east, the lower part of the Beaver valley and, on the west, the remarkable "neot" of drumlins that occupies much of Bighead valley. Below this scarp, both to north and east, lies the lower scarp of the Manitoulin formation.

At "Griersville Rock" the road between Meaford and Kimberley is graded through the upper portion of the Escarpment revealing on each side small but interesting sections of Middle Silurian dolomite—the Amabel formation—and the underlying Lower Silurian shales. These sections (Figures 2A and B) display features not previously recorded in southern Ontario, which have developed as a result of mass wastage of the scarp edge during former colder conditions. These features were observed and measured in September 1965 and it is the purpose of this paper to consider their characteristics and the probable conditions under which they were formed.

Reprinted from *Geographical Bulletin*, Vol. VIII, no. 4 (1966), pp. 369-76, by permission of the Department of Energy, Mines and Resources.

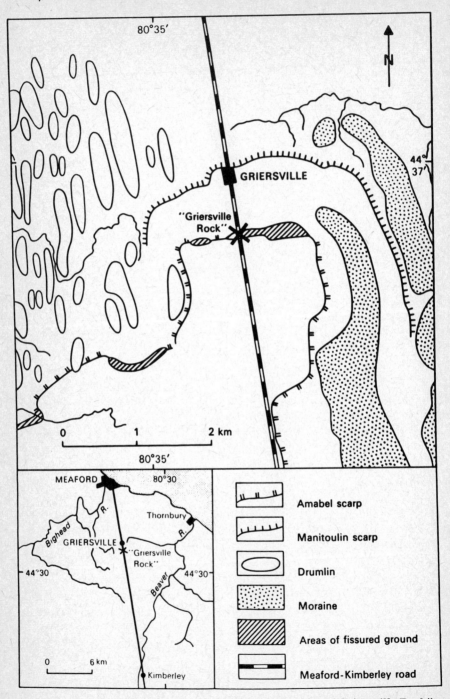

Figure 1. Geomorphological features in the vicinity of the "Griersville Rock".

Description of the Sections

By analogy with descriptions given by Bolton (1957) of other exposures along the Niagara Escarpment, the evenly and thinly bedded underlying shales observed in the "Griersville Rock" sections appear to comprise the upper 5.5 metres of the Cabot Head formation (Lower Silurian) although they are generally of a medium to dark blue-grey colour rather than green as described by Bolton. At the north end of the western section (Figure 2B) they are underlain by just over 2 metres of similarly laminated purplish-red shale whose thin beds are so separated and split by horizontal cracks and so segmented by countless vertical cracks that under the hammer it disintegrates into small blocks and flaky fragments. Generally, this bedded shale is overlain by a layer of pulverized, unstratified shale of variable thickness, consisting of small flakes and platy fragments in which no trace of original bedding remains. Nowhere along the sections on either side of the road was undisturbed shale seen in contact with the base of the cap rock dolomite. Nevertheless, an original sharp, ungraded contact can be assumed.

The pale, yellow-brown dolomite probably includes more than one division of Middle Silurian although no marked change in lithic character was observed. A maximum thickness of 7 metres was measured but, in the southern part of the eastern section (Figure 2A), a thickness of 9 metres was estimated. Extrapolating from Bolton (1957, p. 18), the lower 2.5 metres can be referred to the Fossil Hill formation of the Clinton group and most of the remainder to the Lions Head, lowest member of the Amabel formation. The formational dip is to the south, averaging 2.5 degrees, but the dolomite is somewhat unevenly bedded. It is dense and blocky, with innumerable small vertical and horizontal cracks occurring only a few centimetres apart in the more flaggy beds and near the ground surface. All the beds of the eastern section are consistently jointed along vertical planes which bear 47 degrees and 162 degrees from magnetic north and thus run obliquely to the lines of section; those on the western section bear 342 degrees and 227 degrees from magnetic north. These master joints are spaced 2 to 3 metres apart, penetrate the full thickness of the dolomite and divide the rock into large squared blocks. A number of these joints have been widened, some by a few centimetres but others, like C to I (Figure 2A) by from 15 cm to 60 cm. These latter are now filled with loamy or blocky material.

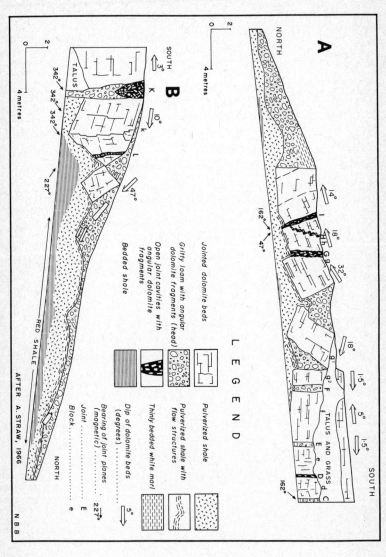

Figure 2. A. A section east of the road at "Griersville Rock".
B. A section west of the road at "Griersville Rock".

South of joint E on the eastern section (Figure 2A) the jointed but fairly massive dolomite dips south at between 1.5 degrees and 3.5 degrees. By contrast, over the northern half most of the blocks dip at high angles to either north or south. On the western section the dip is generally high and to the north. The strongly inclined blocks are underlain by from 50 cm to 250 cm of the pulverized shale. In the sections on both sides of the road this shale shows evidence of flow structure near the lower corners of the blocks and in places is squeezed up for a distance into the joints between them.

Most of the joints are infilled with reddish-brown, gritty, silty-clay loam containing numerous sharp-edged dolomite fragments. These are usually small, between 5 cm and 15 cm in diameter, but in some joints, specifically the open ones in Figures 2 and 3, fragments as much as 35 cm to 45 cm in diameter were observed. In all joints, many of the flatter fragments stand vertically. The infill in joint F is of special interest; bright grey-blue pulverized shale has been forced upward into it for about 45 cm and the remainder of the cavity is filled with red-brown stony, gritty clay, wetter and heavier in texture than the material in other joints, and with a few rounded pebbles in addition to dolomite fragments.

At the north end of the eastern section the dolomite blocks give way to a layer of unstratified brown, stony loam containing angular pieces of dolomite as much as 60 cm across, their long axes declining north parallel to the surface slope. This deposit rests quite sharply on a metre or so of pulverized shale with some mixing observed at the contact. Similar material occurs on either side of block g^1 and overlies block k and the pulverized shale on the western side of the road. Generally, the angular dolomite fragments are 15 cm or less in length, but in troughs in the underlying shale where the layer of brown, stony loam is thicker, large dolomite slabs up to 1.5 metres long and tilted strongly northward are concentrated. Both the pulverized shale and the gritty loam with angular dolomite fragments are derived largely from local bedrock. Some glacial drift may be incorporated into the loam although the only traces readily observed were in joint F in the eastern section.

Discussion

Displacement of the dolomite blocks appears to be a result of mass movement under gravitational influences, and separation and movement of the dolomite blocks was clearly facilitated by failure of the upper part of the Cabot Head shale and by the presence of master joints. Lateral flow of broken shale toward the scarp edge seems to have resulted from the presence of excessive quantities of groundwater which may have penetrated the fissured dolomite from the ground surface above the scarp. The thoroughly mixed character of the shale fragments and the suggestion of involutions in the pulverized shale in the western section, suggest that deeply penetrating frost may also have played a considerable part. Repeated freezing could account for the myriad cracks in the bedded shale and alternate thawing for the periodic concentration of water at and about the dolomite-shale junction. The widening of the joints in the dolomite could have resulted from one or, more probably, both of two processes.

Along the eastern section, at least south of joint F (Figure 2A), the dolomite appears to have been stretched. It would seem likely that near the scarp face the softened shale has been extruded under the weight of the dolomite, allowing joint-blocks to cant and creep slowly forward and thus opening some of the more discrete joints as well as widening most of the master ones. The present open joints and the dolomite slabs that have fallen and lodged in them possibly testify to this process. Yet in addition the joints may have been forcibly widened by the growth of ice veins during former more severe climatic conditions. No direct evidence of this was observed although the sides of all joints, both open and infilled, were jagged and roughened as if by mechanical processes.

Dissolution of the dolomite by percolating water may be discounted as a major factor in joint enlargement at the "Griersville Rock".

The joints in the sections do not appear to have opened simultaneously. Joints $C, E,$ and F (Figure 2A) and K (Figure 2B) are filled with gritty loam, identical to that which intervenes between the displaced joint-blocks, overlies block k and rests on the broken shale at the north end of both the eastern and western sections. These joints were obviously open fissures at the time the gritty loam was produced, if not earlier, but the present open joints, particularly G, H and I which reach to the surface, appear to have widened since it was formed. The unstratified, locally derived character of the gritty loam and the freshness and angularity of its included fragments point to its production by cryoturbation and solifluction processes involving frost-riving of the dolomite as well as repeated freezing and thawing to facilitate the downhill movement of materials lubricated periodically by melted ground ice and snow. Dines *et al.*, (1940) used the term *head* to describe similar deposits in England produced under periglacial conditions.

Ice formerly overwhelmed the whole of the Niagara Escarpment in the vicinity of the "Griersville Rock" and striations on the dolomite surface as well as the orientation of drumlins in the nearby Bighead valley (Figure 1) testify to its movement from north to south. Parts of the upper surface of the dolomite to the north show clear evidence of abrasion by glacier ice but any suggestion that ice was responsible for disturbance of the joint-blocks is refuted by the fact that the blocks have moved northwards, i.e., downslope, under the influence of gravity, whereas the pressure of the ice against the scarp would have kept the joints closed. The ice left little drift along the face and brow of the scarp and, indeed, may have achieved considerable erosion through abrasion. It is unlikely that the *head*, which is quite different from glacial drift, was produced prior to glaciation. It must, therefore, either have survived or followed it, and the inclusion of clayey constituents and erratic pebbles in the filling of joint *F* seems to prove it was formed after deglacierization.

Probable Sequence of the Development of the Present Features

The lithic, structural and stratigraphic details of the road sections permit a reconstruction of the probable general progress of their development.

(1) After prolonged glaciation by south-flowing ice, deglacierization of the area took place. In the succeeding ice-free and relatively temperate period, some joints probably began to open as a result both of solution and of creep of the joint-blocks. In the case of joint *F* this allowed entry of small amounts of glacial drift.

(2) This period was followed by one of intense cold, during which frost penetrated 10 metres or more beneath the surface, freezing any groundwater present in the dolomite and upper layers of the shale. The ice was perhaps segregated into vertical wedges in some of the joints and into thin layers within the shale.

(3) The period of intense cold appears to have been followed by somewhat less severe periglacial conditions during which alternating freeze and thaw pulverized the upper layers of the shale and assisted in the displacement of the dolomite blocks. Saturation of the broken shale during thaw periods would permit creep of the blocks under their own weight which, in turn, would induce flow in the shale and facilitate its limited injection upwards into spaces between the blocks. Such freeze-thaw conditions would also be responsible for the solifluction. That the formation of *head* was in

part contemporaneous with block movement is shown in the eastern section where the backward tilt of block g^1 has "trapped" a large mass of *head*. Accumulation of *head* seems to have continued after the blocks had become immobile for it overlies some of them (e.g., *k*) and, to the north, spreads tens of metres downslope over the shale in front of the outermost blocks. Deep permafrost is not essential to the occurrence of solifluction. In addition to freeze-thaw cycles, the melting of snow and ground ice and the concentration of groundwater near the base of the dolomite due to the general impermeability of the underlying shale would provide the necessary conditions.

(4) This periglacial phase gave way in its turn to the more temperate conditions that have seemingly persisted until the present. With the amelioration of the climate a few fissures developed as both the detached blocks and the main dolomite beds sagged into the pulverized shale or crept minute distances over it. Soil horizon development, though limited, indicates not only the present stability of the *head* but also its fossil character.

Comparisons with Areas of Periglacial Features in England

Features similar to some of those in the sections at the "Griersville Rock" have been observed in the southern Pennines of England which experienced severe periglacial conditions during the last Wisconsinan Glaciation. In numerous locations along the scarps of Millstone Grit and Lower Coal Measures large blocks of gritstone and sandstone have been moved along bedding planes, not only along those that decline toward the scarp but also along some that dip inward from the scarp faces and which must have necessitated some slight upward component in the lateral block displacement. This lateral movement is considered to have resulted from the enlargement of vertical and near-vertical joints—the initial operative force being, presumably, the growth of ice in these joints—which gradually pushed the blocks outward toward the scarp faces where they became more strongly influenced by gravity. Where a critical amount of outward movement occurred they tilted or fell downward, frequently merging into solifluction deposits or landslide debris. It is suggested that the widening of the joints and some movement of the dolomite blocks at the "Griersville Rock" was achieved by similar processes under similar climatic conditions.

The nature of the mass movements invites further comparison with similar phenomena in England that were first described at length by Hollingworth, Taylor and Kellaway (1944). In and around opencast ironstone workings in Northamptonshire, certain more resistant Jurassic rocks (Northampton Sand, Lincolnshire Limestone and Great Oolite Limestone) are bent or curved downward at outcrop toward lower ground. Such vertical displacement may amount to as much as 35 metres and creates considerable difficulty in geological mapping. This bending of the strata is termed *cambering* and is common where incompetent strata outcrop beneath the cap rock as on valley sides and scarp slopes. In section, *cambered* rocks are frequently fissured and faulted and the widened joints or fissures, when wholly or partially filled with material from above, are known as *gulls*. When rocks are cambered toward valleys, *valley bulges* are frequent, having been produced by lateral movement of clay or some other incompetent strata from beneath the cap rock, or even by upward movement in the valley floor which may result in anticlinal structures often with faulted margins. Dury (1959, 1963), following Kellaway and Taylor (1952), regards all these features in association as periglacial phenomena. He considers that joints and bedding planes were wedged open by ice, and that thaw of frozen underlying clays rendered them wet, weak and capable of being squeezed outward by the weight of the cap rock. Sections across *cambers* in Northamptonshire and Lincolnshire show that the curved cap rock often gives way to a number of large detached joint-blocks which have slipped or crept downslope. These may be back-tilted to give what Hollingworth, Taylor and Kellaway call *dip-and-fault* structures, although locally pronounced reversals of dip may occur.

These features of the English landscape clearly have much in common with those at the "Griersville Rock" although by comparison with instances the writer has examined in Lincolnshire and Northamptonshire, the "Griersville Rock" example appears to be a relatively minor occurrence. Nevertheless the Niagara Escarpment would seem to be truly cambered at this point and the debris-filled, partially open joints are firmly regarded as *gulls*.

In the light of these comparisons, it now becomes pertinent to consider whether the "Griersville Rock" features are a manifestation of a prolonged periglacial phase.

For some hundreds of metres east of the road sections, along the Bruce Trail, the scarp face is breaking away. Several large joint-blocks and long slices of rock are wholly or partially detached from the main face, and some are disintegrating under modern winter conditions into jumbles of freshly broken angular fragments. As indicated in Figure 1, areas of fissured ground were noted on the air photographs at other points along the scarp.

Along a narrow zone near the scarp edge, the opening of small joints spaced a metre or less apart, and the rounding of the upper edges of the joint-blocks has led to the production of limited areas of clint-and-gryke formation.

Behind the scarp face, the higher ground surface is creased by a number of linear depressions, up to 1 metre both in width and in depth, which become less distinct and more widely spaced back from the scarp edge, but which are still clearly visible on air photographs (Ontario Department of Lands and Forests, 54-4424, no. 30-50 and no. 30-51) especially east of the road in an area of rough grassland and scrub. These depressions mark the run of enlarged joints or gulls and confirm that cambering has affected more parts of the Escarpment than just the section at which the road crosses it.

The freshly broken rocks along the scarp face are the result of present day frost action but, in front of the scarp, aprons of dolomite blocks spread 150 metres down the shale slopes. These aprons appear to be stabilized, the rocks being partially covered with moss and lichen. Lacking fresh frost-shattered edges, they are regarded as products of an earlier solifluction period. Similar aprons of stable blocks of Amabel dolomite were noted at a point about 1 km east of Kimberley in the Beaver valley, and again about 2 km north-northwest of Eugenia farther to the south, situations in which glacial drift was not plastered against the dolomite edge. Another excellent example of former periglacial mass movement, this time of blocks of Manitoulin dolomite, exists along the south-facing slope of a deep ravine on Blue Mountain east of Loree, 16 km due east of "Griersville Rock".

The periglacial features at "Griersville Rock" are not therefore an isolated occurrence as other fossil features exist around the Beaver valley, and no doubt at other places along the Escarpment and elsewhere in southern Ontario. A fairly prolonged periglacial phase may therefore be suspected, one which was contemporaneous with one of the later substages of the Wisconsinan Glaciation. It is suggested that this locality was too far removed from ice to be greatly affected by climatic deterioration during the Cochrane re-advance (Hughes, 1956), and it becomes tempting to consider these periglacial manifestations as being produced during the Valders substage when ice perhaps invaded the Lake Simcoe and northeastern Georgian Bay areas (Deane, 1950) but did not extend as far as the "Griersville Rock".

References

Bolton, T. E., "Silurian stratigraphy and palaeontology of the Niagara Escarpment in Ontario", *Geol. Surv. Can.*, Mem. 289 (1957), p. 141.

Chapman, L. J. and Putnam, D. F., "The physiography of Southern Ontario," Ontario Research Foundation, Toronto, (1951), 267 pp.

Deane, R. E., "Pleistocene geology of the Lake Simcoe District Ontario", *Geol. Surv. Can.*, Mem. 256 (1950), p. 91.

Dines, H. G., Hollingworth, S. E., Edwards, Wilfred, Buchan, S., and Welch, F. B. A., "The mapping of head deposits", *Geol. Mag.*,Vol. LXXVII, no. 1, (1940), pp. 198-226.

Drury, G. H., "The face of the earth," Pelican Book A 447, Penguin Books Ltd., Harmondsworth, Middlesex (England), (1959), 225 pp.; The East Midlands and the Peak. T. Nelson, London (1963), 299 pp.

Hollingworth, S. E. Taylor, J. H., and Kellaway, J. A., "Large-scale superficial structures in the Northampton Ironstone Field", *Geol. Soc.* (London) *Q.J.*, Vol. C, pts. 1 and 2 (1944), pp. 1-44.

Hughes, O. L., "Surficial geology of Smooth Rock, Cochrane District, Ontario", *Geol. Surv. Can.*, Paper 55-41 (1956), p. 9.

Kellaway, J. A., and Taylor, J. H., "Early stages in the physiographic evolution of a portion of the East Midlands", *Geol. Soc. (London) Q.J.*, Vol. CVIII (1952), pp. 343-66.

11

Notes on Avalanches, Icefalls, and Rockfalls in the Lake Louise District, July and August, 1966

J. Gardner

Most people who frequent the high mountain country have, through intuition or cursory observation, developed many ideas of their own regarding mass movements. Indeed many a mountaineer's actions are partly governed by these ideas. The results of this paper would seem to confirm many of the notions that people have developed. Under the term "mass movements" are being considered such phenomena as snow avalanches, falls of ice, and rockfalls. It should be noted that less perceptible movements, such as solifluction and soil creep, are of some significance to the geomorphologist but they will not be directly dealt with in this paper. Snow avalanches, falls of ice and rockfalls are of varying significance in the movement of earth materials down the mountain slopes, and in the sculpture of the mountain slopes themselves. It is not this geomorphic significance that is dealt with in this paper; rather it is the frequency distribution of mass movements over the period of a day and over the period of the summer, which is taken to include July and August of 1966.

The methods that were used in the actual field observations of avalanches, icefalls and rockfalls are relatively simple, while those that were used for the treatment of the accumulated data are a little more devious. During the time spent in the field an inventory of directly observed mass movements was kept. The notation in this inventory included the data, time, the location, and the type of mass movement observed. If the mass movement was a rockfall the number of fragments, size of fragments, and the distance they moved downslope were noted. The study or observation location was re-

First published in *Canadian Alpine Journal* (1967), pp. 90-95.

stricted almost entirely to the Valley of the Ten Peaks and vicinity. The result of the observations is therefore a great amount of detailed data for a relatively small area of the Rockies.

The collection of data commenced on July 1 and finished about August 25. For obvious reasons, direct observation was limited almost entirely to the daylight hours; however some data were collected for the evening hours. In some cases the data are very weak, particularly in the early and late daylight hours (i.e. 0300-0700 and 1800-2100 hours). This is due to the small number of sample days accumulated over the summer for these time periods, and it is something that should be kept in mind while reading the results. It should be noted that those mass movements recorded are representative of only one location within the study area at any one time. Thus, what is being dealt with here does not represent the total mass movements in the Lake Louise district for the summer of 1966, but is only a sample. Therefore the results are not to be taken as an exact reproduction of reality as it existed at any one time; rather, it can only be hoped that they will focus our image of that reality a little more clearly.

A brief definition of terms is probably in order at this point. The term "avalanche" refers to any rapid or easily perceptible downslope movement of material, the major portion of which is snow. An "icefall" involves the free or rapid downslope movement of ice fragments, normally from a hanging glacier. By "rockfall" is meant the freefall or bounding fall of rock fragments. A rockfall normally originates from a point on a free face. The movement of rock fragments from one part of a scree or debris slope to another part is referred to as a "shift", and is not being considered here.

In treating the frequency characteristics of the mass movements, a number of quantitative summarizing measures were utilized. Of primary importance is the frequency distribution of mass movements on the average day in the summer. For this analysis the "average day" was divided into twenty-four 1-hour intervals, and each observed mass movement was replaced in its appropriate time interval. The total mass movements occurring in each 1-hour interval over the two months was calculated. This resulted in an absolute value for each interval but this was felt to be an unrealistic value in that some intervals had a considerably greater accumulated observation time than others. As a result, the total number of mass movements for each 1-hour interval was divided by the number of hours that that interval had been under observation during the two

TABLE I. THE "AVERAGE DAY".

	Mass Movements	Avalanches	Rockfalls	Icefalls
Mean no./hr. of observation	2.018	1.353	0.364	0.291
Variability from hour to hour	0.985	1.203	1.050	1.083
Concentration in 4 max. hours	56.47% (3.4)*	65.15% (3.9)	58.88% (3.5)	55.76% (3.3)
Concentration between 11 a.m. and 3 p.m.	56.47% (3.4)	65.15% (3.9)	58.88% (3.5)	31.64% (1.9)

*The values in parentheses refer to the number of times greater the percentage concentration is than 16.7% (the expected concentration in four hours if the distribution were even throughout the "average day").

THE SUMMER PERIOD.

	Mass Movements	Avalanches	Rockfalls	Icefalls
Mean no./hr. of observation/day	3.070	1.947	0.617	0.454
Variability from day to day	1.501	2.326	1.088	1.227
Concentration in 4 max. days	42.24% (5.48)*	62.27% (8.09)	29.27% (3.80)	31.76% (4.13)

†The values inparentheses refer to the number of times greater the percentage concentration is than 7.7% (the expected concentration if the distribution of mass movements were even throughout the summer).

months. The resulting value is the mean number of mass movements per 1-hour interval of the "average day" (see Table I). As noted before, the values for the early and late daylight hours are weak because of their small number of accumulated observation hours.

During July and August about 1400 mass movements were observed, the majority of which were avalanches. Figure 1 represents the frequency distribution of all mass movements on the "average day". A cursory glance tells one that there is a marked concentration between 11 a.m. and 4 p.m. What concerns one most when looking at such a distribution is the degree to which the mass movements are concentrated in this time period. Several statistical summarizing measures are available to quantitatively express the degree of dispersion or concentration. Most of these, however, were not found to be strictly applicable to the case under consideration. Therefore, the

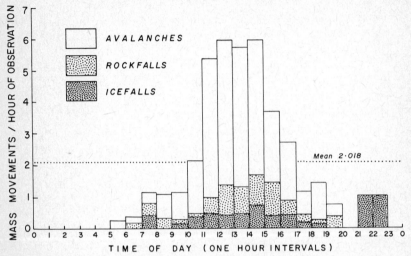

Figure 1. Frequency distribution of mass movements on "average day".

percentage concentration of mass movements in the four 1-hour intervals with the maximum number was computed. This value could then be compared with the percentage concentrated in the four mid-day hours (i.e. 11 a.m. to 3 p.m.) and could also be related to the number of mass movements in the four hours if the distribution had been even throughout the day. In the case of all mas movements, 56.47% were concentrated in the four maximum hours which correspond to the four mid-day hours. This value is 3.4 times the number of mass movements we would expect to be in this time period if the distribution was even throughout the day (see Table 1). On the "average day" the mean number of mass movements per hour was 2.018. The variability in the number of mass movements from one hour to the next was also of interest and is a measure that could give some insight into the nature of the frequency distribution of mass movements throughout the day. For example, if the variability measure is very low one might expect the distribution to be relatively even over the time period being considered. The variability measure being used here is referred to as the "variability coefficient" and is derived from the standard deviation about the mean number of mass movements per hour on the "average day", divided by the mean number of mass movements per hour.

To this point mass movements have been discussed in general, and the types of summarizing measures have been roughly explained. Of more interest however, are the frequency characteristics of the individual components, namely avalanches, icefalls and rockfalls. On the "average day" a mean of 1.353 avalanches per hour was observed with a variability coefficient of 1.203. The concentration of avalanches in the four mid-day hours was very marked with an average of 65.15% occurring at that time or 3.9 times the expected number in an even distribution. In this case, as with mass movements generally, the four maximum hours corresponded to the four mid-day hours. There was only 0.364 rockfall recorded per hour on the "average day" and there was considerably less variability from hour to hour (1.050) than was experienced with avalanches. It was found that 58.88% of all rockfalls, or 3.5 times the expected number if the distribution were even, occurred during the four maximum hours which again corresponded to the four mid-day hours. The pattern of icefall distribution on the "average day" was somewhat different than that exhibited by avalanches and rockfalls. A mean of 0.291 icefall per hour was noted, and a variability coefficient of 1.083 was derived. Again there was a relatively high concentration (55.76%) during the four maximum hours but only 31.64% of all icefalls came during the four mid-day hours. This value is only 1.9 times the value expected if the distribution were even.

To summarize the foregoing analysis it might be said that mass movements during July and August, 1966, showed a marked concentration in the hours between 11 a.m. and 3 p.m. The first significant activity on the "average day" began about 8 a.m. and carried on until about 8 p.m. The most frequent occurrence was between 2 and 3 p.m. Significant avalanche activity started somewhat after 9 a.m. and continued until about 7 p.m. (see Figure 1). The most frequent occurrence of avalanches was between 1 and 2 p.m. but occurrences were almost as frequent during the other three mid-day hours. It was found that significant rockfall activity commenced as early as 6 to 7 a.m. on the "average day" and continued until 8 p.m. The concentration of rockfalls into any one hour was less marked than in the case of avalanches and it was found that they were more or less evenly distributed in the four mid-day hours. The study of icefalls revealed that they were more evenly spread through the day with significant activity continuing until midnight in many cases.

These data collected for one summer period certainly do not provide us with a sound basis for predicting the occurrence of mass movements on any given day in the future. The pattern for another summer could be quite different. The techniques of observation still require considerable improvement and this in itself may alter the frequency distributions as they appear in this paper.

Besides the frequency distributions of mass movements on the "average day", the frequency distribution of mass movements over the two months are also of interest. Many of the summarizing measures used in the treatment of the "average day" are also used in

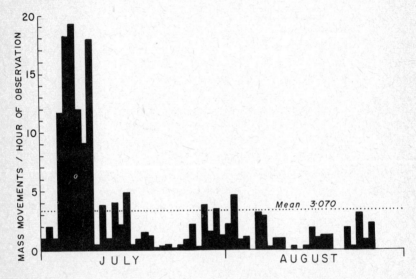

Figure 2. Frequency distribution of mass movements July and August 1966.

the analysis of the two-month period. Figure 2 shows a very marked concentration of mass movements during the first ten days of July. This is largely due to the high frequency of avalanches during that period. The mean number of mass movements per hour of observation per day was found to be 3.070 (see Table I). About two-thirds of these were avalanches which showed a mean of 1.947 as compared with means of 0.617 and 0.454 for rockfalls and icefalls respectively. As a basis for comparing the three types of mass movements with regard to their variability from day to day over the summer period, a variability coefficient was computed. Mass movement in general showed a variability of 1.501 while avalanches showed the greater variability which is summarized in the value of 2.326 (see Table I). Rockfalls and icefalls exhibited variability coefficients of 1.088 and 1.227 respectively. In order to give a somewhat clearer picture of the various degrees of concentration, the percentage of movements represented by the four maximum days was calculated. 42.24% of all mass movement occurrences were recorded on the four maximum days during July and August. This percentage is 5.5 times greater than what would be expected if the occurrences were evenly distributed over the two months. This relatively high concentration into four days is undoubtedly affected by the avalanches, which showed 62.27% of all occurrences on the four maximum days or 8 times the expected figure for an even distribution. Rockfalls and icefalls represented 29.27% and 31.76% of all occurrences during the four maximum days respectively. These concentrations for rockfalls and icefalls are only 3.8 and 4.1 times the expected concentrations had the distribution been even.

In summarizing, it is apparent that avalanches were markedly concentrated at a time of the summer when one would expect a considerable amount of snow to be still lying on the high mountain slopes. This is not to say that avalanches did not occur at other times in the summer. They occurred to a greater or lesser degree all summer in this area. From the analysis it would also appear that rockfalls and icefalls occur with more or less equal intensity throughout the summer.

Analysis of environmental factors such as temperature, precipitation, sunshine, etc. that might be responsible for the occurrence of mass movements is only just beginning. No one factor has yet shown a very high correlation with the distribution of occurrences. This would seem to suggest at this early stage that there are numerous contributing factors, each of which varies greatly over a relatively small area in this high mountain country. One could intuitively suspect that there would be a relatively high correlation between amount of snow and avalanche activity during the summer months. However, this complex system is not to be unravelled as easily as that, and much more detailed field study is required before we can point to any one environmental factor as being especially significant. The "average day" analysis suggests that rockfalls and avalanches, with their marked peaks of occurrence at a certain time of day, are related to some factor that undergoes strong daily or diurnal fluctuations. Icefalls, with their more scattered distribution through the day, would appear to be partly independent of daily fluctuations. The marked concentration of avalanches early in the summer suggests the presence of an over-riding factor such as snow amount, as previously mentioned. Rockfalls and icefalls, with their more scattered distributions over the summer period, are probably independent of such an over-riding factor.

12

Snowcreep Studies, Mount Seymour, B.C.: Preliminary Field Investigation

W. H. Mathews and J. R. Mackay

Snowcreep is a slow, continuous, glacier-like downhill movement of the snowpack (U.S. Dept. Agr., Forest Serv., *Avalanche Handbook*, 1953, p. 145). The quasi-viscous movement takes place under shear stress sufficient to produce permanent deformation but too small to produce shear failure as in snowslides and avalanches (see definition by Stokes and Varnes, 1955, repeated in A.G.I. Glossary, 1957, p. 67).

The movement of the snowpack can be considered as having three components: (1) a vertical settling or compression of the layers under the direct influence of gravity; (2) a shearing within the snowpack usually parallel to the ground surface, higher layers gradually riding downslope over lower ones; and (3) at least locally, a sliding of the snowpack as a whole over the ground beneath. The resulting movement of any point within the pack depends on the magnitudes of all three components, the slope of the ground surface and the distance of the point above the ground.

The settling or compression of the snow layers and the shearing movement within the snowpack have been investigated by a number of research workers (Kojima, 1960; Martinelli, 1960). By means of markers set in the snow it has been demonstrated that the rate of downslope movement diminishes with depth and that in many cases the creep velocity at the base of the snowpack is too small to record. It is possible, however, to demonstrate that an appreciable slip may occur along the interface between the snowpack and the ground beneath, although few experiments have been undertaken to establish the basal rate of movement.

First published in *Geographical Bulletin*, no. 20 (Nov. 1963), pp. 58-75.

The creep of the snow may have an appreciable effect in moving subjacent loose rocks and soil, injuring vegetation and disturbing man-made structures in its path. This study was prompted, in fact, by observations in the alplands north of Garibaldi Lake, B.C., which suggested that basal snowcreep was an important agent in the downslope migration of soil and the formation of sorted soil stripes (Washburn, 1956). Movements at higher levels within the snowpack are also of interest since these can affect the growth and shape of shrubs and trees (Shidei, 1954; Kataoka and Sato, 1959) and exert pressures against poles, avalanche fences and buildings (Furukawa, 1956; Haefeli, 1953; Bucher, 1948).

The present study has been concentrated on the movement of the lowest snowpack layers that might shift stones and soil and that increases the displacement of higher layers of snow. This is the component least investigated to date and indeed, the one that Bucher (1948), in his valuable theoretical study of pressures exerted against avalanche defences, intentionally ignored so as to make the mathematical derivations manageable.

The site for the investigations was selected at Mount Seymour, B.C., which, unlike neighboring Garibaldi Lake, is reasonably accessible during winter and has adequate snowfall and where snowcreep is amply evidenced by the bent trunks and branches of shrubs and even of large trees. The experiments were made at the crest and on the upper slope of a rocky knoll, at an altitude of 3,600 feet, a half mile northwest of the terminus of the Mount Seymour Park road, and 10 miles northeast of Vancouver (Figure 1).

Factors that might affect the rate of the basal snowcreep studied at this site have included snow depth, density, grain size, the liquid-water content of the snow, and temperature. The possibility of adhesion to the ground and the influence of obstacles upslope have also been considered. Field studies were started on Mount Seymour in the fall of 1958 and have been continued each winter since then. Visits have been made to the site about once a week during the period of snow cover, usually from December to April. Much more limited observations of snowcreep have also been made at Garibaldi Lake and, in 1961-62, at Ottawa, Ont.

Instrumentation and Methods

Design of simple, inexpensive and reliable apparatus to record the behavior of the basal layers of the snowpack has so far been the main task. From this has come experience in problems and their solutions that may be the most valuable result of the work. Methods have now been established to measure with reasonable success and

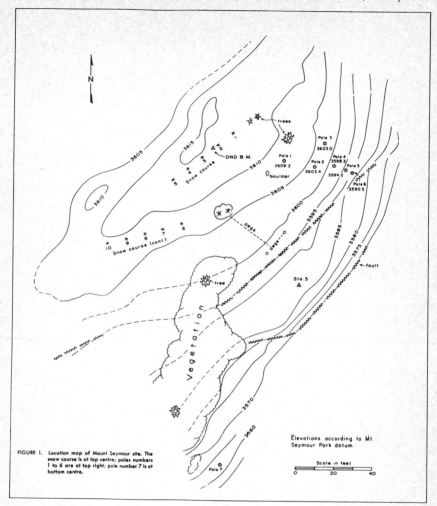

Figure 1. Location map of Mount Seymour site. The snow course is at top centre; poles numbers 1 to 6 are at top right; pole number 7 is at bottom centre.

accuracy the following items: (1) the direction of movement; (2) the rate of movement of the lowermost layer of the snowpack; (3) the total movement of the basal layer during a single winter; and (4) the pressure exerted by the creeping snow against a stationary object at any level.

Direction of movement

An obvious way to establish the direction of movement of the snowpack is to place a loose identifiable object (e.g., a stone with distinctive lithology or a coloured marble) on a fixed point on the ground surface and to locate it when the snow disappears the following spring. A paint spot or chiselled mark can serve as a reference point on a rock surface; a buried ceramic magnet can be used beneath loose soil. Precautions must be taken to ensure that the movable objects are not dislodged by passing animals, are not deflected by irregularities on the surface or do not roll on steep slopes. Replicate tests are desirable to check against such accidents.

Metal ribbons capable of being bent downslope by the winter's movement of the snowpack may also be inserted in the ground or in cracks in bedrock. Such ribbons should be sufficiently strong to escape flexure by wind or freshly falling snow but flexible enough to bend in the creeping snow. They should be sufficiently broad to avoid cutting the moving snowpack but narrow enough to exert no influence on the direction of bending. For the type of snow occurring at Mount Seymour, 2.5-mm-wide metal ribbons that become permanently flexed with a bending moment of 75 gm/cm proved satisfactory, but natural cracks in the rock capable of holding the ribbons were scarce, and wooden dowels split to receive the lower ends of the ribbons were therefore inserted in ¾-inch-diameter holes drilled into bedrock.

As might be expected the direction of movement was downslope. The general slope, however, rather than local declivities establishes the direction of movement, and on undulating surfaces, where lateral stresses may be significant, interesting flow patterns may appear.

Rate of movement of the lowermost layers of the snowpack

The most satisfactory technique tried to date for measuring the rate of movement of the lowermost layers of the snowpack consists in the use of 3- and 6-inch-diameter wooden spheres tethered loosely by means of single-strand Monel wire a few feet downslope from slender steel posts mounted vertically in holes drilled into bedrock. Although the balls are coated with white or aluminum paint to minimize the melting of adjacent snow, a rind of frozen snow found around some of them when they are dug out in the spring suggests that disturbance of the thermal condition of the adjacent snow has not yet been eliminated. As the posts are only an inch in diameter and are at least several feet upslope from the balls, they are believed to be too small to exert any significant influence on the basal snow-creep at the site of the balls. As creep proceeds, the movement of a ball causes the withdrawal of a graduated tape that is mounted at the top of the pole and so is visible above the snowpack. Displacements of the balls can thus be observed from a distance at convenient time intervals.

A similar technique was adopted by in der Gand (1957) at Davos, Switzerland. The wire, however, was attached upslope to a continuous recorder capable of monitoring displacement as small as 0.5 mm within a 15-minute interval. One such unit has been installed at Mount Seymour for the winter of 1962-63.

Instead of wooden balls, spherical plastic fish-net floats and ellipsoidal wooden net floats were also tried. The plastic net floats were found to be subject to serious denting and crushing against the ground. Ellipsoidal floats proved unsuitable because, when they rotated about the vertical axis during creep, they presented cross sections of varying sizes and shapes to the moving snow, and their downslope rotation during movement was no longer free.

Radioactive rock specimens were also placed over a sheet of photographic paper protected from light and water by black paper and pliofilm and from the scoring action of the rocks by a sheet of thin masonite. The separation of the radioactive specimens from the photographic paper was so great, however, that the only result was a diffuse fog.

Another method required the vertical insertion of polyethylene tubing in the snowpack with the object of measuring deformation by means of an electronic strain-gauge probe (Williams, 1957, 1962), but the tubing proved to be too flexible to resist the stress of the wind against the protruding tube. Moreover, the tube curved and stiffened at freezing temperature.

Total movement of the same basal layer of snowpack during a single winter

Some of the methods previously discussed for measuring the rate and direction of movement can be applied to this problem. For the most consistent observations, the writers are using 6-inch and 3-inch diameter fir balls placed on the ground and tethered loosely with Monel wire to slender stakes inserted upslope. The stakes consist of wooden dowelling $1\frac{1}{8}$ inches in diameter drilled horizontally through with $\frac{1}{4}$-inch-diameter holes plugged on the upslope sides with small corks. A slit in the cork holds the wire, and tension can be adjusted by easing the cork into or out of its seat. The distance from stake to ball, not less than 2 feet, is measured at the beginning of the winter season and again on exposure in the spring to determine aggregate movement. Some of the stakes are long enough to measure snow depth.

The methods requiring the use of movable objects (e.g., stones) placed on the ground and metal ribbons have proven less satisfactory. The use of movable objects calls for precautions against movement prior to snow burial; and metal ribbons are not free to move at the ground surface.

Pressure exerted by the creeping snow against stationary objects

This is being measured hydraulically by means of transducers consisting of a square plate mounted at the upslope end of a piston, which, in turn, bears against a flexible brass bellows. The bellows, a length of ¼-inch O.D. copper tubing and a Bourdon gauge form a continuous closed system, which is first evacuated of air and then filled with liquid. Initial experience with 6 feet of tubing between transducer and gauge indicated some temperature sensitivity, which became accentuated as the length of the tubing was increased. A 60:40 water-glycerine mixture, which remains unfrozen down to about $+4°F$ and has a relatively small thermal coefficient of volume expansion, has been selected as the most suitable fluid for the system. Moreover, as the transducer, tubing and gauge are mounted near the ground surface, all parts of the system can be maintained during the winter at or very close to $32°F$ at the Mount Seymour site. The instruments have been calibrated at this temperature, and the calibration permits conversion of Bourdon-gauge readings to loading on the plates and pistons. Transducers have been mounted to record the component of load applied parallel to the ground surface on plates 3 and 5 inches square. Other-size plates can be installed as required providing, however, that the load totals do not exceed the limits of either the bellows or the Bourdon gauges.

It is recognized that scaling of loads from single or clustered plates to large structures is not possible with the data on hand, but information is being sought on the snow conditions contributing to maximum loading at any one site on a given size of plate.

Miscellaneous snow properties

Mean snow depth and density at the Mount Seymour site have been determined by standard methods of snow-surveying with a Mount Rose sampler. Because of marked drifting of snow and variations in depth, however, seven snow poles were set up on the slope where the creep studies were undertaken. These poles have served also for tethering the wooden balls and for mounting pressure transducers. Snow depths were recorded weekly. From time to time a pit was dug several feet away from the poles and densities at various depths were determined by using a sampling tube of known internal volume and a spring scale.

Temperatures for the winters of 1960-62 were measured at depths of 20.75 and 39.25 inches (0.5 and 1.0 meters) below ground surface in a hole drilled into bedrock and later filled with zonolite. A telethermometer that could be read to 0.1°C and thermistor probes were used for ground temperatures. The thermistors were calibrated on the site by using snow-fresh water mixtures. Snow temperatures were also obtained in pits at various depths below the snow surface.

A few attempts have been made to determine the free-water content of the snow by means of simple calorimetric methods (see Yosida, 1959) requiring the use of thermos flasks.

Ramm penetrometer tests were made on a few occasions.

Discussion

Observational data obtained by consistent methods are available for the most part for only one or two winter seasons. As snow conditions vary from year to year and from place to place as well as from month to month, it would be premature to draw firm conclusions. Some consistent results have nevertheless been obtained, and they show that displacements of 2 to 3.5 feet do occur in a single winter within the bottom 3 to 6 inches of the snowpack on the steeper slopes. Movement persists throughout much of the season and reaches a maximum rate of 5 inches a week, especially on the steeper slopes, beneath deep snow that is being subjected to rapid recrystallization during thawing periods. Pressure-plate observations indicate that downslope components of load on 25 square inches (roughly comparable to the cross-sectional area of the 6-inch balls) can reach 25 pounds, which is enough to push loose stones down inclines. Presumably for this reason the bedrock slopes at the Mount Seymour test site are free of soil and boulders. The loads and amount of movement are also adequate to account for the deformation of trees on the nearby timbered slope.

Individual studies are detailed year by year and discussed in what follows.

Winter, 1958-59

On November 29, 1958, five round plastic fish floats, each 6 inches in diameter and weighing 310 grams, were placed on bare rock surfaces of varying slopes. A Monel wire, attached to a ring in each ball, led upslope through a lower and an upper pulley fixed firmly to a tree, and had a counterweight suspended from its free end. The downslope movement of the ball beneath the snowpack was then read by observing the displacement of the counterweight against a scale. As there was a snow cover of 1 to 2 feet when the equipment was installed, trenches were dug to bare rock so that each plastic ball and wire leading from the ball to the tree rested on the ground. After installation, the trenches were filled with snow. Thus the initial movement of the counterweight resulted from the taking up of slack in the wire. The counterweights which weighed less than 100 grams, were too light to exert any uphill pull on the balls, especially since the connecting wires were bent around two pulleys and were covered by snow for at least 15 feet. The displacement of the counterweights might therefore equal, but could not exceed, the basal movement of the snowpack. Ball number 1 was on a 19-degree slope; number 2 was mounted below number 1 on a 35-degree slope; number 3 was on a 13-degree slope; number 4 on a 35-degree slope; and number 5 below number 4 on a 50-degree slope. As the balls had to move downslope slightly to take up slack in the wires, the readings during December were considered unreliable and the movements shown in Figure 2, section a, are therefore calculated from the beginning of January. On February 28, it was found that a vandal had severed the wires connecting ball numbers 1, 2 and 3 to the trees. On March 7, the wire from ball number 2 was repaired, but those from numbers 1 and 3 could not be reattached.

Ball number 5 was exposed by April 25; numbers 2 and 3 were bare by May 10, but both were found partially crushed by snow pressure, and the added friction may have restricted movement; number 5, also bare on May 10, was not crushed. Movement of the balls was obviously fastest on the steepest slopes. The ball on the 50-degree slope moved nearly half an inch a day in early March.

To determine the relative movement within the snowpack, vertical holes made with a snow-sampling tube were filled with sawdust, and numbered ping-pong balls were inserted at measured intervals. The filled holes were later excavated, and the movement was determined. Hole (a), made in 92 inches of snow on March 15 was excavated on May 18 when the snow depth was 42 inches. From the ground surface to a height of 17 inches the sawdust-filled hole was straight; from 17 inches to 42 inches (at the snow surface), it curved gradually downslope, its horizontal displacement being 1.5 inches greater at 42 inches than at 17 inches above ground. On March 30, hole (b) was made in 120 inches of snow; by April 30, when there was 53 inches of snow, the top of the hole was not directly above the bottom but was 5.5 inches farther downslope. On April 5, hole (c) was made in 103 inches of snow; by May 10 there was 60 inches of snow and the top of the hole had moved 4

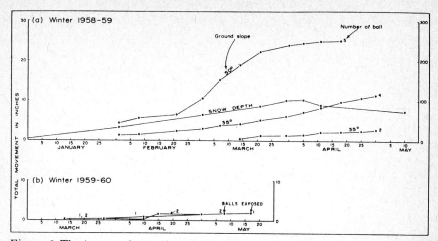

Figure 2. The top graph (section a) shows the total winter movement of balls placed on varying slopes for the winter 1958-59. The snow depth (scale in inches on the right) is for the snow course shown on Figure 1 (see table II). The bottom graph (Figure 2b) shows the total movement for the winter 1959-60.

inches downslope. In all three holes, the upper parts were as much as 5 inches off plumb relative to the bottom, thus indicating more rapid movement at the surface than at depth. The absolute amount off the vertical, however, is not reliable, because it was difficult to make certain the initial holes were vertical, whereas it was easy to measure the slope of the excavated holes by plumb bob in an open pit.

To make it possible to compare snowcreep with snow conditions, a standard snow survey was carried out. It involved sampling at 10 points at 10-foot intervals. The results are shown in the accompanying table.

SNOW SURVEY, MOUNT SEYMOUR, 1959

Date	Day (from Dec. 31)	Depth (in inches)	Water equivalent (in inches)	Density
Jan. 1	.1	6	—	—
" 31	31	35	—	—
Feb. 28	59	,68	27.4	.41
Mar. 21	80	89	32.9	.37
" 30	89	101	38.3	.38
April 5	95	101	—	—
" 11	101	89	40.5	.46
" 30	120	75	36.5	.46

Winter, 1959-60

On December 13, 1959, three balls were installed in a manner similar to that of the previous winter (Figure 2, section b). Ball number 1 was of plastic (as used the year before), whereas numbers 2 and 3 were wooden ovoids 6 inches long, 3.5 inches in diameter and 255 grams in weight. As the movement of number 3 was exactly the same as that of number 2, only the movements of 1 and 2 are shown in Figure 2, section b.

Winter, 1960-61

In the winter of 1960-61, six steel poles were inserted upright in holes drilled in bedrock. The mechanism for reading the displacement of balls was similar to that of preceding years, except that the Monel wire from each ball passed through a hole drilled at the base of its pole, then up through the pole to a tape readable above the snow at the top of the pole. Spherical wooden balls 6 inches in diameter and 1,080 grams in weight were used. Displacement of the balls, snow depth and ground temperatures are shown in Figure 3.

Attempts were also made to determine the free-water content of the snow near pole number 2, by using thermos bottles as calorimeters. On March 25, the free-water content of 2 feet of snow, sampled from the bottom, middle and upper part of the snow core was: bottom 9.5 per cent free water; middle, 12.0 per cent; upper, 12.0 per cent.

On April 2, snow sampled for the full 2-foot depth gave a free-water content of 20.8, 19.8 and 21.5 per cent with a mean of 20.7 per cent. On April 9, when the snow depth was 18 inches the free-water content was 27.1, 27.1 and 25.7 per cent with a mean of 26.6 per cent. Thus in the period from March 25 to April 9, the free-water content rose from 11 to 27 per cent.

Winter, 1961-62

Observations on rate of snowcreep were repeated at the same site with essentially the same equipment as in the previous winter (Figure 4). The most significant changes were the substitution of one 3-inch wooden ball for a 6-inch ball, the addition of two 3-inch balls operating in tandem with 6-inch balls previously installed and the design and installation of four hydraulic gauges for measuring pressure exerted by creeping snow against stationary surfaces. Records of snow depths, temperatures, displacements of the 3-inch and 6-inch balls, and snow pressures were made about once a week and were compared with weather and snow conditions.

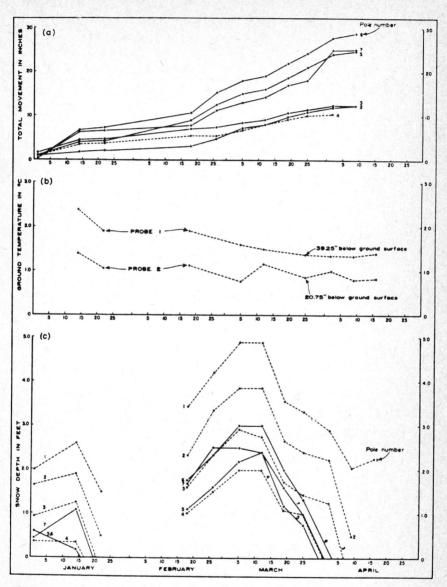

Figure 3. Data for the winter 1960-61. Section a shows displacements of the balls attached to the numbered snow poles. Section b shows ground temperature measured at the site of pole number 1. Temperatures were based on thermistor readings. When the stability of the thermistors was checked in late 1962, some drift was observed; so the readings may be slightly inaccurate. Snow depths are shown in section c. The gap at the end of January and in early February is due to melting of the early January snow cover and exposure of the slope.

The rates of snowcreep in 1961-62 corresponded with remarkable closeness to those of 1960-61 at the same sites; but, during two cold periods, in mid-January and late in February, when the snowpack froze to the underlying rock, snowcreep slackened noticeably (Figure 4).

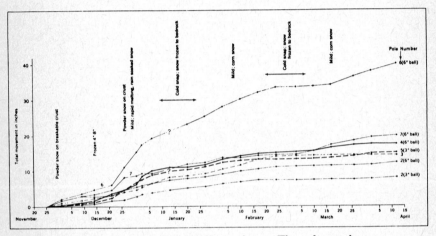

Figure 4. Winter conditions of 1961-62 are shown. The pole numbers correspond to those of Figure 3.

The increase in the rate of movement upward through the snowpack is clearly indicated. Thus the 3-inch ball at pole number 2 moved only 56 per cent of the distance that the 6-inch ball moved at the same site; at pole number 4 the 3-inch ball moved about 70 per cent of the distance travelled by the 6-inch ball. Similarly, the upper surfaces of the balls moved forward more than the centres, thereby contributing to a forward rotation; thus the ball attached to pole number 6 moved forward 2.0 feet but, because of approximately 270° rotation, the upper surface (and the tape) moved 3.5 feet. The ratio of the displacement of the centres of the balls to that of the top surfaces varied from 50 to 95 per cent; for all eight balls the average is 72 ± 18 per cent. If it is assumed that the movement of the tops and centres of the balls accurately reflects snow movement at corresponding levels, then the movement within the snow diminishes in some places virtually to zero at the bottom. Friction between the balls and the ground, however, may restrict free movement.

	Elevation above ground level	Cumulative displacement
		(feet)
Pole No. 2	6″ (top of 6″ ball)	1.15
	3″ (top of 3″ ball)	0.65
	3″ (centre of 6″ ball)	0.58
	1½″ (centre of 3″ ball)	0.32

At other places, movement clearly takes place at the base of the snowpack or within the bottom 1½ inches.

	Elevation above ground level	Cumulative displacement
		(feet)
Pole No. 4	6″ (top of 6″ ball)	1.47
	3″ (top of 3″ ball)	1.08
	3″ (centre of 6″ ball)	1.38
	1½″ (centre of 3″ ball)	0.92

Downslope increase both in rates and in total movement (most notably within 3 inches of the ground surface) without any very marked increase in slope were observed both in 1960-61 and 1961-62.

The snow varied from the fresh powdered kind found early in the season to fine, damp snow and frozen snow; with the alternation of cold and mild moist weather, frozen and wet snow alternated, becoming coarser as the season progressed. The content of liquid water in the snow, as determined by crude calorimetry in thermos bottles, varied from zero in cold, frozen snow to a maximum of 15 to 20 per cent in fine, damp snow subjected to heavy rain. The quantities of water retained by the coarse 'corn' snow that formed during mild periods toward the end of the season were roughly uniform and constituted about 5 to 10 per cent by weight. The increase in grain size and the corresponding change in grain shape presumably led to the reduction in the amount of moisture that could be retained against the pull of gravity in this well-drained environment. Snow densities averaged for the full thickness were not recorded early in the season but later ranged from 0.40 to 0.45 gms/cm.[3]

The conditions most favorable for snowcreep seemed to be the state of mild thaw that prevailed in early January in finer-textured snow with higher water content. Similar snow depths in February and March with wet but coarser snow produced an appreciably slower movement. Below-freezing conditions were least favorable.

Snow pressures exerted against 5-inch-square plates showed a very great range (0 to 45 pounds) and a lack of consistency from week to week. Their maximum occurred at the end of December, early in the first main period of thaw. Later in the season similar snow depths produced much lower pressures. Observed pressures can presumably be correlated not only with depth but with rate of creep; and on two occasions pressures on the higher plate on pole number 4 matched or exceeded the lower plate, then buried under four times the depth of snow. The results, however, are too erratic to prove any firm relationship.

Instrumental problems in measuring the rate of creep were minor. Of greatest concern was the coating of hard frozen snow found around three of the balls still buried when the installation was dismantled. This coating, if it persists through the season and adheres to the rocky substratus, could all result in marked discrepancy between the movement of the balls and the movement of the surrounding snow. The cause of this coating is not at all clear.

Instrumental problems in measuring snow pressures include the disturbing effect of variable air temperature on exposed parts of the tubing and on the Bourdon gauge, and the melting of snow in front of the plates because of the conduction of heat down poles and tubes. All parts of the equipment should remain buried beneath the snow, the gauges being enclosed in a box to permit reading.

References

Am. Geol. Inst., "Glossary of geology and related sciences," *Am. Geol. Inst.*, Nat. Acad. Sci. Nat. Res. Council, publ. 501, (1957).

Bucher, E., "Beitrage zu den theoretischen Grundlagen des Lawinenverbaus," *Beit. z Geol. der Schweiz, Geotech. Ser. Hydrologie*, Lieferung 6 (see also Snow, Ice, Permafr. Res. Est., Trans. 18, 1956, 109 pp.) (1948).

Furukawa, I., "Vertical distribution of snow pressure on slope (in Japanese with English summary)", *Seppyo*, Vol. XVII (1956), pp. 15-17.

Haefeldi, R., "Creep problems in soils, snow and ice," *Proc. Third Internat. Conf. on Soil Mech. and Foundation Eng.*, Vol. III (1953), pp. 258-51.

in der Gand, H.R., "Ergebnisse der Gleitmessung, Winterber," *Eidg. Inst. Schnee u u Lawinenforsch*," Vol. XX (1957), pp. 111-14.

Kataoka, K. and Sato, S., "On the bend of lower part of the afforested cryptomeria by snow (in Japanese with English summary)," *J. Jap. Soc. Snow and Ice*, Vol. XXI (1959), pp. 13-19.

Kojima, K., "Viscous flow of snow cover deposited on slope (in Japanese with English summary," *Low Temp. Sci. Ser. A.*, Vol. XIX (1960), pp. 147-64.

Martinelli, Jr., M., "Creep and settlements in an alpine snowpack," Rocky Mtn., Forest and Range Expt. Sta. Forest Serv. U.S. Dept. Agr., Research Notes no. 43, (1960), p. 4.

Shidei, T., "Studies on the damages on forest tree by snow pressures (in Japanese with English summary)," *Bull. Gov. Forest Expt. Sta.*, no. 73- (1954), p. 89.

Stokes, W. L., and Varnes, D. J., "Glossary of selected geologic terms with special reference to their use in engineering," *Proc. Colo. Sci. Soc.*, Vol. XVI, (1955).

Washburn, A. L., "Classification of patterned ground and a review of suggested origins," *Geol. Soc. Am. Bull.*, Vol. LXVII (1956), pp. 823-66.

Williams, P. J., "The direct recording of solifluction movements," *Am. J. Sci.*, Vol. CCLV (1957), pp. 705-14; "An apparatus for investigation of the distribution of movement with depth in shallow soils layers", *Nat. Res. Council*, Div. Bldg. Res., Building Note (1962).

Yosida, Z., "A calorimeter for measuring the free water content of wet snow (in Japanese with English summary)," *Low Temp. Sci.*, Ser. A., Vol. XVIII (1959), pp. 17-28.

13

Planning Avalanche Defence Works for the Trans-Canada Highway at Rogers Pass, B.C.

P. A. Schaerer

The Selkirk Mountain range in the interior of British Columbia is one of the major obstacles to be crossed by the Trans-Canada Highway. Various routes through this range have been investigated; that through Rogers Pass was selected.[1] The pass lies between the towns of Golden and Revelstoke in Glacier National Park and was chosen by the Canadian Pacific Railway as the route for the first railway link between Eastern Canada and the Pacific Coast. It was in use from 1885 until 1916 when the Connaught tunnel was built and the railway line through the Pass was abandoned.

It was known that avalanches would be one of the major problems for any road built through the Selkirk range (Fig. 1). Accordingly, in 1953 the Department of Public Works of Canada, which is responsible for the construction of the Trans-Canada Highway through Glacier National Park, began reconnaissance work for the highway and organized a preliminary survey of avalanche sites and avalanche activity. These avalanche observations were under the direction of N. C. Gardner. In 1956 an avalanche observation station was established to carry out the more detailed study of avalanche activity required for the design of the defence system. The National Research Council, through its Division of Building Research, co-operated in the organization of the avalanche observation station. Between 1957 and 1960 the author was in charge of this station and was responsible for planning the avalanche defence. This paper describes in brief the

First published in *Engineering Journal*, Vol. XLV, no. 3 (1962), pp. 31-38. Also *Research Paper No. 152* of the Division of Building Research Council, Canada. Reprinted by permission.

[1] Hague, J. P. "Selection of the Trans-Canada Highway Route through the Selkirk Mountains," *The Engineering Journal*, Vol. XLI (June 1958), pp. 57-60.

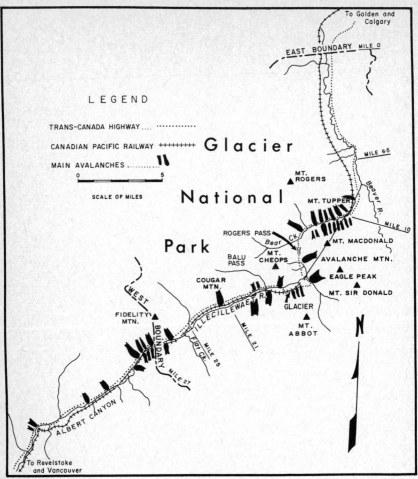

Figure 1. Rogers Pass with main avalanches.

observations which were made and summarizes the defence which was chosen.

Terrain

The summit of Rogers Pass is 4,300 ft. above sea level, and the peaks of the Selkirk range rise to 11,000 ft. The valleys associated with the Pass are short and comparatively steep on the east side, but rise to the summit with a gradual climb from the west. A typical Selkirk valley is narrow and has steep sides.

The mountain sides on the Rogers Pass route have a terrace located between 5,500 and 6,500 ft. above sea level. The terrace goes gradually over into the scree slopes and rockfaces which rise to the mountain ridges. Below the terrace a sharp drop over cliffs leads to talus slopes and alluvial fans into the valley bottom which is between 300 and 1,000 ft. wide. In some places the mountain sides are close together, resulting in a narrow V-shaped valley. The highway is cut into the talus slopes and alluvial fans, and except for two short sections is located on the north side of the river which flows through the valley. The lower mountain sides and valleys are covered with heavy timber and dense brush. Trees become scattered on the terrace, leaving space for alpine meadows.

Climate

Rogers Pass is in the region popularly known as the interior wet belt of British Columbia.[2] High annual precipitation and heavy snowfall are its most distinctive features. The average annual precipitation, measured between 1921 and 1950, at Glacier close to the summit of Rogers Pass, is 18 in. of rain and 342 in. of snow, while the maximum annual snowfall ever observed was 680 in., measured during the winter 1953-54. The magnitude and frequency of the 24-hour snowfalls at Glacier during the winters between 1953 and 1960 are shown in Table I.

TABLE I. FREQUENCY OF SNOWFALLS AT GLACIER.

Winter	Number of days with snowfalls				Total snowfall for the winter, inches
	Less than 4 in.	4 to 12 in.	12 to 20 in.	20 in.	
1953/54	56	11	7	2	680
1954/55	35*	17*	–	–	310
1955/56	42	30	–	–	336
1956/57	47	17	3	1	341
1957/58	74	26	1	–	321
1958/59	85	41	3	–	442
1959/60	84	29	1	–	368

*The records of the winter 1954-55 include only the snowfalls after 1 January 1955.

Storms with a high rate of snowfall are not frequent and are usually of short duration. A day with heavy snowfall is usually followed by one or several days with only light snowfall. During winters of light snowfall three storms may occur yielding more than 16 in. of snow in a three-day period, but during winters of heavy snowfall 10 such storms may occur. Only occasionally, about once in three years, is there a snowstorm that contributes more than 36 in. in a three-day period. Most snowfalls are accompanied

2 Chapman, J. D., and D. B. Turner, "British Columbia Atlas of Resources," B.C. Natural Resources Conference, (1956).

by wind which deposits large amounts of snow on the lee side of the mountain ridges. The maximum depth of snow on the ground in the valley is about 100 in., and on the mountain, at the 6,700-ft. level the greatest measured depth of 160 in. was observed on May 1, 1959.

Temperature during a snowstorm normally ranges between 20 and 32°F. After the storm has ended, it is usual for cloudy weather to continue and the temperature to change relatively slowly. Frequently, however, the temperature rises during a snowstorm and the snowfall in the valleys changes to rain. The temperature falls below 0°F. only a few times during the winter and this cold weather usually does not continue for more than a week.

Avalanche Survey

The avalanche survey had to produce the following information about the average and very large avalanches that occur at each site: their rupture zone, path and terminus; their depth and width; their frequency of occurrence; the prevailing conditions responsible for their occurrence.

Information was obtained by monthly patrols through the area during the first three years, and after 1956 by daily and weekly patrols to record observations on each avalanche that occurred, even on those which terminated far from the highway. The rupture zone and path of important avalanches was sketched on photographs and all avalanches that deposited snow near or on the highway right-of-way were traced on location plans. Information on the conditions that cause avalanches was obtained through snow cover and weather observations at four observation sites in the valley and two mountain observatories located at an elevation of about 6,800 ft. Precipitation and temperature measurements made at Glacier previous to 1953 were available from the Meteorological Branch of the Department of Transport.

The avalanche survey was completed by site studies in summer and winter. Extent, slope angle, exposure, soil conditions, age of trees in the rupture zone, path and terminus were observed at each site. Valuable information was also obtained from the Canadian Pacific Railway Company, who kindly made available their records on all avalanches that had affected railway operation during the period 1910 to 1952.

In 1960, when the final plans were developed for the first stage of defence, there were available detailed observations of avalanche activity during seven winters and partial observations from another 40 years. Records of avalanches that occurred during the years before 1953 revealed that there were periods of more than one year in which avalanche activity was a maximum. These periods of maximum activity are associated with periods of high snow-fall and high temperature. Unfortunately, the present survey was conducted during what the records indicate to be a period of low activity, both as to the size and the number of avalanches. Observations taken do not, therefore, cover the worst possible conditions which may be encountered. The survey did produce a good picture of avalanches that may be encountered during an average year, but maximum conditions had to be deduced from the few large avalanches that occurred during the observation period and from the incomplete observations of earlier years.

Avalanche Sites

Avalanche sites can be easily recognized by the scars on the timbered mountainside. The area where avalanches start to slide and where the bulk of snow originates is called the rupture zone. The previously described terrace at an elevation of about 6,000 ft. divides the rupture zone in many of the avalanche sites. The avalanches rupture in a lower zone on or at the toe of the cliffs below the terrace or in an upper zone on the steep slopes rising to the mountain ridges. Many of the small avalanches which originate in the upper zone stop on the terrace. Large avalanches overrun the terrace and gain a high speed as they fall over the steep slopes below. There is evidence that many lower rupture zones at Rogers Pass were once covered with timber that prevented the occurrence of avalanches. Fires appear to have removed these trees and destroyed a very effective natural defence.

The avalanche path is the track followed by an avalanche during its descent. The majority of avalanches at Rogers Pass are confined in gullies which have been carved in the mountains by running water. A few avalanches reach the valley over open, bare slopes. The terminus is the area where the main body of the avalanche comes to rest. Avalanches slow down on the alluvial fan or the talus slope at the lower part of the mountain. Minor ones will stop there but major avalanches advance into the valley and in some cases reach the opposite side.

The narrow valley in Rogers Pass and the fact that avalanches reach the valley floor from both sides made it impossible to construct a highway that would avoid all dangerous sites. In the 35-mile highway section between the east boundary of Glacier National Park and Albert Canyon on the west side 74 avalanche sites had to be crossed; 61 of these lie within the Park. As the avalanches do not occur under the same conditions at each site, and since the topography and area of the rupture zones and the avalanche paths may produce avalanches of different sizes, the hazard to the highway varies from one avalanche site to the next.

The 74 avalanche sites can be classified as follows:

Nine sites where minor and major avalanches occur frequently and where at least one major avalanche can be expected to reach the highway every year;

Twenty-one sites where minor avalanches occur frequently and usually reach the highway once or more than once each winter; the mass of the sliding snow is small and may not cover the whole width of the highway; large avalanches occur occasionally but not every winter;

Thirteen sites where avalanches occur only under severe conditions and not more frequently than once in two years; the avalanches may deposit a large amount of snow on the highway;

Thirty-one sites where avalanches occur only under severe conditions and not more frequently than once in two years; the snow that would reach the highway from these avalanches would usually be airborne and little would be deposited on the highway.

The most active avalanches are concentrated in a narrow defile, about 2½ miles long, between Mt. Tupper and Mt. MacDonald. Here the concentration of sites is so great that they merge into an almost continuous slide belt (Fig. 1). Another section with active avalanches is just outside the western boundary of Glacier National Park. Between and outside these two areas the avalanche paths are more scattered.

Classification of Avalanches at Rogers Pass

Avalanche classifications have been suggested by various authors since avalanches were first described in the late 19th century.[3] Each classification serves a particular purpose, e.g., a skier travelling in the rupture zone may look at the avalanches from a viewpoint different from that of the engineer who is only concerned with the effect the moving avalanche has on structures. The following classification, which uses the cause of avalanches as a basis, proved to be most practical for the design of the defence system at Rogers Pass.

The frequency of occurrence and the proportion of the avalanches in each class was determined for each site and the defence chosen accordingly.

---

[3] de Quervain, M. R., "Avalanche Classification," International Union of Geodesy and Geophysics, General Assembly of Toronto 1957, Vol. IV, Snow and Ice Commission, Gentbrugge 1958, pp. 387-92.

Dry Snow Direct Action Avalanches

The majority of all avalanches fall into this category. They occur during or immediately after a snowstorm, and avalanche hazard can be recognized by observing the snow cover and the weather. Minor avalanches may reach the highway after snowfalls of 10 in. when accompanied by wind. Experience has shown that snowfalls of two to three days' duration, yielding more than 16 in. of new snow and accompanied by strong wind or 22 in. without wind may cause an avalanche hazard at all known sites. This amount of new snow is only a general guide; an unstable snow cover, low temperatures or other influences may cause avalanches before this amount has been reached. A continuous snowfall, exceeding 30 in. of new snow, may cause unusually large avalanches. Dry snow avalanches move on the surface and are accompanied by a cloud of snow dust. The avalanche may become airborne over cliffs or very steep terrain and the cloud of snow dust obtain a high speed. Some avalanches may deposit large amounts of snow on the highway; others may cause only strong windblast and deposit light snow.

Wet Snow Direct Action Avalanches

These avalanches occur when snowfalls are followed immediately by rain or warm weather. They can be large and reach the highway when the snowfall exceeds 12 in. The wetter the snow the more the avalanches move on the surface, with little or no dust cloud and accompanying windblast. Records of the railway company indicate that most of the big avalanches which have blocked rail traffic for long periods of time were the result of heavy snowfalls followed by warm weather. It would appear that major occurrences of wet snow direct action avalanches would create the most unfortunate situation for the highway, because the avalanches could cover the highway with deep snow simultaneously at many sites.

Dry Snow Delayed Action Avalanches

These avalanches are also called climax avalanches, and are caused by various factors which build up the avalanche hazard over a period of time during which snowfall does not necessarily have to occur. By careful observation of the snow cover and the weather it is possible to evaluate the avalanche hazard and determine the periods when such avalanches may affect the highway. These avalanches have the same effect as dry snow direct action avalanches, but unlike that class appear to occur sporadically and do not normally affect large sections of the highway simultaneously.

Spring Thaw Avalanches

These avalanches occur as a result of the loss of cohesion of snow as it melts. They usually occur in cycles at many sites simultaneously during a few hot days, but single occurrences were observed also. The avalanches can involve large masses of wet, heavy snow and often slide on the ground. They usually have a lower speed and a smaller range than the dry snow avalanches, and tend to flow in channels.

Avalanche Defence Methods

From the observations made on avalanches and their causes, it was necessary to formulate a general defence plan for the whole route, and to indicate the defence method that should be applied for each avalanche site within the general plan. To the greatest possible extent the Trans-Canada Highway was constructed through areas safe from avalanches. When an avalanche path could not be avoided, attention was given to locating the highway in such a way as to reduce the effect of the avalanches, e.g., placing the highway near the tip of the avalanche terminus where it would be reached infrequently by avalanches or high on the mountainside where the avalanche path is narrow and snow clearing to the downhill side easier. Cuts were daylighted in order to avoid deep snow deposits.

Much valuable experience is now available from countries such as Switzerland, Austria and the United States on various methods of defence against avalanches.

Active Defence
Structures and vegetation
  (permanent measures)
  Retaining barriers
  Snowfences and windbaffles
  Braking barriers: Earth
    mounds, Catching dams
  Diverting dams
  Snowsheds
  Reforestation
Explosives (temporary
  measures)

Passive Defence
Avalanche warning (temporary
  measure)
Highway closures
Avalanche detection

All methods may be applied alone or in combination. They are described below and rationalized for the defence at Rogers Pass.

Retaining barriers are constructed in the rupture zone and control avalanches at their source. They can create a completely safe site. It was found that costs for barriers in the remote and large rupture zones which occur at Rogers Pass were prohibitive. Retaining barriers could be built economically only at minor avalanche sites where the rupture zone is close to the highway, e.g., on long, steep banks from which significant amounts of snow may slide and cover the highway.

Snowfences and windbaffles control in the rupture zone the drifting snow that may create an avalanche hazard. It was considered that this defence is not suitable for Rogers Pass, because the structures must be built at high elevations and on exposed mountain ridges where construction and maintenance are difficult. This defence does not prevent dry or wet snow direct action avalanches that are not accompanied by wind. Furthermore, it is possible that deep snow would soon cover the structures.

Braking barriers are obstacles in the avalanche path which slow down or stop the avalanches before they reach the highway. Of the various structures which can be used, earth mounds and catching dams were judged to be suitable.

Earth mounds are built in two or more rows in a checkered arrangement on the flat section of the path where the avalanches would normally have slowed down considerably. At Rogers Pass, the alluvial fans at the terminus of the avalanche sites have low slope angles and are a suitable location for mounds. A series of experimental mounds was constructed in 1957. Observations during the following three winters showed that at locations where the mounds were constructed dry snow avalanches were slowed down while passing between the mounds and terminated just beyond; spring thaw avalanches stopped at the first row of mounds. It was decided that mounds should be built as defence against wet snow avalanches and smaller dry snow avalanches at other sites where not more than three avalanches per winter are likely to occur. Because construction costs are low, they can be built as defence against large dry snow avalanches also. In this case, it is not expected that the mounds will stop the avalanches completely, but they should retain a great part of the heavier sliding snow and reduce the dimension of the avalanche on the highway. The mounds built at Rogers Pass are 15 to 25 ft. high with a distance of 60 to 80 ft. between centres. The number of mounds and their size depends on local conditions such as normal size of avalanches and nature of terrain.

Catching dams are built perpendicular to the avalanche direction. Their efficiency is limited because fast-moving avalanches can jump over them and the spaces behind are often filled with snow after one avalanche has occurred. It can easily be calculated that a catching dam requires about the same earth quantity as two rows of mounds of equal height. Mounds, however, are more efficient in stopping avalanches. It was considered that dams could be built at Rogers Pass immediately above the highway where not more than two avalanches are likely to occur and where there is insufficient space for construction of mounds. Catching dams are between 10 and 15 ft. high. During the two winters between 1957 and 1959, tests were made with catching dams built by pushing snow to the outside edge of a wide terrace. These proved effective against small avalanches, but since snow dams have to be rebuilt each winter earth dams were considered more practical.

Diverting dams divert an avalanche laterally to an area where it can run out harmlessly. Use of small dams with a natural earth slope or a vertical wall on the side facing the avalanche was considered for different avalanche sites where they could be built easily on the long alluvial fans. It was concluded, however, that if these dams were constructed in the terminus zone of the avalanches they would be backfilled by early snow deposits and prove ineffective for later avalanches. Furthermore, it was found that earth mounds produce the same degree of protection as earth dams at lower cost. In two cases, even a snowshed proved to be less expensive than the required long and high dam.

Diverting dams with modified function are used in association with snowsheds. In this case the dams restrict the spreading of the avalanches and keep them in a defined straight path. Such dams were once built on Rogers Pass in association with the numerous snowsheds which protect the railway. Remains of the dams, consisting of stone-filled log cribs, can still be found along the abandoned railway line. But construction methods have changed during the past 70 years, and for the highway it proved more economical to construct high dams with a natural earth slope to reduce the width of the avalanches and minimize the length of the snowshed. These dams are 20 to 25 ft. high. For some sites, where frequent avalanches occur during the winter, it was possible to increase the effective height of the dams by excavating a channel in the avalanche path. The excavated earth was used for highway fill.

Snowsheds divert moving avalanches over the highway. Expensive structures, they were considered only for sites where other active defence methods proved to be inadequate. At Rogers Pass the highway must cross most avalanche sites in the terminus zone, and loads from deposited snow can be very large. For three winters, therefore, snow depth, slope angle and density of deposited avalanche snow were surveyed in sites where sheds are planned. The static load for each snowshed was estimated from these observations. The largest design static load was calculated to be 1100 p.s.f. on the uphill side of the snowshed and 500 p.s.f. on the downhill side. This corresponds to a snow depth of 29 ft. and 13 ft. respectively, with an assumed average snow density of 38 lb./cu. ft. The shed must also be designed to withstand the forces produced by a moving avalanche. For one site this dynamic force was calculated to be 700 p.s.f. vertical load and 350 p.s.f. lateral force due to friction. Lighter sheds, to be built in the path of smaller avalanches, were designed for deposit loads of 450 p.s.f. and 150 p.s.f. and 300 p.s.f. moving load.

Reforestation in the rupture zone serves the same purpose as retaining barriers. Reforestation projects must be associated with temporary retaining barriers, which because of expense were not considered for Rogers Pass.

Explosives are a temporary defence measure that is cheap and versatile. When the avalanche hazard dictates avalanches can be released under control by detonating explosives in the rupture zone. Under certain circumstances, particularly when there is a hazard for dry snow delayed action avalanches, the explosion can create a stable snow cover without producing an avalanche. Much experience on this method of avalanche defence has been gained in the U.S.A. and in Switzerland, where it has proved effective and economical against direct action avalanches and dry snow delayed action avalanches, although not against spring thaw avalanches.

The normal technique is to use projectiles fired from an artillery weapon. Studies and experiments indicated that the 75-mm. and 105-mm. howitzer are most suitable for avalanche control at Rogers Pass. Some rupture zones are too rugged for economical firing and at a few avalanche sites no suitable firing position could be found. Since firing must often be done during snowstorms when maintenance personnel are already busy with snow removal work, and because a large portion of the highway is in the National Park which should not be converted into an artillery range, it was decided that artillery fire should be used only against avalanches that cannot be controlled economically by other methods. Hand-placed and preplanted explosive charges were found unsuitable at Rogers Pass because of the remoteness of the rupture zones.

Highway closure for the duration of dangerous avalanche conditions is a simple protective measure. It requires an organization to study snow cover and weather, evaluate avalanche hazard and order closure and re-opening as necessary. It was considered that although the Rogers Pass route must be open for the whole winter, short closures would be permissible.

An avalanche detection system consists of an electronic device which detects an avalanche as it occurs and operates warning signals. In a long path, an avalanche may be detected some time before it reaches the highway and traffic signals could warn oncoming vehicles before they enter the dangerous area. In short paths the device may only signal avalanche occurrence to the maintenance headquarters. The former use of the system may be feasible at a later stage of defence at a few selected avalanche sites at Rogers Pass, but the latter use of the system was chosen for two sites where frequent small avalanches may cover the highway. The Radio and Electrical Engineering Division of the National Research Council has built an avalanche detection device for Rogers Pass that is now being tested.

Avalanche Hazard Evaluation and Prediction

Defence by means of explosives and highway closure requires the evaluation of avalanche hazard to determine when explosives should be used and when the highway must be closed or reopened. An avalanche is caused by different factors associated with terrain, snow cover, snowfall, wind and temperature, all in close relationship. Certain rules have been established through experience on the dependence of avalanche hazard on these factors. Hazard can be evaluated quite accurately for the time of observation, but the prediction of future hazard is only as good as the weather forecast. In practice, the avalanche hazard forecaster has to assume that weather will follow a certain pattern and his prediction is based on this. The evaluation of the avalanche hazard can be approached by two different methods, called here the testing method and the analytic method.

The testing method has been developed in Switzerland;[4] evaluation of the avalanche hazard is based on snow cover observations. Stability of a snow cover can be tested directly from time to time and weather factors such as snowfall, wind and temperature used to determine its stability between observations. With experience and tests, it has been found which factors may lead to a fracture and which conditions contribute to stabilization.

The analytic method is the technique used in the U.S. Forest Service.[5] Most avalanche sites with which this service was concerned produced direct action avalanches. Observations over a few winters indicated the weather and snow cover factors mainly responsible for creating them. By determining the magnitude of each factor, it was possible to evaluate avalanche hazard. This method requires more accurate weather observations than the testing method, but produces better results.

In practice, both methods are used in combination. Between 1956 and 1959 observers at the avalanche observation station in the Rogers Pass area made snow cover and weather observations and evaluated daily the avalanche hazard. The testing method was used as the basis for evaluation. Weather conditions that caused avalanches at specific sites were analysed and rules for forecasting direct action avalanches were found. Other studies were made to establish a rule of thumb for the hazard evaluation of spring thaw avalanches. In 1959 the responsibility for avalanche evaluation and prediction was transferred to the Department of Northern Affairs and National Resources. The experience gained in combining the testing method with the analytic method has been passed on to this Department.

Defence Plan

Study of avalanche defence for the Rogers Pass route indicated that the cost of a defence that would guarantee a continuously open highway would be unreasonably high. It was realized also that the required structures for such a defence could not be built before the scheduled completion date of the highway. On the other hand it was recognized that a passive defence only would not avoid frequent and long closures, which could not be accepted for the Trans-Canada Highway. Based on these considerations a plan with three stages of defence construction, combining active defence with

---

[4] Bucher, E., and others, "Lawinen, die Gefahr für den Skifahrer," Geotechnische forschende Gesellschaft, (1940), Aschmann and Scheller, Zurich (out of print).

[5] Forest Service, U.S. Department of Agriculture, Snow Avalanches, Agriculture Handbook No. 194, (1961).

passive defence by highway closures, was chosen. The length of time that the highway would probably be closed each winter would be decreased with the completion of each stage.

The first stage of defence will be in operation when the highway is opened, and the other stages can be introduced later when more experience with highway operation and avalanches is available and when shorter closure times are demanded. Studies on the defence required for the three stages were made and costs estimated in relation to prices for 1960-1961. The different stages are described below and their estimated effectiveness shown in Table II. Initial costs for the different stages are those for defence structures and for establishing an avalanche warning service. The annual cost is the estimated amount of money to be spent each year for control of avalanches by explosives and for the avalanche warning service, but does not include any cost for snow removal.

The First Stage of Defence

The following defence plan was chosen for the first stage of defence:

Defence structures will be built that will give full protection at sites:
—where the avalanche survey has indicated that dangerous dry snow direct action avalanches usually occur after snowfalls yielding less than 16 in. new snow when accompanied by wind, or less than 22 in. new snow without wind;—where spring thaw avalanches were observed to have reached the highway location regularly every year; and
—where dangerous dry snow delayed action avalanches usually occurred at least once every year.

This means that defence structures are to be built at sites where avalanches would have affected the highway once or more than once each year during the period from 1953 to 1960. The plan required the construction of eight snowsheds, with a total length of 4,825 ft., a total of 80 earth mounds constructed at two sites, and catching dams at two sites with a total length of 800 ft. Associated with the snowsheds are diverting dams between 100 and 600 ft. long that confine the avalanches.

In addition to these structures it was decided to build a total of 240 earth mounds at seven sites, a total of 700 ft. of diverting dams at two sites and a total of 1,400 ft. of catching dams at three sites where avalanches were observed less frequently, but where the local conditions make inexpensive structures feasible. These secondary defences do not offer complete protection to the highway, but will probably retain smaller avalanches and reduce the size of larger ones.

Artillery fire is to be used against dry snow delayed action avalanches at sites that are unprotected or insufficiently protected by structures.

The highway is to be closed when dry snow direct action, wet snow direct action, or spring thaw avalanches are likely to occur at unprotected sites. This means that the highway would normally be closed until about 12 hours after snowfalls exceeding 16 in. when accompanied by wind or 22 in. without wind. Furthermore, it might be closed when a snowstorm is followed by warm weather or during days when larger spring thaw avalanches are likely to occur.

An avalanche hazard evaluation and prediction service is to be established that will determine when and where avalanches have to be controlled by artillery or the highway closed.

The Second Stage of Defence

The defence measures of the first stage are to be supplemented by structures to provide full protection at sites where wet snow direct action avalanches and spring thaw avalanches reach the highway more frequently than once in 10 years. Additional snowsheds, retaining barriers in rupture zones close to the highway, earth mounds and diverting dams will be built.

The Third Stage of Defence

This stage would guarantee an open highway for the whole winter, except for short intervals when explosives would be used to control direct action avalanches and dry snow delayed action avalanches at sites where no defence structures were built in the first and second stage. Additional defence structures would have to be built at sites where the terrain forbids artillery fire or where artillery fire would be uneconomical, as well as at sites where spring thaw avalanches may reach the highway.

Although defence structures for all stages will be designed for situations which may occur once in 10 years, conditions may be encountered once in 15 to 20 years that produce very large avalanches that follow unusual paths. An avalanche evaluation and prediction service will be necessary to determine the time and location for artillery fire and also to order closure of the highway for the extremely bad situations which may overcome the defence system.

TABLE II. COMPARISON OF THE DIFFERENT DEFENCES BASED ON 1960-61 PRICES.

	Estimated Costs		Closure of Highway in an Average Winter		
	Estimated Initial Cost	Annual Cost	Time	Total Days	Maximum Days
Closures only, no active defence	$ 20,000	$40,000	25	75	10
First Stage: combined active and passive defence	5,300,000	52,000	7	12	6
Second Stage: combined active and passive defence	2,900,000	50,000	4	5	4
Third Stage: active defence only	$4,400,000	$70,000	None	—	—

Concluding Remarks

Avalanche observations and subsequent development of the defence system for the Trans-Canada Highway in the Rogers Pass area is the first project of this kind to be undertaken in Canada. It is to the credit of the engineers of the Department of Public Works that an avalanche survey was initiated with the first highway location studies in 1953. Because of their foresight, observations from seven winters were available upon which recommendations could be based for a defence plan and the specifications for the first stage of defence by the time detailed engineering planning was begun. The defence plan chosen was one that could be completed in successive stages, each stage giving greater protection to the highway. Structures near the highway, explosives and an avalanche evaluation and prediction service form its basis. Structures constructed for the first stage of defence will protect the highway at those sites where major avalanches will affect the highway once or more than once each year. The plan lays down principles to be followed in determining successive stages of defence, but details of this defence require further observations of the avalanches and their effect on the highway when it is in operation.

14
Iron Mining in Permafrost Central Labrador-Ungava

J. D. Ives

Introduction

Decision to prepare for the exploitation of the Knob Lake iron ore in 1950 set in motion one of the largest commercial developments of the Canadian northland. An analysis of the economic and political considerations behind this decision is beyond the scope of this paper, although it can be assumed that a basic consideration was, that if sufficient reserves could be proven, the immense capital outlay would be justified in terms of relatively cheap, large-scale methods of production. In this sense, the amount of proven ore necessary to provide confidence in a long-term venture is a key figure. This was initially set at 300 million tons of ore with less than 20 per cent silica giving an average grade (dry basis) of about 59 per cent Fe., and was increased to 400 million tons in 1949. Of vital importance to the Iron Ore Company of Canada was the acquisition of a "captive market" of American steel companies who would guarantee a minimum purchase, together with company ownership of the Quebec North Shore and Labrador Railway. A crash exploration program between 1948 and 1950 surpassed the target of 400 million tons of proven reserves (417 m. tons) and the vast machinery of railroad, town and mining plant construction was set in motion.

For the purpose of this paper it is assumed that, with 417 million tons of high-grade reserves, a balance between cost of producing and shipping the ore, and sale price had been set, leaving a safe profit margin. It must also be agreed that, in order to estimate the cost of a ton of ore at tidewater, or blast-furnace, as the case may be, some assumptions were made on the type of ore and the manner of handling it, one of the assumptions being that it was normal with respect to temperature; in other words, that it was not

First published in *Geographical Bulletin* no. 17 (1962), pp. 66-77. Reprinted by permission of the Department of Energy, Mines and Resources.

frozen. Iron Ore Company of Canada officials maintain that, based upon Mesabi Range experience, it was closely known in 1950 what it would cost to mine ore with a sufficient profit margin to allow for unknowns such as landslides and washouts on the railroad, and permafrost. However, the fact remains that the extent of permafrost in the mining area was incompletely known in 1950 so that, unless the "margin" was extremely wide, it could hardly be expected to engulf the higher costs of mining in frozen ground in a routine manner.

Ruth Lake mine was the first to be exploited and a token shipment of $1\frac{3}{4}$ million tons was made in 1954. Full operations were begun in 1955, when French mine, discovered in 1948 beneath the site of the exploration camp at Burnt Creek, and Gagnon mine, were opened up. Production for 1955 reached $10\frac{1}{2}$ million tons, slightly less than 5 million tons below the capacity of the existing installation. During mining operations in 1955 small "pods" of frozen ore were encountered in two of the Gagnon pits and caused some local handling difficulties. Also, the Iron Ore Company of Canada became involved with the technical difficulty of successfully mining and shipping "sticky ores," a problem which later became related to the frozen ore difficulty.

Trenching operations in 1955 in the large Ferriman orebodies led to the discovery of "ground frost" a few feet beneath the surface by H. E. Neal and L. Pituley. Permafrost was suspected, although a shaft driven into the Ferriman orebody to a depth of 50 feet in 1948 encountered frozen ground, and the discovery of "ground frost" in the Kivivic and Sunny orebodies 25 to 30 miles farther north in 1949 and 1950, were the first encounters with what has since proved a problem in mining operations.

Following the 1955 trenching experience at Ferriman, a number of exploration drill holes were utilized by H. E. Neal in 1956 for the installation of thermocouple strings according to National Research Council specifications. This was followed in 1957 by widespread thermocouple installation by B. Bonnlander under Neal's supervision and permitted initial ground temperature exploration of the Ferriman, Rowe and Star Creek orebodies. Almost simultaneously with the acquisition of the first results three pits were opened in the Ferriman area. Many of Bonnlander's thermocouple readings gave results of 30 to 33° F. (Bonnlander, 1957, 1958) and, as the limit of accuracy of the method was possibly greater than ± 1° F., doubt remained as to the extent of permafrost. With the opening up of the Ferriman orebodies much of this doubt was rudely removed.

The early cuts pierced frozen ore with a high water content and resulted in production costs considerably higher than those in normal, unfrozen ore.

From this introduction two problems emerge which have a bearing upon the economic production of ore in central Labrador-Ungava. The first is that if the cost of mining in permafrost is considerably higher than that of mining in unfrozen areas, other things being equal, what becomes of the initial estimates that profitable exploitation was possible if based upon 400 million tons of proven and essentially unfrozen reserves? Obviously, the proportion of frozen to unfrozen ore (and technical improvements in mining) become significant. Should a large proportion of the reserves prove to be in permafrost, long-term production and production methods may need careful review unless technical advances can greatly reduce costs. There should at least be a pressing need for permafrost research with a technical objective—that of evaluating the extent and thickness of permafrost in an economic manner. The second problem is the *means* of improving production techniques to reduce the cost of mining in permafrost. It should be emphasized that this second problem has received a large amount of attention by the Company and great improvements have already been made, whereas the first, and possibly more basic problem has received only indirect attention.

It has been argued that, since certain orebodies are definitely frozen, the most logical approach is to disregard the academic problem of estimating the over-all area of permafrost occurrence and to concentrate on improving mining techniques. In view of this, an attempt is made to examine the nature of the technical difficulties involved.

Open-Pit Mining Problems in Permafrost

Problems of large-scale mining in permafrost are numerous and closely related to mining and shipping problems in general so that only some of the main points can be discussed. These may be divided into three groups: (1) blasting and drilling (2) removal of ore from pit face to railway car (3) transportation from mining area to blast-furnace.

(1) *Blasting*—Blasting problems increase in proportion to the frozen water content of the ore concerned. An extreme case is the Wabush ores, situated 150 miles south of Schefferville. Here water content is very low and the ore has less than 1 per cent ice content so that, frozen or unfrozen, the special Schefferville problem does

not occur. Higher water content in the Ferriman ore body, however, results in serious problems. Where water content exceeds 10 per cent, ice segregation, in the form of lenses $\frac{1}{4}$ to 2 inches thick, pipes, and more widely disseminated ice crystals, results in the absorption of a large proportion of the energy generated by each blast, and in incomplete and unsatisfactory rupture of the pit face. Incomplete rupture of the working face sets in motion a vicious circle of events, because this renders much more uncertain the success of each succeeding blast, which in turn can reduce ore removal to a standstill. In practice, far greater amounts of explosives are required in permafrost than in unfrozen ground. Also, as the "toe" becomes unmanageable, secondary blasting becomes necessary with attendant stoppages in mechanical removal and loading of ore. Much research into types of explosives and method of installation and spacing of the charges has resulted in much greater efficiency, greatly reducing the cost of mining. David Selleck, blasting research officer, states that he is "firmly convinced that if we can get the explosive in the correct location, an efficient job of work can be done" (Selleck, 1961). Multiple-row blasting, starting with an "idealized" face, permits the planning of a blast in such a way that the subsequent face, after mucking, is also ideal. This is accomplished by wasting energy in the back row of holes in order to cut down the "back break". However, drilling problems also complicate the locating of explosives. Friction of the drill bit on frozen material causes the sides of the borehole to melt and slump. Thus loss in depth and diameter ensues, rendering correct location of explosives problematical.

(2) *Removal of ore from the pit face*—This problem is intricately associated with the success or failure of the blasting process. A typical blast in frozen ore, as witnessed by the writer, results frequently in the production of large blocks, many of which cannot be handled by the electric shovels and Euclid carriers. The practice in this case has been to bulldoze large blocks to the centre of the pit and (*a*) to await natural thaw and break-down (*b*) to employ secondary blasting or (*c*), to attempt break-down of the blocks by mechancial percussion. If the supply of large blocks is heavy, the pit floor becomes congested, increasing the difficulty of fully effective deployment of large-scale machinery and accordingly reducing production. Another problem is that secondary blasting of the "toe", or only partially successful primary blasting, tends to result in an uneven pit floor. An extreme case is when the shovel cannot

approach sufficiently close to reach ore from the upper part of the pit face.

More successful break-down following primary blasting, but still leaving frozen particles exceeding 1 to 2 feet in diameter, permits successful handling from pit to screening plant, but can result in increased wear and tear on the plant, jams, and further reduction of production and increase in cost.

(3) *Transportation to the blast furnace*—During the 360-mile railroad journey to Sept Iles on warm summer days there is a gradual thawing out of at least part of the frozen ore. This can accentuate the problem of "sticky ore" whereby fine-grain ores with a critical level of water content partially stick to the sides and bottom of the rail car during dumping at the Sept Iles terminal. Continued thawing presumably extends this problem to all transport operations until the blast furnace is reached.

The Extent of Permafrost in Relation to Iron Ore Reserves

As indicated above, the proportion of frozen to unfrozen ore within the 400-million-ton-reserve is a most critical factor in long-term ore production in central Labrador-Ungava, yet so far this proportion is merely a subject for the imagination. Despite this, some comments are in order. Detailed ground temperature measurements and a coordinated research program in permafrost, involving vegetation, climate and meteorology and ground characteristics, were initiated in May, 1959, by the writer in liaison with the Iron Ore Company of Canada and the Division of Building Research, N.R.C.

Ground temperature measurements were taken at numerous sites, both in depth and in area, beginning in September, 1959, and were continued until June, 1960. This period was extended to include a full year by the help of Charles McCloughan, senior observer of the McGill Laboratory, who maintained thermocouple readings during the summer of 1960 while the writer was in Scandinavia. A full report of the program has been published elsewhere and acknowledgments are included there. Much more work remains to be done on the data before a complete report of the research findings can be published. However, preliminary consideration of the data has a significant bearing upon the problem at hand.

1. Over wide areas the permafrost is compatible with the climate of recent decades, it exceeds 150 feet in thickness in many places and probably occurs in thicknesses of more than 300 feet.

2. There is a strong relationship between vegetation type, depth of snow cover and development of permafrost. In this sense the

writer is tempted to suggest that all areas extending above the tree-line, and especially where the wind prohibits the accumulation of deep snow, are areas of potential permafrost. This is placed against a basic mean annual air temperature of 24° F. for Schefferville.

3. Below the tree-line relic patches of permafrost may be expected to occur across wide areas of central and south-central Labrador-Ungava.

4. The southern boundary of contemporary permafrost (discontinuous) probably lies as far south as the Laurentide scarp and should continue into the high mountains of Gaspé and New England which rise above the tree-line.

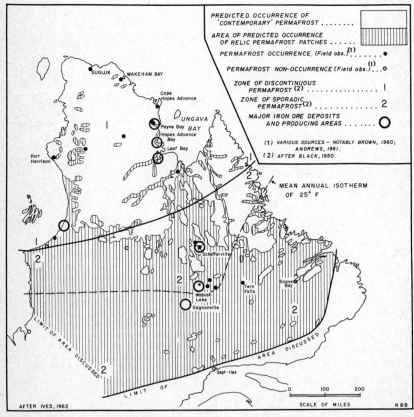

Figure 1. Permafrost distribution and iron-ore reserves in Labrador-Ungava.

Following these preliminary conclusions it is instructive to plot iron-ore distribution against predicted permafrost occurrence (Figure 1). Undoubtedly, the Wabush area is marginal with contemporary permafrost in one orebody and as yet undetected in others at lower elevations, despite many thousands of feet of exploratory drilling. In addition, low water content and the "hard rock" nature of the specularite-haematite ore in this area render the permafrost problem negligible. The Lac Jeanine area is also marginal although scattered patches of relic permafrost should be anticipated. The vast Ungava Bay reserves fall squarely into a zone of extensive contemporary permafrost and mining hazards associated with this location add to the already considerable problems facing any future exploitation (Lloyd and Nutt, 1960).

Within the Schefferville vicinity more than 85 per cent of the ore mined in 1959 and 1960 was unfrozen. Estimated production for 1961 is 7½ million tons, a serious cut-back due to the present economic situation, and this entire quota will be mined from non-frozen orebodies. Thus no permafrost mining is being undertaken during the current year. At this point it is relevant to refer to the problem of the assessment of the proportion of frozen ore to the total known reserves of this general area. Due to incomplete knowledge of permafrost conditions and the need for further research into the relationships between permafrost occurrence and surface factors, only a very subjective estimate of this proportion can be made. It is suggested, therefore, that a considerable amount of the Schefferville reserves, and especially those north of the existing mines, are in permafrost.

The several hundred million tons of unfrozen high-grade ore, plus the huge deposits in the Wabush Lake area that have to be beneficiated before shipment, leave the Iron Ore Company of Canada with a very long-term working reserve. With the passage of time, rising costs and further technical advances might well obliterate the problem. However, it can be maintained that permafrost distribution will prove one of the factors which will influence the pattern of mining development in this area, and the need for detailed permafrost research becomes increasingly apparent.

It is difficult to evaluate the full impact of permafrost occurrence on iron mining, and the purpose of this paper has been restricted to a discussion of some of the theoretical and technical problems. Undoubtedly these are important and it is perhaps correct for the geographer to consider permafrost occurrence as an economic factor in mining operations. It is also hoped that permafrost research in the peninsula, both for its own sake and because of the economic implications, will become one of the applied fields of geomorphology to which the geographer can adapt himself and produce, on the one hand, gratifying pure research results, and on the other, important contributions to the fuller utilization of Canadian resources.

References

Andrews, J. T., "Permafrost in southern Labrador-Ungava," Geography Dept., McGill Univ., Misc. Paper no. 1, Unpub., mimeo, (1961), p. 5.

Black, R. F. "Permafrost," *Smithsonian Inst. Rept. 1950*, (1950), pp. 273-301.

Bonnlander, B., "Progress report on permafrost investigations," Unpub. rept., Iron Ore Company of Canada, Schefferville, P.Q., (August, 1957); "Permafrost research, in Scientific studies in the Labrador peninsula," McGill Sub-Arctic Research Paper no. 4, McGill Univ., Montreal, December 1958, mimeo (1958), pp. 56-58.

Brown, R. J. E., "The distribution of permafrost and its relation to air temperature in Canada and the U.S.S.R.," *Arctic*, Vol. XIII, no. 3 (1960), pp. 163-77.

Hare, F. Kenneth, "A photo-reconnaissance surey of Labrador-Ungava," *Geog. Br., Dept Mines & Tech. Surv.*, Ottawa, Mem. 6, (1959), 83 pp.

Ives, J. D., "A pilot project for permafrost investigations in central Labrador-Ungava," *Geog. Paper no 28*, Dept. Mines & Tech. Surv., Ottawa (1961), 26 pp.

Lloyd, Trevor and Nutt, David, C., "Transportation of Ungava iron ore," *Can'. Geogr.*, no 15, (1960), pp. 26-38.

Selleck, D. J., Personal communication dated 10th May, 1961.

15

Bedrock Channels of Southern Alberta

R. N. Farvolden

Introduction

Over much of the glaciated area of North America the stream valleys that existed prior to glaciation are now partially or completely buried by glacial drift. The channels of these earlier drainage systems commonly contain sand and gravel deposits which originated in the same manner as sand and gravel bars along present-day rivers. These deposits are for the most part exposed only where recent erosion has removed the drift cover. Where extensive sand and gravel deposits are well sorted and occur below the water table, they are excellent aquifers. In an area such as southern Alberta, underlain by shale and impermeable sandstone, the only aquifers capable of yielding large supplies of groundwater, apart from gravel bars along present-day rivers, are the sand and gravel deposits associated with bedrock channels. In present interstream areas these aquifers may be the only possible source of large supplies of water. In semiarid, southern Alberta these aquifers have a special significance for surface supplies of water are often difficult to develop because all but the main trunks of the larger rivers are intermittent streams.

The locations of some of the bedrock channels are known from surface and subsurface data and this report is an attempt to show the pattern of the bedrock channel systems.

Purpose and Scope of the Report

This report presents the results of observations carried out during four years of field work in Alberta. The logs of approximately 8,000 shallow borings have been tabulated and used in compiling the map of bedrock channels. Unfortunately, about 75 per cent of the logs are from borings in the Edmonton and Red Deer areas

First published in "Early Contributions to the Groundwater Hydrology of Alberta," *Research Council of Alberta*, Bulletin no. 12 (1963), pp. 63-75.

(Farvolden, 1963a) and only 25 per cent are from the rest of the area. Nevertheless, it is believed that sufficient information has been collected to allow mapping of the locations, pattern and gradients of the major bedrock channels. In some places the exact positions of the channels are known, but in others it is likely that errors of two or three miles, or more, will be revealed by future work.

The map accompanying this report (Figure 1) has been compiled for use in the exploration and development of groundwater resources in Alberta. It is hoped that by mapping, in a general way, the divide areas and the bedrock channels the development of this type of aquifer will be encouraged.

The pattern of the bedrock channels and the relation of the bedrock surface to the present-day surface is used as evidence to explain the erosional history of the Alberta plains.

Description of the Area and Physiography

The area included in this report is that part of the plains region of Alberta lying east of 115° west longitude and south of 54° north latitude. The Rocky Mountains and the foothills belt form the southwestern boundary of the area but farther north the western edge of the map-area is well within the plains region because of the northwesterly strike of the mountain front.

The surface slopes from an elevation of about 4,000 feet in the plains region of southwestern Alberta to about 1,900 feet at Frog Lake in township 57, range 3, west of the Fourth Meridian. The surface is rather flat and featureless except for several distinctive upland areas, some of which rise hundreds of feet above the surrounding plains, and for the valleys of present-day streams which are deeply incised into the plains surface. The area is almost completely covered by a mantle of glacial drift which is 30 to 100 feet thick over most of the area. Morainal deposits and abandoned meltwater channels relieve the otherwise monotonous plains topography.

The annual precipitation for the area ranges from less than 14 inches in the south to over 18 inches in the north. The January mean temperature ranges from over 15°F in the south to 0°F in the north and the July mean temperature is about 65°F.

The Bedrock Surface

The present-day surface is nearly coincident with the bedrock surface over most of the southern half of Alberta. The major upland areas are underlain by bedrock and the major bedrock channels are, for the most part, coincident with broad depressions in the present

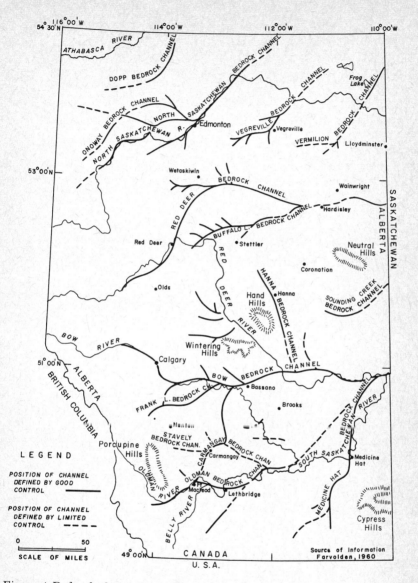

Figure 1. Bedrock channels of southern Alberta.

land surface. A large number of narrow, glacial meltwater channels have been incised into the surficial deposits and in many cases into the underlying bedrock. The majority of these channels are now abandoned except for intermittent streams, but some of them still carry permanent streams. The map of the bedrock channels (Figure 1) does not show these features for it is presented as a map of the preglacial drainage system, or perhaps more correctly, as a map of the drainage system that existed prior to the last glaciation of the area.

Development of the Bedrock Surface

Since the last uplift of the Rocky Mountains, probably in Oligocene time, the history of the western plains has been one of degradation although, during two intervals at least, degrading was halted and aggrading on a large scale took place. It seems likely that in each case aggraded material was deposited on a newly formed peneplain. The older of these two surfaces is the Cypress Plain, and the younger is the Flaxville Plain (Warren, 1939). Both of these surfaces have been nearly removed by erosion and only remnants of them rise above the present prairie level.

Erosion has been more or less continuous since about the end of Miocene time and, because there is no evidence of major post-Miocene tectonic movement in Alberta, the individual valleys must have had the opportunity to approach maturity or possibly old age. The tendency of the bedrock channels to occupy broad lowland areas between widely separated, low and rounded upland areas lends support to the idea of a mature plain drained by mature streams. There seems to be no reason to suspect that the drainage systems were not completely integrated and dendritic, as suggested by Lobeck (1939, p. 484), for the underlying rocks are flat-lying shales and soft sandstones.

The gradients from the interstream uplands toward the bedrock channels appear to be relatively even. If erosion took place by scarp retreat, it progressed to a degree that the valleys were several miles wide and the hills in interstream areas became quite rounded. The net effect is the same as that to be expected from sheet erosion.

Most major rivers in southern Alberta occupy ancient bedrock channels along some portion of their course, and here it is common to find the stream channel to be alternately incised into bedrock and then underlain by alluvium for short distances. The outcrop pattern and the occurrence of alluvium below river level is obviously controlled by the relation between the location of the present channel and that of the bedrock channel. In some instances 50 feet of alluvium have been found below river level but this likely means that the channel had a different course and a lower local base level at one time than it has now. An example of this is found in the Red Deer River valley at Red Deer. It is suggested that the long period of uplift and erosion that began in middle Tertiary time has not changed significantly, and the regional base level of the streams of southern Alberta is as low or lower than it has ever been since the uplift of the Rocky Mountains.

Several of the bedrock channels have been mapped in detail where sufficient subsurface control is available. Where this has been possible, the gradient of the bedrock channel is approximately that of a similarly located valley today. This condition is postulated to hold throughout the area and it is used to help determine possible connections between segments of bedrock channels. The uniformity of the gradients of the bedrock channels is strong evidence that there has been little or no tilting of the land mass since preglacial time.

The block diagrams (Figures 2A to 2G) are an attempt to illustrate a possible sequence of erosion and uplift that can be used as a working hypothesis to explain the features of the present and past plains surfaces and their drainage systems. Figure 2A represents the Cypress Plain. Remnants of this plain are capped with gravel strata for which there is indirect paleontological evidence to indicate an Oligocene age (Warren, 1939). Deposition of the gravels may have been contemporaneous with the last major uplift of the Rocky Mountains. Figures 2B, 2C, and 2D illustrate that, although the land mass continued to rise, the drainage systems and land surface developed to maturity and possibly old age. This cycle must have been repeated several times for the remnants of different erosion surfaces are found throughout Alberta (Warren, 1939). Each time the land was uplifted and another cycle began the streams would continue to occupy the same channel, because that would remain the lowest ground. Thus, the locations of the main features of the present bedrock topography, particularly the main divide or upland areas, and the stream channels, were determined at an early stage in the history of the plains.

Some evidence indicates that, in places, the divides have been breached by headward erosion and that stream piracy occurred. This has caused major changes in the pattern of the bedrock channels but not in the overall bedrock topography. As the individual drainage systems were developing, the controls on local base level, such as lakes and waterfalls, were removed. The gradients of the headwater streams would thereby be alternately decreased and increased, in turn causing changes in the rates of headward erosion in the divide regions. Then, if the relations of elevations and gradients between the streams on either side of the divide were favorable, stream piracy could result. The whole pattern

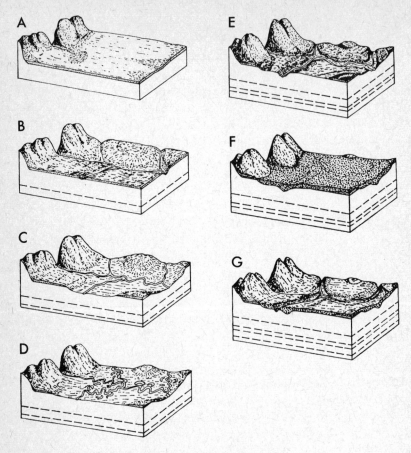

Figure 2. Evolution of a system of bedrock channels.
A. Deposition surface in Oligocene time.
B. Streams dissect the plain as the land is uplifted.
C. Uplift continues but not fast enough to prevent maturing of the
    drainage system and land surface.
D. Uplift is very slow and rivers approach old age.
E. Relatively rapid uplift causes rejuvenation of the drainage systems.
    This sequence may have been repeated several times.
F. The surface is overrun by continental glaciers leaving a mantle of drift
    over all but the highest hills.
G. The drainage system developing since the retreat of the glaciers follows
    the preglacial system in some areas, but deviates from it in others.
    A slight uplift has left the land surface higher than at any time since
    the Oligocene.

of the drainage system might thus be changed. Stream piracy is a
common phenomenon in nature and there is some evidence to indicate it was a factor in determining the pattern of bedrock channels
in Alberta. If it is assumed that the bedrock topography is of

preglacial origin, stream piracy is the most likely explanation for the apparent downstream bifurcation of several bedrock channels.

In several bedrock channels for which there is good subsurface control, there appears to be a deep, narrow gully at the bottom of the channel. This is illustrated in figure 2E. It is interpreted as evidence of a relatively rapid uplift of the land mass similar to minor uplifts that had previously started new cycles of erosion, as illustrated in Figure 2B. In this instance, however, the cycle was interrupted by continental glaciation. Figure 2F shows that, although the drift is not nearly thick enough to obscure the major features of the bedrock topography, segments of some bedrock channels are completely buried by the drift. Most bedrock channels are still coincident with topographic depressions and are now occupied by streams or lakes. In some cases, however, the drainage was diverted by the glacier and the bedrock channels were buried in drift and abandoned (Figure 2G).

The Bedrock Channel Systems
The North Saskatchewan Bedrock Channel System

The North Saskatchewan Bedrock Channel crosses the northwest corner of the map-area following a northeast direction. The channel rises in the mountains but is not shown above Tp. 47, R. 9, W. 5th Mer. on this map. From this point downstream to Tp. 51, R. 3, W. 5th Mer., the present North Saskatchewan River follows the ancient bedrock channel. Then the river leaves the bedrock channel which forms a broad arc that curves to the north through Big Lake to the city of Edmonton. Here the present river reoccupies the bedrock channel and continues to do so downstream to Tp. 58, R. 19, W. 4th Mer. The abandoned portion of the bedrock channel is, in places, partly obscured by a thick drift cover and in other places is easily detectable from the present topography. East of Tp. 58, R. 19, W. 4th Mer. the river likely leaves the bedrock channel once more but there is no proof of this supposition at present.

A tributary bedrock channel that rises in the bedrock uplands area around Tp. 49, R. 3, W. 5th Mer. is occupied by the present North Saskatchewan River from the Fifth Meridian to the city of Edmonton. A major channel joins the North Saskatchewan Bedrock Channel in Tp. 52, R. 2, W. 5th Mer. This channel has been mapped mainly on the basis of present topography and its genetic relation to the main channel is not known.

The Onoway Bedrock Channel swings in a wide arc from Tp. 50, R. 8, W. 5th Mer. around Wabamun Lake and joins the North Saskatchewan Bedrock Channel at the east end of Big Lake. The upper portion of this channel as shown on Figure 1 is occupied by the Pembina River, the valley of which changes markedly where it leaves the bedrock channel. The lower portion of the Onoway Bedrock Channel is well known from subsurface data and the middle reaches coincide with a depression on the present land surface. The relations between this channel and the North Saskatchewan Bedrock Channel system suggest that the Onoway Bedrock Channel may be an early stream course, abandoned due to stream piracy. It is also possible that this bedrock channel has been formed through some process associated with glaciation and should therefore have been omitted from this map.

The Pembina River in Tp. 60, R. 1, W. 5th Mer. occupies the Dapp Bedrock Channel, the tentative location of which is shown, although adequate subsurface data to substantiate this interpretation are not available.

The Vegreville Bedrock Channel System

The Vegreville Bedrock Channel rises in the plains area, in the upland west of Beaverhill Lake and trends northeast. The portion of the channel west of Tp. 54, R. 12 W. 4th Mer. is well established by subsurface data and is a deep, narrow gully in the bedrock surface, in places completely filled with glacial drift. The northern half of the bedrock channel has been located on the basis of the present topography and scattered well logs.

The Vermilion Bedrock Channel is not well established by subsurface data, except in the headwaters region near Birch Lake. However, scattered well logs indicate a headwater tributary pattern as shown on the map, and below the town of Vermilion the present Vermilion River follows the old valley for a short distance. Near Frog Lake the presence of the channel has been confirmed by subsurface data, and there is likely a junction with the Vegreville Bedrock Channel in the vicinity of Cold Lake just outside the map-area.

The Red Deer Bedrock Channel System

The Red Deer Bedrock Channel completely traverses the central Alberta plains. The upper portion of the channel nearly coincides with the valley of the present Red Deer River but in Tp. 39, R. 27, W. 4th Mer. the river leaves the bedrock channel which, east of this locality, is partially buried by glacial drift. The upper Red Deer Bedrock Channel follows a fairly constant northeast direction for about 90 miles and then swings east in a wide arc. It is possible that at an earlier stage of development than that shown on the map, the floor of the upper Red Deer Bedrock Channel was slightly higher than it is now and the channel did not swing east as noted above, but maintained its northeasterly direction to join the North Saskatchewan Bedrock Channel near Edmonton. Stream piracy is a mechanism that may have caused the abandoning of this course. Subsequent erosion has nearly obliterated the former channel. A similar argument could be advanced in favor of an early connection between this bedrock channel and the Vegreville Bedrock Channel but the intervening bedrock topography is not as favorable, although the channel pattern seems suggestive.

West of Hardisty the Red Deer Bedrock Channel is well defined by both surface and subsurface information but there is only scattered subsurface control with which to fix the position of the lower end of the channel. However, logs of wells in Tp. 43, R. 1, W. 4th Mer., the nature of the topographic low that runs through Reflex Lake and Manito Lake just east of the Alberta-Saskatchewan boundary, and the fact that the elevation of the floor of the bedrock at the boundary (1,825 feet) is close to that required by the gradient (3.5 feet per mile), indicate the location for the channel that is shown on the map is the best choice at present.

The Buffalo Lake Bedrock Channel rises in the highland area east of Red Deer and rapidly becomes a major channel because it is joined by several short but large tributaries. It is a major tributary to the Red Deer Bedrock Channel, the confluence likely being as shown on the map. Throughout the upper half of its course the channel is coincident with present-day valleys but the lower half is not readily apparent from surface features. It seems possible that, had headward erosion of this channel not been interrupted, stream piracy would have occurred, causing diversion of the ancestral Red Deer River through the Buffalo Lake Bedrock Channel.

The Bow River Bedrock Channel System

The Bow River Bedrock Channel rises in the Rocky Mountains and, after entering the plains region, follows an easterly course across southern Alberta. Above Bassano the present Bow River occupies the bedrock channel or is diverted only slightly from it. Thirty miles below Bassano the present Red Deer River has a similar relation to the lower portion of the same bedrock channel, following it very closely for the most part. The intervening portion of the channel is mapped on the basis of several water-well logs which indicate the presence of a depression on the bedrock surface between the upper and lower channel segments, with the correct relations with regard to elevation and direction.

The Bow River Bedrock Channel is joined by several tributaries from the north and at least two tributaries from the south. Those

that enter from the north all drain a broad upland area that forms
the divide between the Bow River and Red Deer Bedrock Channels.
West of the present Red Deer River the streams that previously
drained south and had a base level controlled by the Bow River
Bedrock Channel now have a base level controlled by the canyon of
the Red Deer River. Thus the present-day streams, where the
drainage has become sufficiently well integrated, have eroded
through and in some cases removed all the coarse alluvium along
their channels, and bedrock outcrops along these channels are
common. As in the region of the headwaters of the Buffalo Lake
Bedrock Channel, these tributaries are shown on the map in greater
detail than other minor tributaries because they are easily recognized from the present topography.

The upper portion of the Hanna Bedrock Channel is well established from subsurface data and topographic features but the lower
half of the channel is not. Information on the Sounding Creek
Bedrock Channel is scant, but it may be a tributary of the Red
Deer Bedrock Channel.

The present Highwood River, above the town of High River
follows an ancient bedrock channel. At this point the river makes
a turn to the north but the bedrock channel continues to follow an
easterly direction crossing the upper portion of the valley of the
Little Bow River above Frank Lake, before bending northeast to
cut through a bedrock upland area in a broad, deep bedrock valley.
This feature has been named the Frank Lake Bedrock Channel.
It is thought that the ancient Highwood River once occupied the
Frank Lake Bedrock Channel but was diverted from this course to
its present course by stream piracy.

The Oldman Bedrock Channel System

The main branch of the Oldman Bedrock Channel drained part
of the Rocky Mountains and plains of southern Alberta as does
the present-day Oldman River, which follows, more or less, the
ancient bedrock channel. However, at an early stage of development of this bedrock channel, the river followed a northerly course
below Fort Macleod, the channel passing west of Blackspring Ridge.
This segment of the channel is given the name Carmangay Bedrock
Channel in this report. At Carmangay the channel makes a wide
turn to the east, and its course around the north end of Blackspring
Ridge is marked by Travers Reservoir. East of Travers Reservoir
there are four bedrock channels or segments of channels, any one
of which may have carried the main stream flow from the southern
Alberta Rocky Mountains at one time.

The main branch of the Oldman Bedrock Channel follows the same general topographic low as the present river from Fort Macleod to its confluence with the Bow River. Bedrock crops out in places where the present river crosses the bedrock channel (Horberg, 1952); this suggests that the present river may have a base level that is lower than that of the ancient river. The location of the lower half of the Oldman Bedrock Channel is not well established and is drawn on the basis of meager surface and subsurface control.

The Medicine Hat Bedrock Channel formerly drained the southeast corner of Alberta. It is a large tributary of the Oldman Bedrock Channel but the location of the confluence shown on the map is based on conjecture. The gradients at various points in the Medicine Hat Bedrock Channel indicate nearby base level control, for, in the upper half of the channel the gradient is steep (10 feet per mile) but below the city of Medicine Hat the gradient is gentle (less than 3 feet per mile), as required for an accordant junction with the Oldman Bedrock Channel. The location of the Medicine Hat Bedrock Channel is well established by subsurface data above the city of Medicine Hat and is fairly well established below that point (Meyboom, 1963).

The Groundwater Resources of the Bedrock Channels

The permanent streams of southern Alberta are widely separated and, the country being semiarid, large water supplies are often difficult or impossible to develop economically at interstream localities. The bedrock formations that underlie the major part of the area are not sufficiently permeable to yield large quantities of groundwater to wells. The bedrock channels, on the other hand, commonly contain sand and gravel strata that are excellent aquifers, and the mapping of the bedrock channels is therefore a necessary step in determining the groundwater resources of the province. Unfortunately, the bedrock channels do not contain sand and gravel at all places, and exploration thus far has indicated that along some portions of every channel only fine-grained alluvium is present and good aquifers do not exist. In many places the bedrock channels coincide with the valleys on the present surface and erosion has removed most of the alluvium that may have been deposited in the channel at one time. In such a channel there is little hope of encountering coarse alluvium below the water table.

The bedrock channels shown on the map vary with respect to the accuracy with which their locations are known. For instance, in the Edmonton-Red Deer area the courses of the bedrock channels are quite well defined, whereas knowledge of the pattern and location of parts of the Oldman Bedrock Channel leaves much to be desired.

The distribution of gravel along the course of any stream channel may depend upon the availability of coarse clastic material and the regimen of the stream. The rocks of the area weather to fine-grained material and were not the source of the gravel found in the bedrock channels. However, the detritus from the more indurated rocks of the mountains has been carried into the plains region and deposited in stream channels since early Tertiary time. The only source of coarse sediments for bedrock channels that rise in the plains is the gravel that was deposited during an earlier stage of the development of the present surface (Figs. 2A to 2G) and that has since undergone another cycle of erosion and deposition. Thus, of the bedrock channels that arise in the plains, those that pass near upland areas capped by gravel should be the most likely to contain coarse alluvium. All the streams in figures 2A to 2G are surrounded by gravel-capped hills. All bedrock channels that rise in the mountains may be expected to contain coarse alluvium along some parts of their courses.

The regimen of a stream controls the sedimentation in the stream bed and one of the most important factors in this regard is the gradient of the stream. Coarse, well-sorted sediments are most likely to occur where the gradient decreases suddenly, which makes knowledge of the gradients of the bedrock channels of great importance. Deposition can also be expected if the load of a stream is increased suddenly as, for example, at the point where a tributary with a higher gradient joins the main channel.

At the intersections of present-day stream valleys with the ancient bedrock channels the coarse sediments of both the modern and ancient streams may be present and have sometimes been resorted and redeposited as thick permeable strata by the younger stream. At such sites ground-water recharge of these permeable strata by induced infiltration may be possible if the bedrock channel floor is lower than the present stream level (Meyboom, 1963). This is an ideal situation for the production of large amounts of groundwater and such locations are worthy of close examination where large supplies of water are required.

There is evidence that the aquifers of the Alberta plains region are being recharged by local precipitation, although the rate of recharge is certainly low. The determination of this rate of recharge is a necessary step in the proper evaluation of the groundwater resources of the bedrock channels.

It is obvious that the map presented in this study can serve only to indicate the areas most favorable for the occurrence of groundwater in abundance. It is also obvious that one is not justified in depending on supplies of groundwater from aquifers in bedrock channels until a thorough testing program has been conducted.

References

Farvolden, R. N., "A farm water supply from quicksand," *Res Coun. Alberta Prelim. Rept. 61-3*, (1961), 12 pp.; "Groundwater resources, Pembina area, Alberta," *Res. Coun. Alberta Prelim. Rept. 61-4* (1961), 26 pp.; "Bedrock topography, Edmonton-Red Deer map-area, Alberta," *Research Council of Alberta Bulletin No. 12*, (1963), pp. 57-62; "Bedrock channels of southern Alberta," *Research Council of Alberta Bulletin No. 12* (1963), pp. 63-75.

Lobeck, A. K., *Geomorphology*, McGraw-Hill Book Co. Inc., New York, (1939), 931 pp.

Meyboom, P., "Induced infiltration, Medicine Hat, Alberta," *Research Council of Alberta Bulletin*, no. 12 (1963), pp. 88-97.

Warren, P. S., "The Flaxville Plain in Alberta," *Trans. Roy. Can. Inst.*, Vol. XXII, no. 48, pt. 2 (1939), pp. 341-45.

Flow, Channel and Flood Plain Characteristics of the Lower Red Deer River, Alberta

H. J. McPherson

Introduction

Several authors have demonstrated that a river may be considered an open system in dynamic equilibrium (Hack 1960; Chorley 1962; Chorley and Hagget 1967). However, in studies to date, attention has been focussed on the "channel system" rather than on the "river system". The channel was treated as a static element and researchers analyzed only the linkages between the "in channel" variables. (Lacey 1930; Inglis 1949; Leopold and Maddock 1953; Wolman 1955; Leopold and Miller 1956; Blench 1957; Colby and Scott 1965; Simons et al. 1965). The principal channel variables were listed by Lane (1957) as: (1) discharge (2) sediment load (3) velocity (4) bed roughness (5) nature of bed and banks (6) bed slope (7) vegetation (8) water temperature and (9) human interference.

Concentration on the channel factors and the channel system is a restrictive and misleading approach. The channel of a river is not static. Rather it constantly migrates. Therefore, the channel and floodplain must be viewed as a dynamically interlinked unit, the dynamic link being the process of channel shift. If the behaviour and characteristics of a river are to be understood, a model must be developed which recognizes the interrelations between the channel and floodplain. Further, in developing equations to express the relations between the variables controlling river behaviour, the channel and floodplain factors must be taken into account as well as the "in channel" variables described by Lane.

In this paper the results of an analysis of the flow, channel and floodplain characteristics in a selected reach of the lower Red Deer River, Alberta are presented. Specifically the aims of the project were to: (1) determine how channel and floodplain characteristics changed over the study length (2) establish if there is a recognizable pattern to the variations and (3) propose an explanation for the alterations.

The following variables are identified in a river channel and floodplain system: (1) discharge (2) sediment load (3) velocity (4) channel dimensions (width, mean depth, cross-sectional area) (5) bed slope (6) bed configuration (7) nature of bed and banks (8) channel pattern (9) channel shift (10) vegetation (11) water temperature and (12) human interference.

Discharge, sediment load and human interference are considered the independent variables and the remainder, dependent variables. Bed slope has been classed as independent (Lane; 1957), dependent (Leopold and Maddock; 1953), semi-dependent (Leopold, Wolman, and Miller; 1964) and partially independent (Brice; 1964). The present writer believes that bed slope is a dependent variable when the river flows on a thick cover of alluvial fill and does not scour to bedrock.

Not all the variables listed above were investigated. Bed configuration and velocity were not analyzed as data were not collected for these. The vegetation assemblage does not change noticeably. Further, water temperatures during the summer of 1963 indicated that this factor also remains constant.

Location and General Description

The Red Deer River rises in the Front Ranges of the Rocky Mountains and flows into the South Saskatchewan River a few miles east of the Alberta-Saskatchewan border some 100 miles northeast of the city of Medicine Hat.

In this paper the lowest 130 miles of the river's length are studied. Here the river is a sand bed stream, flowing on an alluvial fill varying from fifty to several hundred feet in thickness.

Climatically, the area is semi-arid with a mean annual rainfall of about 13″. Seasonal variations in temperature are marked, with winter readings falling below 0°F., and summer maximum temperatures rising well into the 90's.

The vegetation consists of prairie grass, sagebrush and cactus. Stands of cottonwoods with areas of dense low brush are found along the modern floodplain.

The bedrock formations outcropping at the surface, or occurring immediately below the surficial deposits, are Late Cretaceous in age and consist of series of marine and non-marine sandstones and shales. The strata dip northeastward between 4 and 10 feet per mile.

Flow Characteristics

Twenty-four years of discharge data for the Empress-Bindloss* gauging station at the lower end of the study length, were analysed for mean annual flow, the magnitude of the bankfull discharge value and for flood frequency.

The mean annual flow is 2414 CFS, the values ranging from 1500 CFS to 4500 CFS in individual years. The bankfull discharge is estimated at 35,000 CFS to 45,000 CFS. A more precise value could not be assigned as the rating curve did not plot as a straight line. (Figure 1)

The return period for bankfull discharge is from twelve to thirty years.

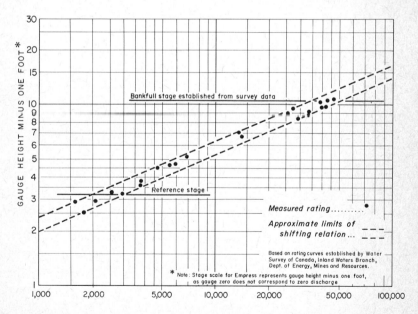

Figure 1. Stage – Discharge Curve, Red Deer River at Empress, Alberta.

---

* The gauge was installed in 1917 at Empress Bridge. In 1931 recording was discontinued and a new gauge was established at Bindloss Bridge, 17 miles upstream, in 1944.

In order to compare the flows at the upper and lower ends of the study length, an automatic recording gauge was installed at Duchess Bridge in 1963. This yielded three years of discharge records. Bankfull discharge at Duchess Bridge, as estimated from the extrapolated rating curve, is between 30,000 and 40,000 CFS. The Duchess Bridge curve is only considered reliable up to a flow of 12,000 CFS as the part of the curve above 12,000 CFS was not substantiated by current metering. Consequently, the estimates of bankfull discharge must be regarded with caution.

Comparisons of the discharge hydrographs of the Duchess and Bindloss stations for the period May to July, 1964 reveal (Fig. 2) that the Bindloss hydrograph reflects almost precisely the variations in discharge experienced at Duchess Bridge with a lag period of approximately 48 hours. This indicates that discharge for low to moderate flows is essentially constant throughout the length.

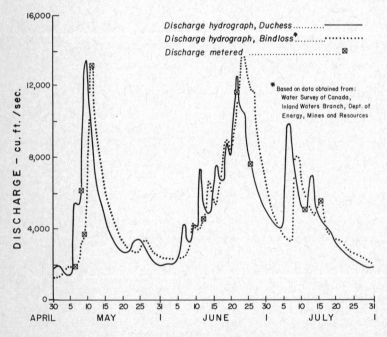

Figure 2. Comparison of Discharge Hydrographs for the Duchess and Bindloss Gauging Stations, April 30th to July 31st, 1964. Data for the Duchess discharge hydrograph obtained from Bobey (1965).

Geomorphology and Sediment Supply

A major problem in most fluvial studies is the accurate measurement of sediment load. A United States Government Inter-agency Committee on Water Resources (1963) stated that no satisfactory sampler for measuring bed load discharge has yet been developed. Blench (1964) noted that calculations of rates of bed load transport using bed load formulae are little better than academic exercises. In this study, no attempt was made to measure bed load discharge or to compute it using any of the well known bed load formulae. Instead, an effort was made to evaluate qualitatively the sediment supply factor in different parts of the valley, based on an understanding of the valley geomorphology. While this approach does not allow figures to be presented for sediment load values in different reaches it does permit a division of the valley into segments, feeding differing amounts of sediment to the stream.

The Red Deer occupies a well-defined valley incised some 200′ to 300′ below the general level of the prairie. A distinct change in valley morphology occurs at approximately 72L/14 (8)*. (Fig. 3). Upstream from this location, the valley is narrow and is only one-half mile to a mile wide at any position. The valley walls are composed of bedrock, veneered with 20′ to 30′ of glacial drift and rise steeply above the floodplain. The major physiographic units within the valley are the modern floodplain, alluvial fans and areas of "badlands" topography.

Downstream of 72L/14 (8) the valley is very much wider although there are local narrowings (Fig. 3). The valley walls are composed predominantly of glacial drift, are not as high, and have gentler slopes than upstream of 72L/14 (8). A diversified assemblage of landforms is present within the valley. The floodplain is wide and well-preserved, fine-grained alluvial terraces occur above the floodplain. Terraces composed of glacial sand and gravel occupy considerable areas of the valley bottom, together with landforms composed of till and extensive sand dune fields.

Sediment supply is influenced in two ways by valley morphology. Between 72L/13 (16) and 72L/14 (2) (Fig. 3) an extensive badlands area developed in the post-glacial period. The badland gullies were eroded into the soft Cretaceous formations. These gullies have contributed, and are contributing, tremendous amounts of fine-grained alluvial material to the channel. As the valley at this

---

*For convenience in presentation the area was divided into four map sheets. The locations at which cross-sections were surveyed are shown on each sheet and a combination of map sheet number and cross-section number is used for referencing.

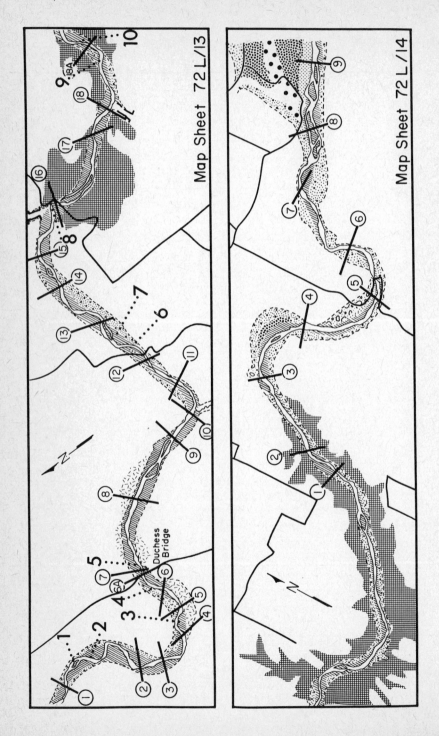

Map Sheet 72 L/13

Map Sheet 72 L/14

Duchess
Bridge

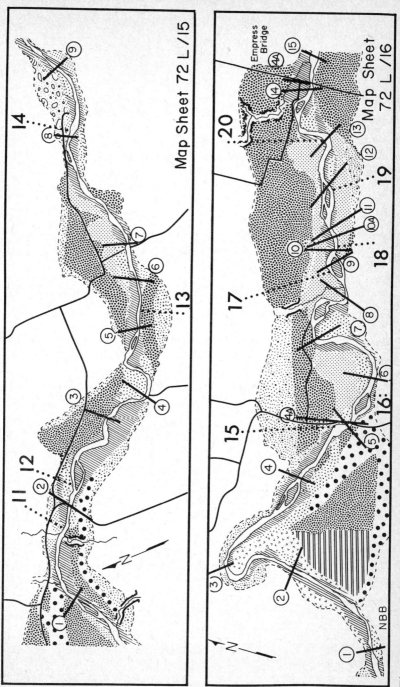

Figure 3. Physiographic Units of the Lower Red Deer River Valley.

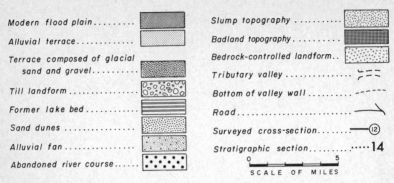

L E G E N D

Modern flood plain.........

Alluvial terrace.............

Terrace composed of glacial
   sand and gravel..........

Till landform................

Former lake bed............

Sand dunes ....

Alluvial fan................

Abandoned river course......

Slump topography ....

Badland topography..........

Bedrock-controlled landform..

Tributary valley ....

Bottom of valley wall..........

Road..........................

Surveyed cross-section.......

Stratigraphic section.........·····14

SCALE OF MILES

Figure 3. Physiographic Units of the Lower Red Deer River Valley.

location is very narrow, most of the sediment eroded from the badlands is fed directly into the channel and little accumulates in the valley bottom to form alluvial fans.

Downstream of 72L/14 (8) where the valley is much wider and the valley walls lower, most of the sediment derived from the wasting of the valley walls was not added to the river, but has accumulated to produce extensive alluvial fans overlying the glacial landforms in the valley bottom. Further, the few tributary streams which join the Red Deer in this stretch are dry for most of the year; even during the spring melt period they carry minor discharges and little sediment. Consequently, little fresh sediment is supplied to the river and the river is largely dependent upon erosion of the floodplain for its sediment.

Channel Dimensions

Channel cross-sectional dimensions were measured at 51 sites along the river (Figure 3). Values for the width, mean depth and cross-sectional area of the channel were measured at each cross section. These data, together with the distance between the cross section and cross-section number 72L/13 (1) (as measured along the centre line of the channel), are presented in Table I.

Visual examination of the data suggested that channel shape does not change to any appreciable degree in the length represented on map sheet 72L/13 but changes progressively downstream of 72L/14 (1). To test if and how channel shape alters, linear regression equations were derived expressing the relationships between downstream distance and the channel dimension parameters for (a) the entire study length, (b) the length represented on map sheet 72L/13 and (c) the reach shown on sheets 72L/14, 72L/15 and 72L/16.

TABLE I. CHANNEL DIMENSIONS OF THE LOWER RED DEER RIVER.

Cross Section No.		Downstream Distance (miles)	Width (feet)	Mean depth (feet)	Cross sectional* area (sq. feet)
72 L/13	(1)	0	800	11.4	9128
	(2)	4.48	550	13.9	7640
	(3)	5.52	790	12.2	9600
	(4)	7.04	700	10.5	7360
	(5)	8.24	450	12.6	5680
	(6)	9.28	950	12.3	11640
	(7)	10.16	560	14.3	8000
	(8)	13.2	1000	12.6	12640
	(9)	16.00	1150	11.1	12760
	(10)	16.88	570	13.8	7840
	(11)	18.4	550	14.5	8000
	(12)	19.84	710	13.6	9680
	(13)	22.48	790	11.5	9120
	(14)	24.8	820	12.0	9840
	(15)	26.32	600	11.3	6800
	(16)	28.72	430	13.2	5680
	(17)	32.0	600	14.0	8400
	(18)	32.12	790	13.4	10560
72 L/14	(1)	48.16	610	12.3	7520
	(2)	49.52	500	13.4	6720
	(3)	53.28	460	10.6	4880
	(4)	56.08	490	10.9	5360
	(5)	59.12	780	9.2	7200
	(6)	61.0	560	9.4	5280
	(7)	64.64	760	9.3	7040
	(8)	67.6	460	9.7	4480
	(9)	70.32	880	8.4	7360
72 L/15	(1)	74.0	700	9.7	6800
	(2)	76.88	600	9.7	5800
	(3)	80.88	660	8.8	5800
	(4)	82.4	590	13.0	7680
	(5)	85.84	440	13.6	6000
	(6)	87.44	700	9.5	6640
	(7)	88.8	700	9.4	6600
	(8)	94.64	1370	7.7	10520
	(9)	96.6	750	9.8	7360
72 L/16	(1)	99.2	590	10.3	6080
	(2)	103.68	830	9.3	7760
	(3)	106.8	590	9.7	5720
	(4)	111.36	730	9.1	6640
	(5)	114.16	530	9.4	4960
	(6)	116.4	590	9.6	5680
	(7)	119.44	900	8.8	7920
	(8)	122.0	820	9.8	8000
	(9)	123.44	1030	8.3	8560
	(10)	124.24	780	6.9	5360
	(11)	125.04	610	8.7	5280
	(12)	126.56	1000	7.1	7120
	(13)	128.72	850	6.7	5680
	(14)	131.68	660	8.8	5800
	(15)	132.8	1190	8.3	9880

* Note cross sectional area was planimetered.

The results of the regression analyses are presented in Table II. These support the visual interpretation and indicate that channel shape remains essentially unchanged in map sheet 72L/13, but becomes progressively wider and shallower throughout the remainder of the length. No change is cross-sectional area and thus in channel size occurs over the whole study length.

Bed Slope

Using a series of bench marks spaced at 10 to 15 mile intervals along the river bank as datum points, a water surface profile (Fig. 4) was constructed for a reference stage discharge of 2500 CFS at Duchess Bridge. Water surface slope is accepted as representing the average bed slope.

Slope is gentlest in map sheet 72L/13 (average slope of 1.45 feet per mile), becomes progressively steeper from the beginning of map sheet 72L/14 to 72L/15 (4) and then flattens out downriver of this point. In the vicinity of Empress Bridge, the average slope is 1.42 feet per mile. By far the most intriguing feature of the slope profile is the local steepening which begins at the end of map sheet 72L/13.

Nature of Bed and Bank Materials

Mechanical analyses were carried out on 19 samples collected from the stream bed with a drag bucket. The positions at which the samples were taken, together with the results of the analyses are indicated in Table III. The bed sediment is mainly sand and the median grain size of the sand does not change systematically downstream.

Bank stratigraphy and thus floodplain stratigraphy, was analyzed at 20 sites along the channel. All the sites selected were in actively eroding high banks and their locations are plotted on Figure 3. Eroding banks were examined because: (a) the individual sedimentary units were clearly defined and thus easily mapped and (b) discrete beds could be traced upstream and downstream for considerable distances.

Criteria for differentiating between vertical accretion sediments and lateral accretion sediments have been described by several authors (Fenneman 1906; Sundborg 1956; Wolman and Eiler 1958; Adler and Lattman 1960). Examination of these criteria suggests that vertical accretion sediments: consist mainly of silt and clay size fragments; occur in relatively thin compact beds; exhibit horizontal stratification; and display no cross-bedded structures. However,

TABLE II LINEAR REGRESSION ANALYSIS OF DOWNSTREAM DISTANCE, MEAN DEPTH, WIDTH AND CROSS SECTIONAL AREA OF CHANNEL FOR DESIGNATED RIVER LENGTHS

River Length	Number of Observations	Shape Parameter	Equation[1]	Correlation Coefficient	Value of Statistic t	Significance Level
All Cross Sections In Study Length	51	Mean Depth	$Y=13.10-0.04X$	0.71	7.04	Significant at 1% level
All Cross Sections From 72L/13 (1) to 72L/13 (18) inclusive	18	Mean Depth	$Y=12.22+0.03X$	0.22	0.92	Not Significant
All Cross Sections From 73L/14 (1) to 72L/16 (15) inclusive	33	Mean Depth	$Y=12.07-0.03X$	0.40	2.44	Significant at 2% but not at 1% level
All Cross Sections In Study Length	51	Width	$Y=656.13+0.90X$	0.19	1.38	Not Significant
All Cross Sections From 72L/14 (1) to 72L/16 (15) inclusive	18	Width	$Y=750.87-2.38X$	0.12	0.49	Not Significant
All Cross Sections From 72L/16 (1) to 72L/16 (15) inclusive	33	Width	$Y=387.5+3.5X$	0.45	2.85	Significant at 1% level
All Cross Sections In Study Length	51	Cross Sectional Area	$Y=873.7-19.34X$	0.43	3.30	Significant at 1% level
All Cross Sections From 72L/13 (1) to 72L/13 (18) inclusive	18	Cross Sectional Area	$Y=9049.86-8.25X$	0.04	0.16	Not Significant
All Cross Sections From 72L/16 (1) to 72L/16 (15) inclusive	33	Cross Sectional Area	$Y=5791.51+9.19X$	0.18	1.02	Not Significant

(1) In the equation X = Downstream distance in miles.
Y = Shape parameter in feet or square feet.

(2) Relationship is considered not significant at levels greater than 10%.

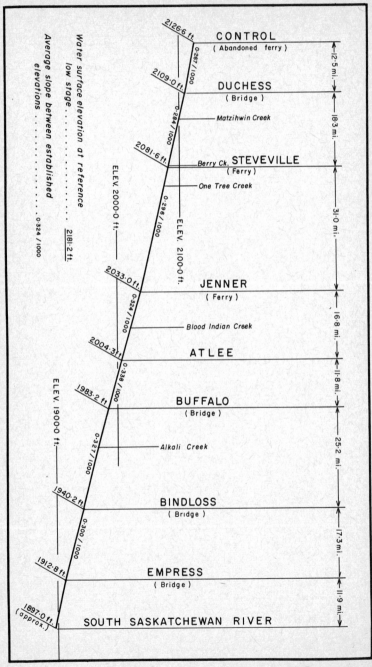

Figure 4. Longitudinal Profile of the Lower Red Deer River (modified after Neill, 1965).

TABLE III ANALYSIS OF BED SAMPLES

Cross Section Location	Channel Position of Sample	10	20	40	60	100	200	Median Size (mm)	Classification
72 L/13 (4)	10' from right bank	91	72	29	8	2	1	0.61	⅜" gravel and sand
72 L/13 (4)	20' from left bank	100	100	100	99	60	80	0.13	Sand
72 L/13 (10)	10' from left bank	100	99	86	53	35	27	0.23	Silty sand
72 L/13 (10)	7' from right bank	99	93	49	8	3	1	0.43	Sand
72 L/14 (4)	Midstream	100	97	60	15	7	6	0.38	Sand
72 L/14 (8)	Midstream	99	97	56	11	1	1	0.38	Sand
72 L/14 (9)	50' from left bank	100	93	75	14	2	1	0.31	Sand
72 L/15 (1)	100' from left bank	94	87	55	16	1	1	0.38	Sandy gravel
72 L/15 (5)	Midstream	99	98	82	16	1	1	0.30	Sand
72 L/15 (6)	Midstream	99	98	85	28	3	2	0.30	Sand
72 L/15 (9)	20' from right bank	100	99	78	30	10	5	0.30	Sand
72 L/16 (1)	Midstream	98	96	87	37	4	3	0.28	Sand
72 L/16 (3)	50' from left bank	100	100	94	29	2	1	0.29	Sand
72 L/16 (4)	50' from left bank	100	99	96	65	11	4	0.12	Sand
72 L/16 (9)	20' from left bank	95	85	94	86	82	79	0.0067	Silty clay and gravel
72 L/16 (11)	30' from left bank	98	93	73	30	3	2	0.30	Sand
72 L/16 (12)	20' from right bank	43	40	29	6	2	1	7.0	Sand and gravel
72 L/16 (13)	Midstream	100	99	74	22	3	1	0.31	Sand
72 L/16 (13)	150' from right bank	100	100	98	62	16	4	0.21	Sand

* Classification used: U.S. Army Corp of Engineers, Lower Mississippi Valley Division.

Sand 2.0 to 0.05 mm

Silt 0.05 to 0.005 mm

Clay <0.005 mm

lateral accretion deposits are composed of sand sized and coarser material, form relatively thick non-cohesive beds, and display a variety of structures. The structures reflect the bed configuration at the time of deposition (Harms and Fahnestock 1965).

At each of the stratigraphic sites the thicknesses of the vertical and lateral accretion materials were determined (Table IV). The vertical accretion deposits are thicker in the length shown on sheet 72L/13 (median thickness 63″ and arithmetic mean 61″) than in the remainder of the study length (median thickness 32″ and arithmetic mean 30″).

As discussed previously, the channel is narrowest and deepest in the section portrayed on map sheet 72L/13. The finding that the vertical accretion deposits are thicker there and that the banks contain a higher proportion of silt and clay, tends to support the contention of Schumm (1960) that a channel will become narrower and deeper as the percentage of silt and clay in the wetted perimeter increases.

Channel Shift

Analyses of variations in rates of channel shift were carried out by comparing the position of the river channel, as shown on the original 1883 township survey plans, with its location as plotted on national topographic map sheets constructed from air photographs flown in 1958. This technique permitted the overall migration of the channel during the 75-year period to be computed.

Changes in the positions only of the eroding banks were studied. There were two reasons for this: (a) a change in the location of the eroding bank reflects a shift in the entire channel, providing average channel dimensions remain the same, because erosion of one bank is balanced by concomitant deposition on the opposite bank. There is no evidence to indicate that the size of the Red Deer's channel has changed within the recent past, (b) the writer was not able to determine from the reports of the early surveyors the criteria applied to fix the limits of the depositional banks. Therefore the position of the depositional banks in different years could not be compared.

The positions of the channel in 1883 and in 1958 were plotted on a map at a scale of two inches equals one mile and the areas of bank erosion outlined. A portion of this map is shown in Figure 5.

The channel was divided into 25 reaches, varying in length from 4 to 6 miles and the area of bank erosion in each reach was planimetered. To facilitate the comparison of erosion rates between

Table IV.

THICKNESSES OF VERTICAL AND LATERAL[1] ACCRETION DEPOSITS COMPOSING THE
FLOODPLAIN AND TERRACES OF THE LOWER RED DEER RIVER

Location of Stratigraphic Section		Thickness of Vertical Accretion Deposits (inches)	Thickness of Lateral Accretion Deposits (inches)	Total Bank Height Above Reference Stage (inches)	Landform
Site No.	Map Sheet No.				
1	72 L/13	67.0"	0.0"	67"	Floodplain
2	72 L/13	63.5"	86.5"	158.0"	Floodplain
3	72 L/13	96.5"	52.5"	148.5"	Floodplain
4	72 L/13	56.0"	92.0"	148.0"	Floodplain
5	72 L/13	39.0"	115.0"	154.0"	Floodplain
6	72 L/13	73.0"	46.0"	119.0"	Floodplain
7	72 L/13	66.0"	18.0"	84.0"	Floodplain
8[2]	72 L/13	60.0"	72.0"	216.0"	Alluvial Fan
9[3]	72 L/13	31.0"	96.0"	154.0"	Floodplain
10	72 L/13	63.0"	73.0"	136.0"	Floodplain
11	72 L/15	32.0"	84.0"	116.0"	Floodplain
12	72 L/15	32.0"	82.0"	114.0"	Floodplain
13	72 L/15	33.0"	70.0"	103.0"	Floodplain
14	72 L/15	122.0"	28.0"	150.0"	Terrace
15	72 L/16	27.0"	41.0"	68.0"	Floodplain
16	72 L/16	33.0"	70.0"	103.0"	Floodplain
17	72 L/16	63.0"	122.0"	185.0"	Terrace
18	72 L/16	20.0"	83.0"	103.0"	Floodplain
19	72 L/16	54.0"	109.0"	163.0"	Terrace
20	72 L/16	33.0"	65.0"	98.0"	Floodplain

1 Thickness of Lateral Accretion Deposits measured above reference stage elevation.
2 At this site the material from bank top to a depth of 84.0" is colluvial.
3 Material from bank top to a depth of 29.0" is aeolian.

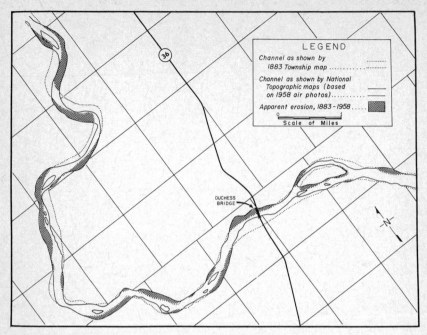

Figure 5. Channel Shift Map of part of Sheet 72L/13.

the different lengths, an erosion index expressing the average amount of bank erosion per mile in each segment was computed. The erosion indices were calculated according to the formulae:

$$E_i = \frac{A_b \times 5280}{L_c}$$

where E_i = Erosion Index

A_b = Area of Bank Erosion in square miles

and L_c = Channel length measured in miles along the centre-line of the channel

Erosion indices for the various reaches are presented in Table V.

Visual analyses of the erosion indices data suggest that a marked change in the rate of bank recession occurs at the end of map sheet 72L/13. The rate of bank recession appears uniformly high through-out map sheet 72L/13, and sharply decreases at the beginning of map sheet 72L/14. Downstream of this point bank recession progressively increases attaining values comparable to those in map sheet 72L/13 in the latter third of map sheet 72L/16. To test this, linear regression equations relating the bank erosion indices and the downstream distance were computed: (a) for the entire study length, (b) for the reach from the beginning of map sheet 72L/13 to 72L/14 (1) and (c) for the length from 72L/14 (1) to the end

TABLE V. DOWNSTREAM CHANGES IN INDICES OF EROSION

Channel Length (miles)	Area of Bank[1] Erosion (sq. miles)	Erosion Index[2]	Downstream Distance (miles)	Location
0.000	0.00000	0.00	0.000	Beginning of
6,500	0.37752	306.60	6.500	sheet 72L/13
5.125	0.20904	215.50	11.625	
5.500	0.22464	215.60	17.125	
5.000	0.22776	240.51	22.125	
3.750	0.2278	320.70	25.875	
5.375	0.18086	178.00	31.250	
4.875	0.13728	148.90	36.125	End of sheet
5.750	0.23088	212.00	41.875	72L/13
6.000	0.17784	156.40	47.875	72L/14 (1)
6.250	0.13416	113.30	54.125	
4.875	0.0624	67.60	59.000	
5.062	0.0652	68.30	64.062	
5.250	0.2652	268.60	69.312	
5.437	0.17472	169.90	74.749	
7.000	0.16224	122.40	81.749	
5.500	0.18096	173.70	87.249	
5.875	0.19344	174.00	93.124	
6.812	0.16224	125.70	99.936	
5.925	0.22464	200.30	105.861	
5.500	0.29016	278.50	111.316	
4.187	0.13728	173.40	115.543	
5.375	0.27456	269.90	120.923	
4.750	0.27768	308.60	125.675	
5.750	0.19032	174.70	131.423	End of sheet 72L/16

1 Bank erosion is computed for the period of 1883-1886 to 1958.

2 Erosion Index= $\dfrac{\text{Area of Bank Erosion X 5280}}{\text{Channel Length}}$

$$E = \dfrac{Ab \ X \ 5280}{Lc}$$

of sheet 72L/16. Results of the analyses are presented in Table VI. It was found that the rate of shift does not change significantly throughout the entire river length. However, it was also discovered, if the lengths upstream and downstream of 72L/14 (1) are considered independently, that the rate of shift significantly increases downstream of 72L/14 (1) but does not vary throughout map sheet 72L/13. It is worthwhile noting the very low correlation coefficient,

Channel Pattern

Along the Lower Red Deer, recognition of lengths with differing channel patterns is difficult as pattern changes are subtle rather than obvious. It could be argued that the pattern over the entire length should be classified as braided, since bars and islands are present everywhere in the channel and the sinuosity is very low. Sinuosity indices calculated for a number of lengths suggested by visual examination, are presented in Table VII. With the exception of the short reach downstream of 72L/16 (5) no length approaches the critical sinuosity index of 1.5, proposed by Leopold and Wolman (1957), to differentiate meandering from straight channels or the value of 1.3 suggested by Brice (1964).

Analyses of the mechanics of channel shift, from the map constructed to show changes in channel position during period 1883-1958 (Fig. 5), floodplain morphology on air photographs, and ground inspection indicate that the pattern in the river length depicted on sheet 72L/13 is very different from that of the remainder of the length. The river segment shown on sheet 72L/13 flows for the most part in a series of regular curves. The migration of the channel is accomplished by point bar formation and meander sweep. Further, floodplain topography is dominated by a series of discrete meander scroll ridges indicating that meander creep is the principal mechanism of floodplain construction. On these grounds, in spite of the low sinuosity index, the pattern is best classified as meandering. Since the meanders are deformed where they impinge against the bedrock valley walls, they might be termed "confined", following Lane (1957).

Throughout the remainder of the length, no systematic migration can be traced. Floodplain construction was accomplished by the successive abandonment of anabranches and the incorporation of these and adjacent bars and islands into the floodplain. This process has been described by several writers (Glenn 1925; Wolman and Leopold 1957; Brice 1964).

TABLE VI. LINEAR REGRESSION ANALYSIS — DOWNSTREAM DISTANCE AND
EROSION INDICES FOR DESIGNATED RIVER LENGTHS

River Length	Number of Observations	Linear Regression Equation[1]	Correlation Coefficient	Value of Statistic t	Significance Level[2]
Entire Study Length	25	Y=171.94+0.23X	0.12	0.57	Not Significant
Beginning of Study Length to 72L/14 (1)	10	Y=186.41+0.54X	0.01	0.26	Not Significant
72L/14 (1) to End of Map Sheet 72L/16	15	Y=8.08+1.84X	0.62	2.86	Significant at 2% level

1 Equation has the form Y=a+bX where Y=Erosion Index
X=Downstream Distance in
Miles, and a and b are constants.

2 Relationship is considered "not significant" at levels greater than 10%.

TABLE VII. SINUOSITY INDICES FOR SELECTED CHANNEL LENGTHS
OF THE LOWER RED DEER

Selected Channel Length	Sinuosity Index[1]
Beginning Sheet 72L/13 to end sheet 72L/14	1.16
72L/13 (1) to 72L/13 (7)	1.2
72L/13 (7) to 72L/13 (12)	1.04
72L/13 (12) to beginning map sheet 72L/14	1.2
Beginning 72L/14 to 72L/14 (8)	1.06
72L/14 (8) to 72L/15 (9)	1.07
72L/15 (9) to 72L/16 (5)	1.10
72L/16 (5) to end of map sheet 72L/16	1.34

(1) Sinuosity Index=$\dfrac{\text{Channel length measured along center line}}{\text{Valley length measured along valley axis}}$

The difficulty experienced in applying existing classifications to the Red Deer points to the necessity of developing a more satisfactory, preferably quantitative, scheme for describing channel patterns.

Discussion and Conclusions

In preceding sections of this paper, the principal variables operating in the channel and floodplain system of the Lower Red Deer were analyzed in succession, and changes in their values over the study section established. This question may now be asked: Is there an identifiable pattern to the changes? and if so, how may this be explained?

The answer to the former question is clearly yes. The results of the analyses indicate that the river system exhibits very different characteristics upstream and downstream of the end of map sheet 72L/13.

In the upstream part, the channel is narrower and deeper than further downstream and does not significantly change shape over the segment. Bed slope values are lower; this is contrary to what might be expected, since slope usually decreases in a downstream direction. The channel pattern is meandering or "confined" meandering. Rates of channel shift are high, higher than elsewhere throughout the study segment and they do not vary. Channel migration is accomplished by the meander sweep process. The morphology of the floodplain is dominated by a series of arcuate meander scrolls, reflecting the mode of channel migration and the channel pattern. The floodplain is higher in this portion of the river's course and the banks are composed of thicker sequences of silt and clay vertical accretion deposits. Perhaps the most striking attribute of the upstream section is that values for the individual variables do not change throughout the length. This is in marked contrast to the remainder where the variables characteristically exhibit a progressive change in value.

Downstream of 72L/13 the channel is shallower and wider, and increases in width and decreases in depth. Bed slope increases sharply at the beginning of the length and then decreases downvalley. However, bed slope is still steeper at every position in the downstream as opposed to the upstream reach. Braiding is the dominant channel pattern, but if the sinuosity index is applied as the criterion to differentiate between braiding and meandering sections, then there is a tendency for the river to return to a meandering habit between Bindloss Bridge and Empress Bridge.

The rate of channel shift is low at the beginning of the reach but increases noticeably downstream. Floodplain height reduces downstream. In addition, the vertical accretion sediments are much thinner.

How can we account for this change in the character of the channel and floodplain? Earlier it was stated that in a river system there are three independent variables—discharge, sediment load and human interference. The remainder are dependent variables and adjust to changes in the independent variables. It has been demonstrated that discharge does not alter and human interference is not believed to be significant. Consequently, variations in these elements cannot explain channel and floodplain variations. That leaves the sediment factor and as this does alter, it is suggested that changes in this element are responsible for the differing character of the river system upstream and downstream of sheet 72L/13. The badland areas, between 72L/13 (16) and 72L/14 (1) are believed to be the key control. Their gullies feed tremendous volumes of sediment to the river and the input of material to the channel and floodplain system triggers a very complex series of adjustments. This results in a change in the river's characteristics. One readily apparent effect is the local steepening of the bed slope and the change from a meandering to a clearly braided pattern downstream of the badland regions.

This study shows how a number of factors in a river system alter in an interrelated manner, underscoring the point that if we are to understand river behaviour we must appreciate that the channel and floodplain are dynamically interlinked. We cannot include only

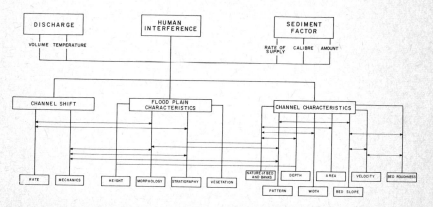

Figure 6. Model Showing Relationships between the major factors in a river channel and floodplain system. Links without arrows indicate equivalence.

the "in channel" variables in our predictive equations; we must also include floodplain and channel measures.

In Figure 6 a model showing the relationships between the major factors in a river system is depicted. Future river research should concentrate on identifying, tracing and evaluating the complex linkages between these variables. Such work will involve the development of quantitative measures for variables, for example, channel pattern, currently described qualitatively. Statistical techniques such as factor and multiple regression models will also prove valuable tools in this research.

References
Adler, A. A. and Lattman, L. H., "Floodplain Sediments of Halfmoon Creek, Pennsylvania," *Penn State College Mineral Industries Experimental Station*, 77, 1 (1960).
Blench, T. "Regime Behaviour of Canals and Rivers," Butterworths Scientific Publications, London (1957); "Discussion on Sediment Transport Capacity in Erodible Channels by Mao and Rice," *Proceedings, American Society of Civil Engineers*, 90, no. HY2 (1964).
Bobey, N., "A Field study of bed activity in the lower Red Deer River," Unpub. Msc. Theses, University of Alberta (1965).
Brice, J. C., "Channel Patterns and Terraces of the Loup Rivers in Nebraska," *U.S. Geol. Surv. Prof. Paper 422-D*, Washington, (1964).
Chorley, R. J., "Geomorphology and General Systems Theory," *U.S. Geol. Surv.*, Prof. Paper 500-B, Washington (1962).
Chorley, R. J. and Haggett, P., *Models in Geography*, Methuen, London, (1967).
Colby, B. R., and Scott, C. H., "Effect of Water Temperature on Discharge of Bed Materials," *U.S. Geol. Surv.*, Prof. Paper 462-G., Washington (1965).
Fenneman, N. M., "Floodplains Produced Without Floods," *Bulletin, American Geographical Society*, 38 (1906), p. 88.
Glenn, L. C., "Geology and Physiography of the Red River Boundary between Texas and Oklahoma," *Pan American Geologist*, 43 (1925), p. 375.
Hack, J. T., "Interpretation of Erosional Topography in Humid Temperate Regions," *Am. Jour. Sci.*, 258A (1960), p. 80.
Harms, J. C., and Fahnestock, R. K., "Stratification of Bed Forms and Flow Phenomena," *Society of Economic Paleontologists and Mineralogists*, Special Publication 12 (1965), p. 84.
Inglis, C. G., "The Behaviour and Control of Rivers and Canals," *Central Board of Irrigation*, Research Publication 13, Simla, India, (1949).
Lacey, G., "Small Channels in Alluvium," *Proc. Inst. of Civil Eng.*, 229, 259, (1930).
Lane, E. W. "A Study of the Shape of Channels formed by Natural Streams in Erodible Materials," *U.S. Army Corp. of Engineers*, Sediment Series, no. 9, Vicksburg, (1957).
Langbein, W. B. and Leopold, L. B., "Quasi-Equilibrium States in Channel Morphology," *Am. Journ. Sci.*, 262 (1964), p. 782.
Leopold, L. B. and Maddock, T., "The Hydraulic Geometry of Stream Channels and Some Physiographic Implications," (1953), *U.S. Geol. Surv. Prof.*, paper 252, Washington, (1955).
Leopold, L. B. and Miller, J. P., "Ephemeral Streams; Hydraulic Factors and their Relation to the Drainage Net", U.S. Geol. Surv. Prof., paper 282-A., Washington, (1956).

Leopold, L. B. and Wolman, M. G. and Miller, J. P., "Fluvial Processes in Geomorphology," Freeman, San Francisco, (1964).

Lustig, L. K., "Sediment Yield of the Castaic Watershed, Western Los Angeles County California—A Quantitative Geomorphic Approach," *U.S. Geol. Surv. Prof.*, paper 422-F., Washington, (1964).

Neill, C. R., "Channel Regime of the Lower Red Deer," Unpub. Report, Research Council of Alberta, Edmonton, (1965).

Schumm, S. A., "The Shape of Alluvial Channels in Relation to Sediment Type," *U.S. Geol. Surv. Prof.*, paper 352-B, Washington, (1960).

Simons, D. B. *et al.*, "Bed-load Equation for Ripples and Dunes," *U.S. Geol. Surv., Prof.*, paper 462-H., Washington, (1965).

Sundborg, A., "The River Klaralven — A Study of Fluvial Processes," *Division of Hydraulics, Royal Institute of Technology, Bulletin 52*, (1956), Stockholm.

U.S. Inter-agency Committee on Water Resources, "Determination of Fluvial Sediment Discharge," Report No. 14 in series Measurement and Analysis of Sediment Load in Stream, Washington, (1963).

Wolman, M. G., "The Natural Channel of Brandywine Creek, Pennsylvania," *U.S. Geol. Surv. Prof.*, paper 271, Washington, (1961).

Wolman, M. G. and Brush L. M., "Factors Controlling the Size and Shape of Stream Channels in Coarse Non Cohesive Sands," *U.S. Geol. Surv. Prof.*, paper 282-G., Washington, (1961).

Wolman, M. G. and Eiler, J. P., "Reconnaissance Study of Erosion and Deposition Produced by Flood of August 1955 in Connecticut," *Trans. Amer. Geophys. Union*, 39 (1958), p. 1.

Wolman, M. G. and Leopold, L. B., "River Flood Plains: Some Observations on their Formation," *U.S. Geol. Surv. Prof.*, paper 282-C., Washington (1957).

17

The Shifting Sands
of Sable Island

H. L. Cameron

Sable Island, a narrow sand ridge some 180 miles east of Halifax, Nova Scotia (Fig. 1), has been known to European sailors since 1497, when it was sighted by John Cabot. Over the centuries it has acquired such an evil reputation as a navigational hazard that it has become known as the "Graveyard of the Atlantic."[1] There have been 204 recorded shipwrecks on the island, and probably as many unrecorded disasters.

The island is the emergent top of a huge sand lens that extends for about one hundred miles along the edge of the continental shelf. The north-south width of the lens is irregular but averages about thirty miles. The visible island is twenty-one miles long and is extended underwater by a bar at each end; the whole forms a fifty-mile-long menace to shipping.

This mass of sand is believed to represent the reworking of the glacial debris deposited on the continental shelf during the low water level of the maximum glaciation in Wisconsin time.[2] The post-glacial rise of sea level submerged the deposits and subjected them

First published in *Geographical Review*, Vol. LV, no. 4 (1965), pp. 463-67.

[1] G. Kobbe, "Graveyard of the Atlantic," *Current Literature* (April, 1900), p. 24; Edward Rowe Snow, "Island of Lost Ships" *Colliers* (Jan. 7, 1955), pp. 66-69.

[2] For discussions on the origin and structure of Sable Island see: Simon D. Macdonald, "Sable Island: Its Probable Origin and Submergence," *Proc. and Trans. Nova Scotian Inst. of Nat. Sci.*, Vol. VI (1882-86), pp. 265-80; J. W. Goldthwait, "Physiography of Nova Scotia," *Canada Geol. Survey Memoir 140* (1924), pp. 141-47; Charles H. J. Snider, "Sands of Sable," *Canadian Mag.*, Vol. LXIX, no. 30 (1928), pp. 8-9; P. L. Willmore and R. Tolmie, "Geophysical Observations on the History and Structure of Sable Island," in "Ocean Floors Around Canada" (papers from a symposium sponsored jointly by the Canadian Committee on Oceanography and the Royal Society of Canada, Section IV), Ottawa (1956), accompanying *Trans. Royal Soc. of Canada*, Ser. 3, Vol. L, sect. 4 (1956), pp. 13-2 (reprinted as *Contribs.* from the Dominion Observatory, Ottawa, Vol. I, no. 31.)

to wave and current action. After the sea had attained approximately its present level, about five thousand years ago, the interaction of the Labrador Current, the Gulf Stream, and the St. Lawrence Current created an eddy on the shelf (Fig. 1), which gathered the sandy parts of the glacial debris into the Sable Island bank.[3] It is probable that the exposed part of the bank was much larger than the present island and that the gradual rise in sea level since has progressively submerged it. Geophysical work indicates that the sand of Sable Island is about 1200 feet deep over bedrock.[4]

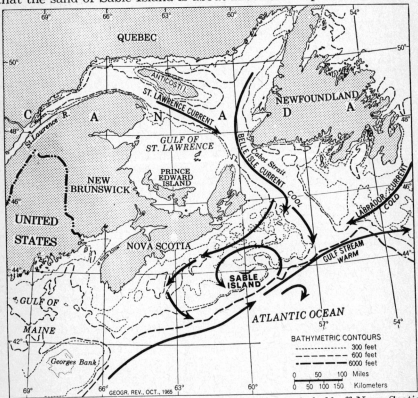

Figure 1. Location of Sable Island on the continental shelf off Nova Scotia and its relation to the principal currents.

---

[3] H. B. Hachey, "Water Transports and Current Patterns for the Scotian Shelf," *Journ. Fisheries Research Board of Canada*, Vol. VII, no 1 (1947), pp. 1-16; R. W. Trites and R. E. Banks, "Circulation on the Scotian Shelf as Indicated by an Experiment using Drift Bottles, *ibid.*, Vol. XV, no. 1 (1958), pp. 79-89. See also Dean F. Bumpus and Louis M. Lauzier, "Surface Circulation on the Continental Shelf off Eastern North America between Newfoundland and Florida," *Amer. Geogr. Soc. Serial Atlas of the Marine Environment*, Folio 7, New York (1965).

[4] Willmore and Tolmie, *op. cit.* (see footnote 2), pp. 17-20.

This would suggest that the lens accumulated in a shallow bowl-like depression, since the average water depth to rocky areas around the island is 350 to 400 feet.

Throughout the long history of Sable Island many reports have been circulated concerning its gradual destruction by the sea. The present study was undertaken for the twofold purpose of checking these reports as a matter of scientific interest and evaluating the potentialities of the island as a base for navigational aids.

The Historical Backgrounds[5]

Sable Island "firsts" date from the sixteenth century. In 1518 the Baron de Lery, after an unsuccessful attempt to colonize New France, is said to have left half of his cattle at Canso, in Nova Scotia, and the rest on Sable Island; this is the first record of domestic animals on the island. The first chart showing the location of the island was one of New France made by Jacopo Gastaldi about 1550. It delineates the fishing banks that extended from off LaHave, Nova Scotia, to Newfoundland, and on it Sable appears as Isola della rena, "Island of Sand." The first recorded shipwreck was in 1583, when Sir Humphrey Gilbert's second-in-command was lost with some ninety-five of his crew. Twelve men escaped, and after spending some time on the island, eventually reached Nova Scotia with their tale of the disaster.

In the seventeenth century stories abound of mariners wrecked on the island, where they lived a Robinson Crusoe existence, subsisting on berries and wild cattle until they could build boats or rafts from the timbers of their broken ships and succeed in reaching the mainland. Many old tales are known, and are confirmed in the archives, of people who wished to settle on the island but were prevented by government niggling over quitrents. However, not all colonists of the time were "willing." In 1598 the Marquis de La Roche brought forty convict settlers to New France. These unhappy wretches were landed on Sable Island until he could establish them on the mainland. Unfortunately, the marquis himself was driven home to France by a gale, and there was thrown into prison by an enemy. It was not until five years had elapsed that the convicts were remembered and rescued. By then only twelve remained; one wonders what happened during those five years.

Little authentic information is available about the island from 1700 to 1800. Sea commerce increased, and Sable reputedly became

---

5 The historical information in this section is derived from Simon D. Macdonald, "Notes on Sable Island," *Proc. and Trans. Nova Scotian Inst. of Nat. Sci.*, Vol. VI (1882-86), pp. 12-33.

the haunt of pirates and wreckers. Dark tales are told of these desperadoes robbing the dead—and the not yet dead—washed ashore from vessels wrecked on the shoals, tales of battles, murder, and sudden death. In 1802 the transport *Princess Amelia* was lost with some two hundred officers and men, and—worse still—with the household effects of the Duke of Kent. This calamity, together with reports that many of the crew had reached shore, only to be murdered by the pirates, brought action by the Nova Scotia authorities. They appropriated funds to set up a lifesaving and assistance station on the island, and the station was built in that same year. The first punitive ship dispatched, the gun brig *Harriet,* in 1803, was also lost. She was probably a victim of the navigational hazards of the bank, but her loss nonetheless impelled the provincial government to proclaim severe penalties for anyone residing on the island without a license.

The toll of wrecks went on, even after the advent of steam, owing mainly to the unknown strength of the currents around the island and the occurrence of heavy fog banks along the border between the Gulf Stream and the cold northern currents. In 1873 two new lighthouses were constructed, one near the west end of the island, and the other near the east end. But "near" is a relative term; the East Light, for example, had seven miles of exposed sandbar to the east of it. There was a great deal of discussion concerning the usefulness of the lights, in which both Mr. Joseph Howe and Mr. Samuel Cunard took a prominent part. The two lights have been continuously maintained ever since, though the West Light has had to migrate eastward. Today the island is beaconed by radio waves in a way that would have appeared sheer black magic to Sir Humphrey Gilbert and his men. Among the aids used for navigation and for oceanographic work on the continental shelf are wireless beacons, radiosonde, and three-lane Decca.

Recent Investigations: Aerial Photography

In 1954 an investigation was undertaken by the writer to check the validity of reports that Sable Island is gradually being destroyed by the sea. The method used was sequential air photography at intervals of three to five years. From the photographs a series of maps was produced by photogrammetry and photointerpretation. This technique proved highly successful and is strongly advocated for studies of other coasts and islands. The National Air Photo Library in Ottawa offers a great reservoir of untapped knowledge, which awaits exploitation not only by scientists for studies of this kind but also for practical applications.

The Royal Canadian Air Force obtained photo coverage of Sable in August, 1952. Three flight lines were required to cover the island at the scale of 1 inch equals 1050 feet, using a 5-by-5-inch format. The first map, showing terrain analysis, was prepared in 1954 from this photography. The Royal Canadian Navy became interested in the study and in 1955 requested the R.C.A.F. to rephotograph the island, as nearly as possible at low tide. Planning for this operation brought out the interesting fact that low tide on the south side of Sable was three-quarters of an hour behind low tide on the north side—direct evidence of the length of the underwater extensions. The photographs were taken in August, 1955, at the scale of 1 inch equals 1000 feet, using a 9-by-9-inch format. In April, 1959, a third coverage was obtained by the R.C.A.F. at the same scale and size, and in June, 1964, a fourth, comparable, survey was flown.

The photography has all been standard panchromatic with a minus blue filter, though it is recognized that infrared cover would have aided in the mapping of the numerous ponds and the sparse vegetation. In all cases Logotronic (automatically dodged) prints have been used for photointerpretation. The Department of Mines and Technical Surveys obtained color photography of the island in 1963, but the pictures were not available for this study.

Geomorphological Interpretation

The terrain on the island can be divided into distinct types: Stabilized sand, unstabilized sand, shallow water, deeper water, grass cover and sand ripples. The loose beach sand and the sand in wet areas are moved only by the sea, as is shown by the sand ripples on the beach between the west end of Lake Wallace and the south shore. Elsewhere on the beaches, beach cusps are well developed. Where wind action has breached the vegetation mat, large crater-like blowouts form rapidly, usually the result of one particularly severe storm. Often they reveal relics of old wrecks or the camps of unknown castaways.

The stabilized dunes are the major topographic feature of the island. None are more than eighty-five feet high, and most are capped with the coarse grass that is the principal vegetation. Those which are not grassed on top are found to contain layers of roots from former grass cover. The winter winds drift the sand over the grass, and the summer growth, if any, must come up through the sand.

The ponds and lakes have a surprising amount of water surface. All the smaller water bodies are fresh, and even Lake Wallace is brackish only because of blown spray. This brings us to one of the most startling features of this desert island. If one digs anywhere in the sand a reasonable distance above the high-tide mark, say twenty-five feet, fresh water will flow into the excavation. Tests indicate that the whole island is underlain by a bubble of fresh water, formed by accumulation of the abundant rainfall and hardly tapped by the inhabitants, human or otherwise. The saltwater-freshwater interface is probably distinct, and the constant renewal of the fresh water will resist anything but a complete breach of the island.

The terminal bars were delineated from the photography, and it may be seen that the East Bar has a break or channel a quarter of a mile wide about a mile from the east end. Several wrecks showed up in the photographs. The most spectacular is that of the ocean freighter *Alfios*, about eight thousand tons, ashore on the south beach opposite the East Light. The most interesting wreck—and later a useful one—is a dragger aground head on to the East Bar about nine miles east of the East Light. This dragger was spotted again in the 1959 photography; if it was in the same place as before, as seems likely, then the East Bar had shifted its width to the northwest, a conclusion first suggested by a comparison of the 1959 map and the 1952 map.

A series of seven outline maps was prepared, based on early sources and on the maps drawn from the recent aerial photographs, to indicate the changes that have occurred in the size, shape, and position of Sable Island in almost 200 years (Fig. 2). Ground checks and a careful check of records established the east end of Lake Wallace as a probable invariant point. With this as a fixed point, the maps were all plotted on a single sheet to facilitate comparison.

The accumulated changes have been superimposed on the 1959 map. They show a loss of nine miles from the west end of the island and a gain of eleven miles on the east end. The sequence demonstrates, then, that the island is gradually being eroded in the west, and elongated in the east, by wave and current action. Serious erosion losses are attributable mainly to severe storms and have therefore occurred spasmodically. Erosion may be accelerated by the recent trend in hurricane tracks, but even under these conditions the island should survive for hundreds of years. The principal threat to its existence is the rise in sea level, which, if continued at the present rate of half an inch in a century, will reduce the island to base-level in about three thousand years.

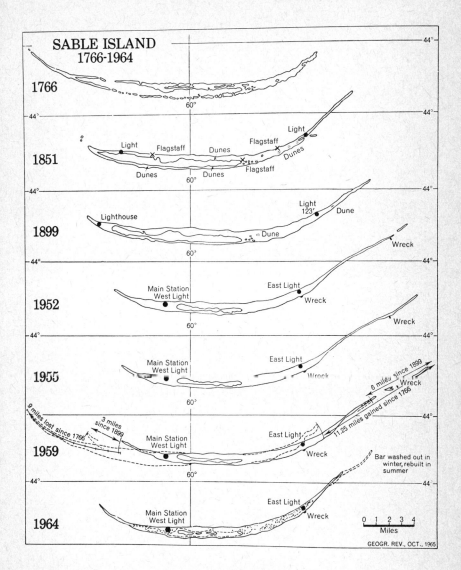

Figure 2. The migration of Sable Island, 1766-1964.

18

Wave Transports of Beach Materials on Long Point, Lake Erie

H. A. Wood

It is generally accepted by geomorphologists that in the development of shorelines the most important modifications are effected by waves. The winds which generate them are only directly operative above water level, and in most cases the currents which are significant geomorphologically are those generated by the waves themselves. Nevertheless, exact information concerning the effect of waves upon a coast has been slow in coming, for quantitative studies of shore morphology are extremely difficult to carry out. The direct measurement of waves is too costly a procedure for most research programmes, and wave tank experiments are of limited value as they cannot duplicate the great storm waves which are the most significant agents of shore modification.

It is the purpose of this paper to demonstrate the application of an alternative quantitative approach to the problem of shifting shorelines based solely on data readily available in wind records and on hydrographic charts. Admittedly the accuracy of the method is only approximate. Now and then, as in many investigations, progress has had to wait upon the making of assumptions which may not be entirely true. Nevertheless, the approach does permit the study of shoreline processes without elaborate equipment, and even in the single application here presented, it points to several important conclusions.

The area selected for study is Long Point, a sand spit on the north shore of Lake Erie. Long Point is probably the most spectacular feature of shore deposition in the Great Lakes and offers ideal conditions for an investigation of this type. Not only is Lake Erie

First published in *Canadian Geographer*, no. 16 (1960), pp. 27-35.

free from the tides which complicate studies of marine coasts, but its surrounding weather stations provide more complete wind records than would normally be available for an oceanic site. Furthermore, surveys made in 1853, 1907 and 1951 show Long Point to have been subject to extremely rapid growth. In the approximately 4000 years of its existence, the spit has attained the remarkable length of twenty-five miles. This rate of growth is a result, in part, of the coincidence of the prevailing southwest winds and the maximum length of fetch on Lake Erie, and in part of the abundance of beach-building materials available in sand cliffs to the west. Even though the spit is now being extended into water about 150 feet deep, it has been increasing in length over the past century by an average annual amount of 20 to 25 feet.

Long Point has not only been growing; it has been changing in other ways as well. The major alteration to have occurred was a breaching of the neck of the spit early in the nineteenth century. Maps published in 1798 and 1817 show the point to be continuous, and early travellers tell of portaging across the isthmus. However, by 1839 a channel a mile wide and fourteen feet deep had been formed across the western end of the Point. This channel was regularly used for navigation, and its position was marked by a light ship. Fourteen years later, in 1853, it was still five-eighths of a mile wide.

It is not known exactly when the gap was completely filled, but by the beginning of the present century the spit was once again joined to the mainland. A shallow bay, however, still marks the position of the former breakthrough, and in this bay deposition is still occurring. In Long Point there is, therefore, a departure from the classical situation in which a spit is divided simply into two sections, a prograding tip and a retrograding neck. Clearly Long Point contains at least two prograding and two retrograding sections, a fact which adds considerably to its value as a subject of investigation. The fact of the breakthrough also adds a note of immediate practical significance to the study, for the area is undergoing rapid recreational development. If Long Point is once again to be broken in two by the waves, it is highly desirable to know approximately where the breach is likely to occur, and when.

The starting point in this inquiry is to ascertain the magnitude of the prime force producing shoreline changes: the energy of the waves. Theoretically this should pose few difficulties as the energy of a deep water wave is known to be proportional to the product of its period and height, both squared,[1] while a graph has been constructed permitting the determination of these wave characteristics when wind velocity and duration and the length of fetch are known.[2] Unfortunately, however, wind records are not normally compiled in such a fashion as to reveal directly the velocity and duration of individual gales or breezes. In the records winds are divided into different velocity groups, but durations are only indicated in terms of total hours per month or year. A further problem is that the weather stations around Lake Erie[3] differ considerably in the winds which they experience, and no single station's records can be taken to the exclusion of the others as the waves which affect Long Point come from all parts of central and eastern Lake Erie, and are developed under the influence most stations. The results ars summarized in Table I.

The latter difficulty was met by accepting, as far as possible, the highest velocities (for winds of each direction) recorded at any of the stations, a procedure aimed at minimizing the influence of local reduction of wind velocities due to friction over the land. Average wind velocities for the different velocity groups were then easily determined using the records of wind mileage available for most stations. The results are summarized in Table I.

TABLE I. WIND FREQUENCIES, LAKE ERIE

(Showing per cent of total time during which wind of each class blows)

Wind direction	Wind velocity		
	0-12 m.p.h. (Average 7.45 m.p.h.)	12-24 m.p.h. (Average 16 m.p.h.)	Over 24 m.p.h. (Average 27.7 m.p.h.)
North	4.5	1.9	0.48
Northeast	9.03	4.45	1.2
East	4.66	1.8	0.18
Southeast	5.57	0.80	0.10
South	8.40	1.96	0.40
Southwest	13.63	7.7	2.4
West	8.15	5.73	1.9
Northwest	7.31	4.47	1.60

[1] From $E = LH^2$ (D. W. Johnson, *Shore Processes and Shoreline Development*, New York, 1919) and

$$L = g \quad T^2$$

$$\overline{}$$
$$2x$$

(O, Krummel, *Handbuch der Ozeanographie, Stuttgart*, 1911) where E is the wave energy, L the length, H the height and T the period; g is the acceleration due to gravity.

[2] C. L. Bretschneider, "Revised Wave Forecasting Relationships," Proceedings of the Second Conference on Coastal Engineering (1952). Bretschneid-

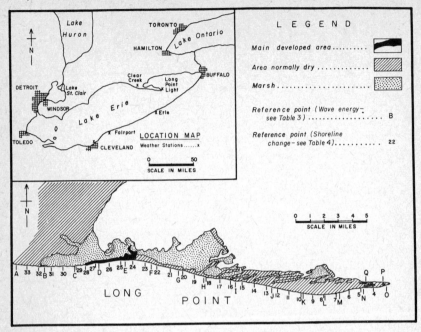

Figure 1.

The duration of individual gales is more difficult to estimate without running through the day to day wind records, a process which would usually be prohibitively time-consuming. In the present instance, a short cut was taken by assuming that winds of over 12 m.p.h. will occur during the passage of low pressure systems while those of under 12 m.p.h. will be associated with highs. From a recent analysis of the occurrence and frequency of cyclones in the northern hemisphere[4] it may be calculated that the average time required for the passage of a cyclone over Lake Erie is about 60 hours, while the corresponding figure for an anti-cyclone is 200 hours. When these amounts of time are divided among winds of various directions in accordance with the frequencies shown in Table I, it is possible to obtain values for the average durations of individual gales and breezes by wind strength and direction. These are shown in Table II.

er's graph is reproduced in "Shore Protection Planning and Design," Technical Report No. 4, Beach Erosion Board, Corps of Engineers, Department of the Army, Washington D.C. (1954), p. 16.

3 Locations of stations, the wind records of which were consulted, are shown in Fig. 1.

4 William H. Klein, "Principal Tracks and Mean Frequencies of Cyclones and Anti-cyclones in the Northern Hemisphere," Research Paper No. 40, U.S. Department of Commerce, Washington, D.C. (1957).

The way was now clear for the evaluation of wave heights and periods, using Bretschneider's graph, and hence of the energy of the waves corresponding to wind groups of Table II. These figures, multiplied by the frequency values of Table I, were plotted for the eight points of the compass and for various lengths of fetch.

TABLE II. WIND DURATIONS, LAKE ERIE.
(in hours)

Wind direction	Wind velocity		
	0-12 m.p.h.	12-24 m.p.h.	Over 24 m.p.h.
North	14.0	5.9	3.5
Northeast	28.0	13.8	8.7
East	14.5	5.6	1.31
Southeast	17.3	2.48	0.73
South	26.1	6.1	2.9
Southwest	42.3	23.9	17.4
West	25.3	17.8	13.8
Northwest	22.7	13.9	11.6

The graphs were completed by the drawing of smooth curves, with the results shown in Figure 2. Upper limits to the energy

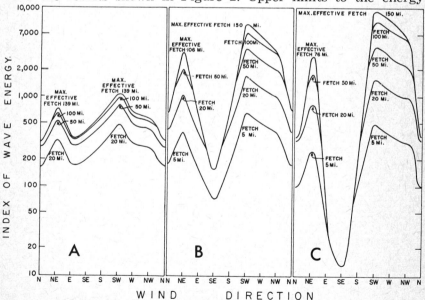

Figure 2. A. Wave energy generated by winds of under 12 m.p.h. on Lake Erie.
B. Wave energy generated by winds of 12-24 m.p.h. on Lake Erie.
C. Wave energy generated by winds of over 24 m.p.h. on Lake Erie.

values were set in general by wind durations, but in a few cases by the length of fetch on Lake Erie. Using these graphs it was possible to obtain an index of the wave energy for winds of each velocity group for any wind direction and any length of fetch.

This energy, however, is not uniformly distributed along the shore. Because of wave refraction, there will be a concentration of wave energy in certain sections, and a diffusion in others. The effect may be evaluated by the drawing on a chart of orthogonals, or lines perpendicular to the wave crests, from deep water to the shore using standard techniques of refraction diagram construction.[5] The interval between orthogonals in deep water divided by the interval at the shore provides a refraction coefficient which can be multiplied by the wave energy in deep water to give the amount of energy per unit of shore length. Values of this coefficient were worked out for winds coming from each 10° of the compass,[6] and selected results are shown graphically in Figure 3. The reference points may be located on the map of Long Point, Figure 1.

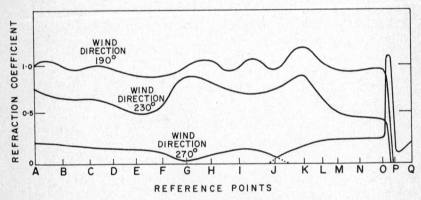

Figure 3. Refraction coefficient for selected wind directions.

Concerning the disposal of this energy, the first distinction to be made is between that part of the energy which is exerted perpendicular to, and that part which is exerted parallel to the shore. The effect of the former is to move materials directly both

---

5 For a description of procedures in refraction diagram construction, see Bretschneider's, "Shore Protection Planning and Design," pp. 31-6.

6 As this procedure is a rather lengthy one, it was not carried out separately for the three wave lengths corresponding to the three wind velocity groups. For each section of the shore a single wave length was used which was the average of the longer waves, those generated by the stronger winds. Since the shorter waves contribute relatively little energy, the error in using the calculated coefficient for them also will be a small one.

landward and seaward. In general, large waves remove materials from the shore, while small ones restore it,[7] but clearly, the two movements would not normally cancel each other out, as in most areas there must be a net loss of substance from the land. Unfortunately, it is not easy to determine directly how much energy is used to bring in sand, and how much to carry it away, yet it would not appear that on Long Point there are large local variations in the energy expended in this way. Consequently, the geomorphic significance of energy expended at right angles to the shore will be assessed indirectly below.

With energy expended parallel to the shore, the situation is different. In this case, the net direction in which the energy is applied remains essentially constant. The effect of this is to move materials in one direction, which, for clarity, will be referred to as "downshore," and as the quantity of material transported increases or decreases downshore, erosion or deposition will occur. The influence of this energy upon shore development will therefore be the more readily evaluated.

The fraction of the total energy which is applied parallel to the shore is, of course, represented by the sine of the angle between wave and shore. However, since this angle becomes continually smaller as any wave approaches the shore obliquely, it becomes necessary to decide where it should be measured. Since most of the wave energy is released after it breaks, it seemed reasonable, in the present study, to measure the angle at the breaking point and again at the shore and to average the two sines together.[8] The results, in a graphical form which permits the ready evaluation of this factor for any wave which might occur on Lake Erie, appears in Figure 4.

Using the procedures outlined above, a calculation was made of the energy expended parallel to the shore at each of seventeen reference points, for winds from each 10° of azimuth and of each of the velocity groups. For each reference point the sum of the energies directed westwards was subtracted from the sum of those

---

[7] A. Guilcher, "Coastal and Submarine Morphology," London (1958), p. 85. Also W. C. Krumbein, "Geological Aspects of Beach Engineering," *Berkey Volume, Geol. Soc. Amer.*, New York (1959).

[8] A graph showing the breaking depth for waves of different heights and lengths is given in Bretschneider's, "Shore Protection Planning and Design" p. D-42. The angle between wave and shore for different depths and angles of wave approach may be taken from "Wind Waves at Sea Breakers and Surf," Publication 602, Hydrographic Office, Washington, D.C., Table 36 (1947), p. 157.

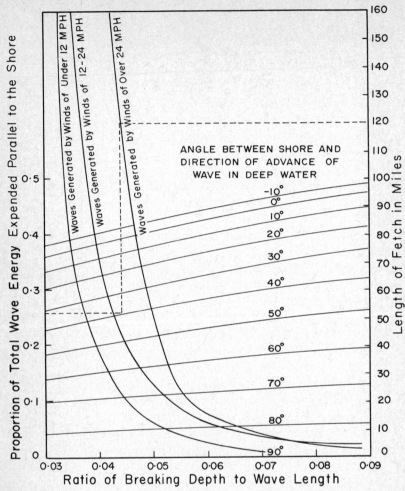

Figure 4. Proportion of total wave energy expended parallel to the shore on Lake Erie; *e.g.*, for length of fetch 120 miles and winds of over 24 m.p.h., with deep water waves advancing at an angle of 40° with the shore, the proportion is 0.256.

pushing eastwards, the former being always the greater except in the case of the two reference points on the north side of Long Point. For convenience each figure was then expressed as a percentage of the value at B, the point where the spit connects with the mainland, with results as summarized in Table III. If it is assumed that there will be no significant local variation in the proportion of energy used in the transport of material, the last column of Table III becomes in actuality an index of downshore transportation.

TABLE III. WAVE ENERGY EXPENDED PARALLEL TO THE SHORE.

Reference point	Easterly component	Westerly component	Net flow of energy		Energy flow as % of value at B (Transportation index)
A	11719	548	11171	Easterly	94.8
B	12349	585	11764	"	100.0
C	11890	553	11347	"	96.4
D	11064	459	10605	"	90.2
E	9063	328	8735	"	74.3
F	6229	642	5587	"	47.5
G	8893	764	8129	"	69.1
H	10949	683	10266	"	87.2
I	10205	620	9585	"	81.5
J	9267	627	8640	"	73.5
K	10628	710	9918	"	84.4
L	10167	789	9369	"	79.7
M	8736	655	8081	"	68.7
N	5834	629	5205	"	44.3
O	5859	874	4985	"	42.4
P	1284	1800	516	Westerly	4.4
Q	1083	2419	1336	"	11.4

This index, in turn, may be expressed in terms of the actual volume of materials moved, a most fortunate circumstance which derives from the very rapid growth of the spit, and the depth of water into which it is being built. Long Point is advancing so rapidly that, as indicated above, its annual increase in length is at least twenty feet. As it is also being extended at a large angle with the bottom contours, it is a simple matter to fix the original position of these contours by linking them across the tip of the spit. This permits the drawing of sections across the spit showing the original as well as the present lake bottom. In this way the cross section of the spit at its tip was found to have an area of 228,000 sq. yds., from which it follows that the annual average addition of material to the tip of the spit is approximately 1,500,000 cu. yds. This figure, however, corresponds to a net transport index of 38.0, a value representing the difference between additions to the tip on the south and removals on the north. Consequently, each index point is equivalent to the transport of approximately 40,000 cu. yds. of beach materials per annum.

Before attempting to relate the volume of materials transported along the shore and the actual rate of shoreline advance or retreat, three other factors should be considered: the offshore gradient, the amount of material in the offshore section of the spit, and the size of particles being transported.

The offshore gradient would appear to have a bearing on whether erosion or deposition will be concentrated at the beach or out in the lake, though on Long Point the significance of this factor must be quite limited, for it is related primarily to the past. Where the shoreline has retreated, the water offshore is shallow. Where it is advancing, the water is deep. Nevertheless, the average gradient has been measured as far as the 7-fathom contour, the depth which agrees most closely with the limit of offshore wave deposition as shown by the contours themselves, as well as by theoretical considerations.

The amount of material extant in the offshore parts of the spit deserves consideration, for this represents the present status of Long Point as much as does its actual shoreline. In a dynamic study, the change in this volume as one proceeds downshore will be the most significant element, and this has been determined from hydrographic data with the results summarized below.

Finally, the size of particles being transported might be significant inasmuch as larger ones would be moved only by larger waves. No sampling was made, but in view of the fact that all materials in the spit originated in the sand bluffs to the west, it was assumed that variations in particle size downshore would be small, and due primarily to abrasion in transit. The distance from the neck of the spit can therefore be expected to give a reasonable inverse index of the size of particles.

The results of all these calculations for thirty stations on the south shore of Long Point are given in Table IV below.[9]

The relationship between the five elements of Table IV has been worked out statistically, with the following results:[10]

Equation: $A = -10.13 + .0225\ B + 16.32\ C + .00895\ D - .327\ E$

Multiple correlation coefficient: $r_{A \cdot BCDE} = 0.727$

Simple correlation coefficients: $r_{AB} = 0.656$; $r_{AC} = 0.186$;

$r_{AD} = 0.352$; $r_{AE} = 0.033$

The standard error of estimate: $s_e = 7.73$

---

[9] Values for materials transported parallel to the shore appearing in Table 4 are derived by interpolation from data in Table 3, each unit of the transportation index being equated with the movement of 39,500 cu. yds. of material per annum.

[10] The author is indebted to Dr. J. D. Bankier, Professor of Mathematics, McMaster University, for effecting the statistical analysis.

TABLE IV. SHORELINE CHANGE AT LONG POINT.

NOTATION: A = Average annual shift in position of shore 1907-1951 in feet.
(positive=aggrading shore; nevative=retrograding shore)
B = Net additions of material produced by downshore movement, in 1000 cu. yds. per mile of shore per annum.
C = Offshore slope in percent.
D = Downshore change in volume of offshore portion of spit, per mile of shore, in 1000 cu. yds.
E = Distance from neck of spit in miles, downshore.

Point No.	A	B	C	D	E
4	− 4.5	19.75	1.26	−380	24.34
5	− 5.7	158	.969	−425	23.49
6	2.3	592.5	.795	−230	22.66
7	− 2.3	473	.764	− 65	21.77
8	3.9	296	.736	−150	20.94
9	2.3	79	.685	−145	20.08
10	4.5	−138.2	.652	−230	19.26
11	− 6.8	−276.1	.576	−380	18.38
12	−11.4	−150	.513	−350	17.51
13	− 6.1	59.3	.467	−395	16.66
14	− 6.8	197.5	.418	−330	15.80
15	0	256.5	.391	−355	14.94
16	−18.2	185.5	.361	−245	14.05
17	−31.8	86.8	.350	−190	13.23
18	−18.2	−197.5	.337	−140	12.33
19	−15.9	−414.5	.328	−160	11.51
20	−22.7	−477	.314	−600	10.64
21	−28.4	−481	.282	−150	9.76
22	−26.6	−355	.279	−150	8.88
23	− 3.4	197.5	.268	−205	8.03
24	3.4	592.5	.263	− 5	7.15
25	9.1	426	.267	130	6.33
26	6.8	296	.274	225	5.46
27	2.3	189.5	.285	320	4.61
28	6.8	138.1	.302	305	3.75
29	8.4	118.5	.317	155	2.90
30	0	98.6	.320	40	2.06
31	− 7.3	19.75	.324	190	1.20
32	− 9.3	− 59.3	.342	385	0.35
33	− 6.8	−130.3	.375	500	−0.48

It will be noted that the correlation is by no means perfect. There are many reasons for this. One is the limitation of the number of factors considered; another, the inaccuracies of the assumptions made in evaluating them. The hydrographic charts may not be entirely accurate, and, in addition, the average yearly change in position of the shore from 1907 to 1951 is not in every case equal to the present rate of change. It must further be borne in mind that that ideal equation may incorporate the elements of Table IV in powers other than the first, but time has not permitted the examination of this possibility.

Although the correlation is only approximate, it is still significant, and several points are worthy of note. First, both the offshore slope and the distance from the neck of the spit have a low degree of correlation with the shift in position of the shore. As indicated above, this is not entirely unexpected. The most significant variable in explaining shore aggradation or retrogradation is the addition or removal of materials by downshore movement, while of intermediary importance is the variation in the volume of the offshore slope.

If C and E in the equation of Table IV are assigned their average values and treated as constants, the following approximation is reached:

$$A = -6.45 + 0.0225 \, B + 0.00895 \, D.$$

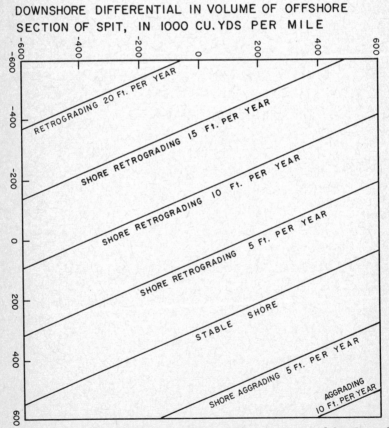

Figure 5. Amount of material annually added or removed by downshore transportation, in 1000 cu. yds. per mile.

This expression lends itself to graphical representation, and is shown thus in Figure 5. It points, furthermore, to a number of important relationships which are here advanced as tentative conclusions:

1. The shore of Long Point may be retrograding even though considerably more material is being brought in by lateral wave transport than is removed in the same fashion.

2. Where the volume of the offshore section of the spit is uniform, approximately 290,000 cu. yds. of beach materials per mile must be brought in by lateral movement each year to maintain a stable shore. Each change of five feet per year in aggradation or retrogradation of the shore is equivalent to a further addition or removal of 220,000 cu. yds. of materials annually per mile.

3. Retrogression of the shore tends to be considerably greater where the volume of the offshore part of the spit decreases downshore, and less where it is increasing. However, only when the annual increase is greater than 700,000 cu. yds. per mile will the shore tend to aggrade even though materials are being removed from it by wave transportation downshore.

Finally, since the discrepancies between the change in volume of the offshore part of the spit and the amount of materials added or removed by wave transport downshore must be accounted for by transportation at right angles to the shore, the data shown in Figure 5 give at last a means of evaluating this elusive quantity. It is shown graphically in Figure 6, on which the following additional conclusions are based.

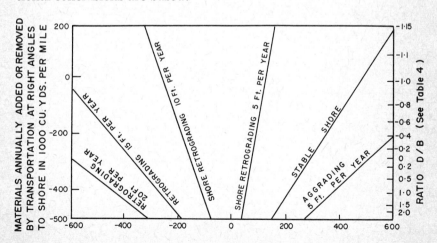

Figure 6. Amount of material annually added or removed by downshore transportation, in 1000 cu. yds. per mile.

4. The amount of material moved at right angles to the shore is a function of the ratio D/B, where D is the downshore change in volume of the offshore part of the spit, and B is the amount of material added or removed by lateral wave transport. Only where the ratio is less than −1.0 is there a net addition of materials to the shore by movement at right angles to it. In most sections there is a large annual net loss of material by direct seaward movement.

5. The net amount of material moved at right angles to the shore is comparable to that moved parallel to the shore.

6. Additions of material to the shore by direct landward movement occur mainly where the shore is retrogressing at a slow or moderate rate. Rapid removals of material from the shore may occur on aggrading shores as well as on rapidly retrograding ones.

Conclusion

Wave energy, the most significant factor in effecting shoreline change, can be evaluated from point to point along a shore using information obtained from wind records and hydrographic charts. Under favourable circumstances, it is also possible to ascertain the amount of material being transported in a given period of time by a given amount of energy expended parallel to the shore.

On Long Point the rate of shoreline change is correlated with the amount of material being added or removed by lateral wave transport, as well as with the downshore variation in volume of the offshore part of the spit. However, most parts of the shore are retrogressing due to large annual removals of material seaward.

Although it is beyond the scope of this paper to do so, the relationships discovered will permit a prediction of future changes on the shore of Long Point. For example, even a cursory examination indicates that the next breakthrough by waves across the spit will occur in the vicinity of reference point I and in about 200 years. It would appear likely, moreover, that the results of the study might well be applicable to other sand spits as well, provided the materials of which they are composed are similar to those of Long Point.

Finally, in pursuing a study of this type, one is struck by the limitless possibilities for further refinement in the method employed. This should not be taken, therefore, as the final word on the subject, but simply as the result of one stage in the search for a more complete understanding of beach morphology.

19

Notes on the Shoreline Recession Along the Coast of the Yukon Territory

J. R. Mackay

The 150-km.-long stretch of the coast between Herschel Island and the Mackenzie Delta terminates in bluffs cut into Pleistocene silts, sands, and gravels. No bedrock has yet been observed; if present, its occurrence must be very local. The coastal bluffs, which are up to 50 m. high, are constantly being undermined by waves and by the melting of numerous thick tabular ground-ice sheets lying close to sea-level. As the ice sheets are found only in fine-grained sediments, coastal retreat is especially rapid along silty to clayey bluffs. Similarly rapid recession has taken place along the northern coast of Alaska[1] and from the Mackenzie Delta east to Langton Bay[2]. It is the purpose of this note to describe some of the geomorphological and historical evidence for recession of the coast of the Yukon Territory. The field observations were made while the writer was carrying out studies for the Geographical Branch, Department of Mines and Technical Surveys, Ottawa.

King Point

In the late summer of 1905 Roald Amundsen with his crew in the *Gjøa* were hopeful of completing the Northwest Passage by sailing west around Alaska, but ice stopped them near King Point, where winter quarters were established. The residence was built an estimated 10 to 15 m. inland from the beach, the magnetic observatory about 5 to 10 m. During the winter a member of the expedition, G. Wiik, died and was buried in the magnetic

First published in *Arctic*, Vol. XVI (1963), pp. 195-97.

1 E. de K. Leffingwell, "The Canning River region, northern Alaska," *U.S. Geol. Surv. Prof.*, Pap. 109 (1919), p. 251; G. R. MacCarthy, "Recent changes in the shoreline near Point Barrow, Alaska," *Arctic*, 6 (1953), pp. 44-51.
2 J. R. Mackay, "The Anderson River Map-Area, N.W.T.," Ottawa: Geographical Branch, Mem. 5, pp. 137.

observatory. In 1908, 2 years after Amundsen's departure, Stefansson[3] photographed the ruins of the winter residence, which by then was nearly engulfed by the sea; by about 1911 every vestige had vanished. Furthermore, the Royal North West Mounted Police, at the request of the Norwegian Government, moved Wiik's grave inland in 1908, because it was in danger of falling into the water[4]. The R.N.W.M.P. moved the grave 230 m. inland and stated in its report that there was "now no danger of the sea encroaching on the present position of the grave". The respite gained was brief, because in about 1923 the grave marker was again moved inland, this time by some 140 m., and by another 100 m. in 1955. On July 13, 1957 the marker was only 7 m. from the edge of the coastal bluff and 20 m. from high tide level. According to these figures, cliff recession from 1906 to 1957 exceeded 450 m. As several of the distances were visually estimated, and some overlap in measurement may have occurred, the actual retreat may have been considerably less than the sum of the individual measurements. Even so, an annual rate of at least several metres may be safely inferred. A comparison of early photographs with the modern terrain adds corroborative evidence. A photograph taken in the winter of 1905-06 by Amundsen[6] shows the *Gjøa* anchored off King Point with Kay Point clearly visible in the background. This viewpoint could not be attained in 1957 until a position was occupied several hundred feet out to sea. In 1913 K. G. Chipman of the Canadian Arctic Expedition took a photograph at King Point.[7]. A comparison with identifiable features in 1957 showed that the total recession for the 50-m.-high sand and gravel hills in the background exceeded 20 m. The foreground of the photograph shows extensive slumping from the melting of ground-ice sheets 2 to 5 m. thick. It is quite apparent that Amundsen's winter residence of 1905-06 was built over a seaward extension of the ice sheets, the same holding true for the area from which Wiik's grave marker was removed in 1908 and again in subsequent years. At King Point coastal recession since 1913 in the area underlain by ground-ice sheets has been at least ten times as rapid as in the higher sand and gravel area without ice sheets.

3 V. Stefansson, "My life with the Eskimo," The Macmillan Co. (1951), figure facing p. 36.
4 G. L. Jennings, Royal North West Mounted Police Reports, Fort McPherson, 16th February 1910, Ottawa.
5 Personal communication from Supt. W. G. Fraser, R.C.M.P., Ottawa.
6 R. Amundsen, "The North West Passage," London: Archibald Constable Co., Vol. II (1908), p. 138.
7 Geological Survey of Canada, Photo. No. 39587.

Sabine Point

When Captain John Franklin explored the coast in 1826, 6 days were spent near Sabine Point, which formed the eastern extremity of a wide though not deep bay[8] (Fig. 1a). Amundsen had no difficulty in locating Sabine Point in 1905. Fig. 1b shows that there is now only a smooth coast with nothing even remotely resembling a "point"; it is also impossible to distinguish any "point" in sailing along the coast. Although Franklin's map may exaggerate the depth of the bay and the prominence of the adjoining points, major smoothing of the coast in the past 150 years is evident.

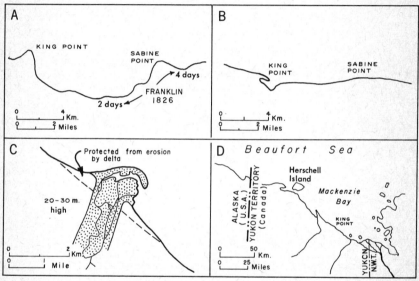

Figure 1. (a) Copy of Franklin's map of 1826 showing Sabine Point with the sites where he spent 4 and 2 days ashore. (b) The present coast of the area shown in (a) — note the absence of any promontories at Sabine and King points. (c) The seaward growth of the delta is believed to have partly protected the adjacent mainland from excessive erosion, while the unprotected areas have been cut back. Recession has been greater on the stormy exposed northwest side.

Kay Point

The earliest map[9] of Kay Point, dating from Franklin's survey in 1826, shows a 4-km.-long island about 5 km. northwest of Kay Point. Today only a shoal under several metres of water marks the site of the former island.

---

8 Capt. J. Franklin, "Narrative of a second expedition to the shores of the polar seas, in the years 1825, 1826, 1827," London: J. Murray (1828), p. 122.
9 Scott Polar Research Institute, Cambridge, England. Lefroy bequest, MS. 248.

Delta protection

Midway between Blow River and Shingle Point, just west of the Mackenzie Delta, an unnamed river has built its delta for 1 km. seawards from cut bluffs, which form the coast (Fig. 1c). The shape of the "mainland" coast suggests that the bluffs on either side of the delta have retreated about 300 m. since delta building began. Greater recession on the northwestern side is in agreement with the direction of storm waves.

Herschel Island

There is strong evidence to support the view that Herschel Island is a glacier ice-thrust feature whose profile, broken only by wave-cut bluffs, forms a continuous curve with a submarine basin[10]. The projected land profile, as shown in Fig. 2, reaches the sea floor at the 15-m. submarine contour. The total amount of postglacial coastal recession is estimated at 2 to 3 km. along the southeastern coast, with an average recession of about 2.5 km.

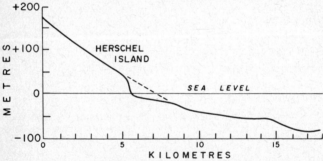

Figure 2. The wave-cut bluff at Herschel Island suggests a recession of 2 to 3 km. The generalized profile runs from northwest to southeast.

Conclusion

From the proffered evidence and additional examples that could be cited, it is apparent that recession of relatively great magnitude has taken place along the coast of the Yukon Territory in postglacial and historic times. The most rapid recession has been in low bluffs of fine-grained sediments with a high ice content; in such areas a rate exceeding 1 m. per year may occur. Coastal retreat in sand and gravel bluffs tends to be less rapid. A collation of Alaskan data with those from the Yukon Territory and the District of Mackenzie shows that much of the coast between Point Barrow, Alaska and Langton Bay, District of Mackenzie is receding rapidly.

---

10 J. R. Mackay, "Glacier ice-thrust features of the Yukon coast," *Ottawa: Geogr. Bull.*, 13 (1959), pp. 5-21.

20

Dynamics of Wind Erosion: 1. Nature of Movement of Soil by Wind

W. S. Chepil

The energy relationships between the air currents, commonly known as wind, and the soil are of great importance in the problem of wind erosion and its control. But, unfortunately, much of the information available in the past on the nature of air currents near the ground and on the importance of wind as a geologic agent in soil formation is too fragmentary and inadequate for wide application to the practical aspect of wind erosion control. Consequently, when research work on wind erosion of soil was begun at this laboratory, little aid was obtainable from the records of previous work, and it was found necessary to investigate the fundamental aspects of the problem, namely, the energy relationship between wind and soil.

The factors influencing wind erosion are numerous and hence add considerably to the complexity of the problem. The most important of these factors may be listed as follows:

I. Air
1. Velocity
2. Turbulence
3. Density, affected by
   a. Temperature
   b. Pressure
   c. Humidity
4. Viscosity

II. Ground
1. Roughness
2. Cover
3. Obstructions
4. Temperature
5. Topographic features

III. Soil
1. Structure, affected by
   a. Organic matter
   b. Lime content
   c. Texture
2. Specific gravity
3. Moisture content

First published in *Soil Science*, Vol. LX, no. 4 (1945), pp. 305-20. Research carried out at the Dominion Experimental Station, Swift Current, Saskatchewan. Reprinted by permission.

It is evident that the wind erosion problem depends on the mutual relationship of a combination of many factors. The influence of any factor that is involved in any condition may be negative or positive with regard to erosion, and in fact, one factor may counteract the influence of another in virtually any situation. For example, although it has been found[4] that wind turbulence increases erosion, yet the degree of erosion of a rough surface, where turbulence is more developed, is much less than that of a smooth surface over which the mean velocity of wind is greater. In both cases it is the net effect of the two opposing trends which determines the actual amount of erosion of the soil. Thus, in order that any condition may be understood and properly interpreted, it is essential that the individual factors involved be known and their relative significance accurately evaluated.

Briefly, the immediate problems for the soil conservationist are to determine the relative significance of the various factors influencing the movement of the soil material by the wind and to appreciate the physical nature of wind translocation. In order to answer some of the elementary questions connected with these problems, it was necessary to run a large series of experiments. The first of these had to do with the physical nature of soil drifting, the results of which are herewith reported.

Review of Literature

The movement of soil particles by the wind has been studied by a number of investigators. Free[6] asserted in his review of the literature on the problem up to 1911 that the largest proportion of the soil carried by the wind is moved in a series of short bounces called "saltation". He reported that sand never bounces high above the ground and in the desert cannot be felt by a person mounted on a camel. The smaller the soil particles, the greater is the influence of the wind upon it and the closer the approach of the path of saltation to a line parallel with the direction of the wind. The smallest quartz particles carried in saltation are about 0.1 mm. in diameter. Particles smaller than this have a velocity of fall lower than the upward velocity of the turbulent wind. Such particles are carried more or less parallel with the general direction of the wind, and form what has been termed a "suspension movement". They may be carried through the atmosphere for long periods of time and will fall to earth only with rain or after the wind has slackened considerably. Fine dust is thus often carried great distances from its original location.

In addition to flow in saltation and suspension, there is still another type of movement. Udden[7] observed that quartz grains larger than about 0.5 mm. in diameter and smaller than 1 mm. are too heavy to be transported through the air, but roll and slide along the surface of the ground. This type of movement is termed, by Bagnold[1], "surface creep". On the other hand, grains greater than 1 mm. in diameter are too large to be moved by ordinary erosive winds.[5]

Although the manner of the transport of sand by wind was understood in a general way many years ago, it was not until recently that a comprehensive study was made by Bagnold[1]. He asserted that dune sand is carried by wind mainly in saltation and surface creep and to a minor degree in true suspension. Bagnold concluded from theoretical calculations that with average dune sand the suspension flow, even under a relatively strong wind, does not exceed one twentieth of the flow in saltation and surface creep.

For arable soils, the only information available is a preliminary report from this laboratory[2], concerning the proposition of soil carried in different types of movement. Chepil and Milne[3] made measurements of the relative amounts of soil carried by wind over cultivated fields. The soil catchers used at that time were satisfactory for measuring rate of flow in saltation and surface creep, but were not entirely effective in trapping fine dust carried in suspension. It was observed that much soil was moved in saltation, but a very substantial proportion constituting the finest fraction was carried in true suspension.

In addition to information on the general mechanism of transport of soil by wind, data are herewith presented on the relative proportion of different types of movement on different soils with varying degrees of roughness of surface, and on the relative nature and intensity of soil movement as influenced by some of the major types of tillage treatments.

Experimental Procedure

In the development of experimental technique to study the quantity of soil transported by wind, it was found necessary to recognize the fundamental differences that exist between the transport of particles by saltation and surface creep and by movement in suspension. A fine grain of quartz dune sand when shot into the air at a speed of 9 miles per hour travels 20 cm. before its speed is reduced 50 per cent. Because of this continuing and directed motion, the rate of the flow of particles by saltation is determined easily by trapping the particles in narrow containers, open to windward. On the other hand, the measurement of the movement of fine dust is rather more complicated, for such particles are seriously affected by the wind's internal movements, that is, by the continued and instantaneous changes in its velocity and direction. Consequently, small particles do not enter the volume of still air contained in the trap, but instead are deflected from it and carried along with moving air. The procedure for measuring the rate of suspension flow is therefore much more complicated. In the first place, the volume of air in which dust is suspended must be determined accurately, and in the second, the dust itself must be filtered off completely and removed into a receptacle in which it may be weighed.

The method used in measuring the quantity of soil moving in saltation and suspension is as follows: Soil material carried by the wind enters a narrow box through a rectangular nozzle, ½ inch wide and 2 inches high, facing into the wind. The particles carried in saltation are trapped in the box and may be removed and weighed after detachment from the rest of the apparatus at a stoppered joint and removal of a tight-fitting cover. The particles in suspension, however, are carried into the trap by suction created by a vacuum pump and collected in a 6-inch column of distilled water in a cylinder. Air intake through the apparatus is measured with a gas meter and its volume corrected on the basis of atmospheric pressure with the aid of a mercury barometer. More than one dust-catching unit can be connected by rubber tubing to a vacuum flask. This type of arrangement facilitates simultaneous measurement of soil flow at various positions. The vacuum pump was driven by a small gasoline motor.

The amount of dust trapped was determined by evaporating the water and weighing the residue. The relative concentration of particles carried in suspension at any height was determined by dividing the weight of the residue by the corrected volume of filtered air. Suspension flow was determined by multiplying concentration per unit volume by velocity of the wind. This was done on the assumption that the velocity of particles carried in suspension was equal to that of the wind. Photographs and indirect measurements substantiated this assumption.

In addition to measurements of soil flow at different heights, the total amounts of soil carried in different types of movement were determined. Total saltation and surface creep were determined by the method of Bagnold,[1] and the suspension flow was ascertained by subtracting the flow in saltation and surface creep from the total flow as determined by the difference in the weight of thoroughly air-dry soil before and after exposure to an erosive wind. These measurements were made both in the open field and in a portable field tunnel, described previously[2], and on different types of soil with varying degrees of roughness of surface.

In order to determine the nature of soil movement under different conditions of soil and wind, photographs were taken of the paths of moving soil particles. Sunlight admitted vertically through a lens in the ceiling of a darkened wind tunnel was used for illumination. The lens was 25 cm. long and 12 cm. wide and had a focal length of approximately 2 feet. It produced a very intense vertical beam of light, which illuminated an area, 25 cm. long and 1 cm. wide, parallel with the direction of the wind. Particles flying through the illuminated space reflected light distinctly, and their paths appeared on a photographic plate as silvery threads, or as a series of dots or dashes against a black background. Photographic exposures of 1/25 and 1/10 of a second were made for the purpose of indicating the shape and length of grain paths and 1/100 and 1/50 of a second for determining the speed of the grains through the air.

Results

Nature of wind translocation

Photographs and direct visual observation indicated that all soils were carried by wind in the three types of movement already mentioned. The relative proportion of each type of movement varied greatly for different soils.

The greatest proportion of the movement in all cases examined was by particles in saltation. After being rolled by the wind, the particles suddenly leaped almost vertically to form the initial stage of the movement in saltation. Some grains rose only a short distance, others leaped 1 foot or more, depending directly on the initial velocity of rise from the ground. They also gained considerable forward momentum from the pressure of the wind acting upon them, and acceleration of horizontal velocity continued from the time grains began to rise to the time they struck the ground. In spite of this acceleration, the grains descended in almost a straight line invariably at an angle between 6 and 12 degrees from the horizontal. On striking the surface they either rebounded and continued their movement in saltation, or lost most of their energy by striking other grains, causing these to rise upward and themselves sinking into the surface or forming part of the movement in surface creep. Irrespective of whether the movement was initiated by impact of descending particles or by impact of rolling grains, the initial rise of a grain in saltation was generally in a vertical direction.

The cause of this vertical rise was not at all apparent. It was found that soil grains jumped vertically off a smooth surface, such as a wooden floor, after rolling for as short a distance as 2 cm., although there were neither soil grains nor other obstructions against which they could strike and rebound into the air. The probable cause of the vertical rise was thought to be a direct impact of a facet of an irregularly shaped grain against the tunnel floor. From a theoretical point of view, however, if no other impact forces are involved, the angle with which a descending grain would rebound from a smooth horizontal surface should be equal to the angle of descent, which was 6 to 12 degrees. Actually, the angle of ascent was between 75 and 90 degrees in the majority of cases. This seemed incredible unless it was presumed that the vertical rise was due to some force other than the force of impact of the grain against the surface.

The only logical explanation of the vertical rise of particles in saltation appears to be by the theory of the Bernoulli effect. This effect is apparently due to two causes, the spinning of the grains and the steep velocity gradient near the ground. Actual photographs indicate that grains carried in saltation rotate at a speed of 200 to 1,000 revolutions per second. The photographs show clearly that about 50 per cent of the grains carried in saltation are spinning, while another 25 per cent exhibit relatively indistinct rotation. It is possible, however, that more than 75 per cent of the grains were rotating but that this could not be indicated without greater intensity of illumination. Nor was it possible to indicate rotation for grains approaching a spherical shape, since photographs merely show the variation in the intensity with which the reflecting facets of the grain were illuminated. The more angular the grain, the more distinctly rotation appeared on the plate.

On account of a rapid clockwise spinning of the grain, the air near the grain surface is carried around it. On the lower side, assuming the wind to be traveling from left to right, this air is moving against the direction of the wind; on the upper side, it moves with it. The velocity of the air at any point near the grain is thus made up of two components, one due to the wind and the other to the spinning of the grain. On the upper side these components have the same direction, whereas below they have opposite directions. Thus, the velocity is greater at the top surface than at the bottom, and according to the Bernoulli theorem, the pressure is decreased at the top and increased at the bottom, and so the grain tends to rise.

The Bernoulli effect is further intensified by virtue of the fact that a steep velocity gradient exists near the ground. At a threshold wind velocity, that is, a velocity just high enough to initiate the movement in saltation, the following is a typical case of average velocities encountered near a smooth ground surface:

Height, $mm.$	0.02	0.1	0.2	0.3	0.6	0.9	1.2	1.5	1.8
Velocity, $cm./sec.$	0	104	152	189	234	254	272	297	304

The variation in air velocity near the ground causes a substantially higher rate of air flow at the upper than the lower surface of the grain of, say, 0.2 mm. in diameter, and this difference in velocity may be expected to produce a similar and additive effect to that caused by the spinning of the grain in a current of air. Consequently, if the total difference in the pressure between the upper and the lower surfaces is greater than the force of gravity acting downward, the grain will rise in a vertical direction.

As no experimental data are available concerning a sphere the size of a sand grain spinning in an air current near the ground, it is impossible to confirm the existence of a vertical component of wind force as postulated above. Experiments are now being undertaken to obtain definite information on this problem.

After being shot into the air, the grains rose to various heights and, because of force of gravity, fell at an accelerating velocity. There was at the time a horizontal acceleration of the falling particle due to the forward pressure of the wind upon it. Photographs indicate that the downward and the forward accelerations were approximately equal, and the inclined path of the falling grain was therefore almost a straight line. Only a slight curvature downward was observed in most cases. The angle of descent varied but little and, as already pointed out, was between 6 and 12 degrees from the horizontal. Smaller grains descended at a somewhat smaller angle than the larger grains. The angle of descent did not vary greatly with wind velocity, but the higher the wind the greater the height to which some grains rose in the air, and hence the corresponding longer average path.

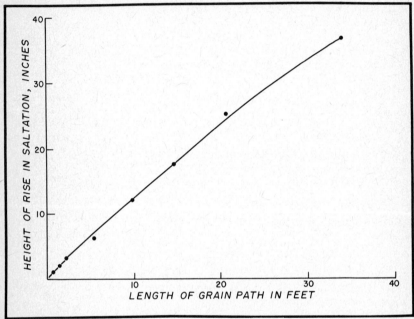

Figure 1. Relation of height of rise in saltation to length of grain path (horizontal plane).

On the whole, the horizontal distance through which the grain continued to rise was about one fifth to one fourth of the total horizontal length of a single leap in saltation.

On a smooth surface consisting only of erosive grains, there was a remarkable consistency in the shape of grain paths. The shape of the paths was much less regular over a rough surface, but the general character of the movement in saltation remained the same. Irrespective of the degree of surface roughness, a close relationship existed between the average height to which the grains rose in the air and the horizontal length of grain leap. This relation is indicated in Figure 1. The ratio of height of rise to the horizontal equivalent of length of grain leap was about 1:7 for a rise up to 2 inches in height, 1:8 for a rise of 2 to 4 inches, 1:9 for 4 to 6 inches, and 1:10 for heights above 6 inches. The results were the same for all soils used in the investigation.

The surface creep of soil grains could not be recorded photographically but was plainly visible to the naked eye. The grains in surface creep were too heavy to be moved by the direct pressure of the wind, but derived their kinetic energy from the impacts of smaller grains moving in saltation. Grains in saltation, on the other hand, received most of their impact energy from the direct pressure of the wind. It is evident that the movement of soil by wind is dependent, not so much on the force of the wind acting on the surface of the ground, as on the velocity distribution to such height as the grains rise in saltation. This height is definitely limited, and it may be concluded, therefore, that wind erosion is mainly a surface phenomenon and is not directly dependent on the condition of the wind above that restricted distance.

Movement of dust in suspension

The results of measurements made in the open field showed that the mechanism of transport of soil by wind is very similar to that of dune sand described by Bagnold[1], except that in addition to movement in saltation and surface creep there was, in some cases, a substantial proportion of the soil carried in true suspension. The presence of fine dust, even in very appreciable amounts, did not seem to affect the nature of the movement in saltation or surface creep, but it greatly influenced both the threshold wind velocity and the intensity of erosion for a given wind.[5]

Once lifted off the ground, the particles in suspension were completely borne up by the wind. In the open country they usually reach great heights and do not fall to the ground except with rain or after the wind has slackened considerably.

The mechanism by which fine dust is lifted off the ground is entirely different from that of saltation. A previous study[5] showed that samples of soil composed only of fine dust particles were extremely resistant to erosion by wind. In fact, quartz particles less than 0.05 mm. in diameter could not be moved by wind velocities as high as 37 miles per hour at a 6-inch height. In mixtures with coarser grains ranging up to 0.5 mm., however, these particles moved readily, and the threshold velocity of the mixtures was lowered very considerably. It may be said, therefore, that movement of fine dust in an air current is mainly the result of movement of grains in saltation; hence, without saltation movement, dust clouds would not arise, except on a relatively limited scale as a result of disturbance by moving vehicles, animals, etc.

The relative quantities of fine dust particles blown off different soils was observed from the photographs. The suspension flow, though clearly visible to the naked eye, was indicated by very faint lines on a photographic plate. The difficulty is attributed to the relatively small diameter of the particles and to their high speed, which is approximately that of the wind. A very sensitive film and light much brighter than sunlight are essential for a clear indication of suspension flow.

In spite of these difficulties, the relative concentration of suspended dust is plainly indicated for different types of soil. Photographs indicate almost no suspension flow over Sceptre heavy clay, an appreciable concentration over Hatton fine sandy loam, and a concentration so dense over Haverhill loam as to mask the appearance of movement in saltation. Wind velocity in all cases was about 17 miles per hour at a 12-inch height.

In contrast to the movement of grains in saltation, the movement of fine dust in suspension, after it has been lifted off the ground, is completely governed by the characteristic movement of the wind. The influence of an eddy on suspended dust was indicated clearly in the photographs, which were taken at the time of a gentle movement of air following a stronger wind that initiated the movement of saltation. The back-eddy of air currents was shown plainly by the characteristic paths of suspended particles and almost duplicated a diagrammatic representation of wind structure previously recorded over the same type of surface with the aid of sensitive oscillating plates.[4]

Concentration of wind-borne particles at different heights

Measurements of the concentration of wind-borne particles in a portable field tunnel and also in the open field indicated that most of the soil movement in saltation was carried below the height of 2 or 3 feet. In fact, over 90 per cent of the soil was transported below the height of 12 inches, and this was found to be true for several widely different soils chosen for investigation.

The results of this study obtained in the open field under various conditions of soil, wind velocity, and surface roughness are given in Table I. In general, the movement in saltation was the same on all soil types investigated, but some variation was found in the proportion of grains carried at different heights. Coarsely granulated soils, such as Sceptre heavy clay, drifted closer to the ground than the more pulverized Haverhill loam, but the difference was not appreciable.

TABLE I. RELATIVE QUANTITIES OF SOIL CARRIED IN SURFACE CREEP AND SALTATION UNDER DIFFERENT CONDITIONS OF SURFACE ROUGHNESS AND WIND.

WIND VELOC-ITY AT 12-INCH HEIGHT	SOIL TYPE	SMOOTH SURFACE							RIDGED SURFACE*						
		Surface creep	Saltation carried below the height of						Surface creep	Saltation carried below the height of					
			1 inch	3 inches	6 inches	12 inches	24 inches	36 inches		1 inch	3 inches	6 inches	12 inches	24 inches	36 inches
m.p.h.		mgm./cm. width/sec.	%	%	%	%	%	%	mgm./cm. width/sec.	%	%	%	%	%	%
17	Sceptre heavy clay	15.9	27	66	91	98	....	....	1.9	20	54	74	96	99+	....
	Haverhill loam	3.1	24	63	86	97	99+	....	0.4	18	47	72	95	99+	....
	Hatton fine sandy loam	2.8	30	65	88	98	99+	....	0.5	21	50	75	97	99+	....
25	Sceptre heavy clay	142.1	28	56	77	91	99	99+	4.5	25	55	76	93	98	99+
	Haverhill loam	9.2	24	54	75	90	97	99+	1.2	18	45	72	95	98	99+
	Hatton fine sandy loam	3.3	29	63	83	95	99	99+	0.6	20	47	73	95	99	99+

*Ridges were 2.5 inches high, 9 inches wide, running at right angles to the wind.

There were large differences in the rate of surface creep, which was particularly high on coarsely granulated Sceptre heavy clay and least on finely granulated Hatton fine sandy loam. Roughly, the amount of surface creep depended on the quantity of erosive grains greater than 0.5 mm. in diameter for cultivated soils and over 0.25 mm. for dune sand.

The relative concentration of wind-borne particles at different heights above a rough surface differed widely from that above a smooth one. Figures 2 and 3 show typical differences in relative concentration of drifting soil over the two types of surface — wind velocity in both cases being the same at 12-inch height. The ridges had a marked effect on lowering the total rate of soil flow and virtually eliminated surface creep. Furthermore, many of the coarser granules that moved in saltation over a smooth surface were apparently trapped by the ridges, thus lowering appreciably the relative concentration of particles near the ground.

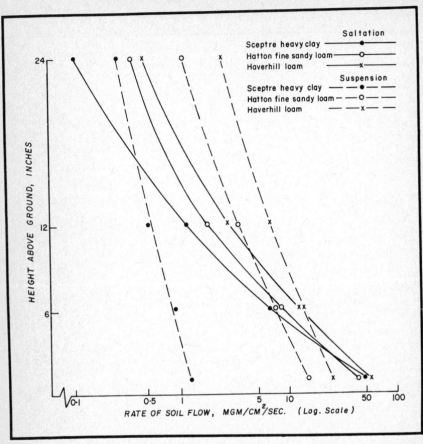

Figure 2. Distribution of wind-borne particles carried in saltation and suspension at different heights above a smooth surface.

The relative concentration of soil particles at different heights remained the same under a wide range of wind velocity; hence the graphs presented in Figures 2 and 3 give only averages for a range of wind velocity from 13 to 30 miles per hour at 12-inch height. The data show that the ratio of saltation to suspension decreased rapidly with height above the ground, as would be exacted. As grains in saltation do not generally rise higher than several feet above the ground, the soil carried above this height is in true suspension and capable of being carried to great heights and over long distances from the original location.

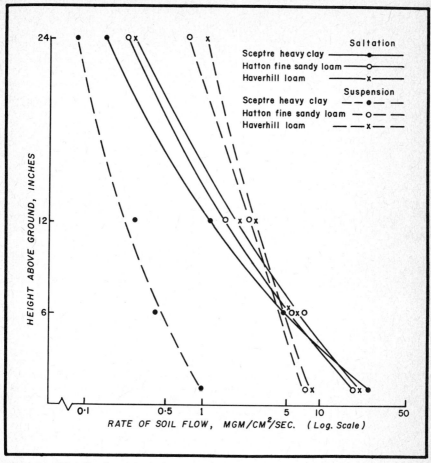

Figure 3. Distribution of wind-borne particles carried in saltation and suspension at different heights above a ridged surface.

Figures 2 and 3 indicate further that virtually a straight-line relationship exists between height and the logarithm of the rate of flow in saltation and suspension over all soils investigated. There seems to be some functional relationship between the rate of soil flow and height, for nearly all of the experimental values fall very closely to curves of the same characteristic shape.

Proportion of different types of flow on different soils

The relative rates of soil transport in each of three types of movement varied widely on different soils (Table II). The proportion moved in surface creep constituted 7 to 25 per cent of the total flow, the lowest rate being on highly pulverized Haverhill loam, the highest on coarsely granulated Sceptre heavy clay. Flow in saltation varied much the same way as surface creep and composed 55 to 72 per cent of the total flow, depending on soil type. The proportion of flow in suspension constituted about 3 per cent of the total flow on Sceptre heavy clay and as much as 38 per cent on typically dusty Haverhill loam and was approximately equal to the proportion of particles smaller than 0.1 mm. found in the soil. The data in Table II indicate that some particles larger than 0.1 mm. must have been carried in suspension, probably those of lowest specific gravity. As the percentage of particles smaller than 0.15 mm. in all soils was substantially greater than the percentage carried in suspension, it is evident that only a small proportion of the size above 0.1 mm. was carried in true suspension.

TABLE II. RELATIVE PROPORTION OF THREE TYPES OF FLOW ON DIFFERENT SOILS.*

SOIL COVER	SOIL REMOVED IN		
	Saltation *per cent*	Suspension *per cent*	Surface creep *per cent*
Sceptre heavy clay	71.9	3.2	24.9
Haverhill loam	54.5	38.1	7.4
Hatton fine sandy loam	54.7	32.6	12.7
Fine dune sand	67.7	16.6	15.7

*The size distribution of erosive grains in the soils was as follows:

	DISTRIBUTION OF PARTICLES				
SIZE OF PARTICLES .... *mm.*	0.83-0.48	0.42-0.25	0.25-0.15	0.15-0.1	0.1
	per cent	*per cent*	*per cent*	*per cent*	*per cent*
Sceptre heavy clay	33.5	46.1	14.9	4.0	1.5
Haverhill loam	13.3	17.3	15.1	22.3	32.0
Hatton fine sandy loam	1.1	6.0	26.4	40.5	26.0
Fine dune sand	0.1	2.1	54.2	35.6	8.0

There was a constant proportion of the three types of flow throughout the wide range of wind velocity used in the investigation, and on this account the values in Table II indicate only the average results obtained for many individual cases. Measurements were made on a level surface on which the surface projections in the form of clods and surface ripples did not exceed 0.75 inch in height.

Increasing the roughness of surface caused a proportional reduction in the rate of movement in surface creep. A rough surface, such as that composed of cultivator ridges extending at right angles to the direction of wind, trapped most of the surface creep but failed to reduce the movement in saltation and suspension to quite the same degree.

Relative efficiency of cultural treatments in trapping drifting soil

The data in Table II gives some idea of the relative amounts of soil that are carried away by the winds. Dust in suspension is transported far and wide; hence the regions in which it is deposited benefit but little in the way of additional soil, but the much more limited eroded area loses a great deal. In saltation and surface creep, on the other hand, soil is not usually carried far and is deposited in or near the vicinity of the affected area. Many cultural and cropping methods are devised to trap the grains in saltation and surface creep to prevent the spread of erosion to surrounding unaffected areas. It has been pointed out that surface creep may be almost eliminated and saltation greatly reduced as a result of ridging a highly erosive soil. The higher the ridges the more effective they are in stopping surface creep and saltation, but as movement in suspension and surface creep is dependent wholly on movement in saltation, the elimination of saltation will eliminate all other forms of movement. The whole program of wind erosion control is therefore based on reduction or elimination of movement in saltation.

Soil ridges may be used to reduce or eliminate saltation. It is often preferable to ridge the whole of the affected area, but this treatment is not always possible, and ridging narrow strips at regular intervals across the field is often resorted to. Stubble and crop strips may be used for a similar purpose.

The width of trap strip that may be required depends partly on the length of jump of grain in saltation and partly on the trapping capacity or receptiveness of the surface. Standing grain stubble is probably the most effective form of trap, for it will trap all the soil moving into it in saltation. It can therefore be considered as fully receptive. The minimum width of stubble strip, however, would have to be not less than the maximum horizontal length of a single leap of the grain in saltation. The data in Table I and Figure 1 combined supply complete information on the approximate percentage of soil grains that may be trapped by any width of a wholly receptive trap strip.

Thus, supposing the width of a strip of short stubble to be 10 feet, Figure 1 indicates that to jump the distance the grain in saltation would have to rise to a height of approximately 12 inches. Furthermore, it is indicated in Table I that between 90 and 98 per cent of the flow in saltation is below this height; hence a 10-foot strip of short stubble may be expected to trap between 90 and 98 per cent of the flow in saltation, depending on soil type and wind velocity. It can be found in like manner that a 2-foot stubble strip will trap about 50 per cent of the flow in saltation and a 30-foot strip over 99 per cent. These values corroborate the results obtained from actual practice in the field.

The data indicate further that the effectiveness of a trap varies somewhat with wind velocity. Thus, it is shown that in order to trap 99 per cent of the movement in saltation a 20-foot strip of totally receptive surface would be required for a 17-mile-per-hour wind, and a 35-foot strip for a 25-mile-per-hour wind.

There are other angles to be considered in deciding on the width of trap strip that would be most effective. The next of these considerations is the height of the stubble. A 3-foot width of 6-inch stubble, for example, will trap on an average about 85 per cent of the total flow in saltation, but a 3-inch stubble will trap about 60 per cent.

The minimum width of trap required depends also on the capacity to hold the blown soil. Long stubble, in addition to being more effective in trapping the encroaching drift, has a greater holding capacity. Other factors being equal, the holding capacity varies directly with the height of the stubble. The effectiveness of the trap is reduced to zero as soon as the trap has reached its holding capacity. Hence, to be fully effective, the minimum width should be that required to store the encroaching drift, plus that width to leeward that will remain relatively free to act as an effective trap. The width that may be required to store the encroaching drift cannot be estimated with any degree of accuracy, for the amount of erosion is dependent to a large measure on the conditions of the weather, which cannot be predicted. Hence, a considerable margin over and above the minimum requirement must be allowed.

Ridges were found to have a lower trapping capacity than the standing grain stubble, for many of the grains bounced off the ridges one or more times before they were finally trapped in the furrows. The trapping capacity of ridges depends on their size. Ridges 2.5 inches high and 9 inches wide were about 50 per cent as effective as a 6-inch wheat stubble, whereas cultivator ridges 5 inches high and 18 inches wide and lister furrows 12 inches deep were 77 to 83 per cent and 85 to 92 per cent as effective, respectively.

The foregoing results were obtained on a highly nonerosive clay soil that was exceedingly resistant to the grinding action of flying grains. Many soils do not exhibit such marked resistance, for the ridges may wear down rapidly and lose much of their sheltering effect. The trapping capacity of the grain stubble, on the other hand, is not at all affected by abrasion.

Conclusions

Because of rapid spinning of the grains moving in saltation and a steep velocity gradient, there appears to be a considerable vertical component of wind force near the ground. On account of these effects the grains rise steeply and descend very obliquely toward the surface. As the downward acceleration, due to gravity, and the forward acceleration, due to wind pressure, are approximately equal, the grains fall in almost a straight line. They strike the surface at an angle of 6 to 12 degrees.

Movement in suspension and in surface creep is a result of movement in saltation. The whole program of wind erosion control should therefore depend on methods designed to reduce or eliminate saltation.

The intensity of soil movement depends not so much on the force of the wind acting on the ground, as on its pressure against the grains as they leap in saltation. Soil movement is therefore dependent not on velocity at any fixed height but on the velocity distribution to the height of saltation.

Dust in suspension does not affect the general character of the movement in saltation or in surface creep, but the presence of dust in the soil increases the minimum velocity required to initiate erosion and decreases the intensity of erosion for a given erosive wind. Once lifted off the ground, fine dust is carried to great heights and distances from its original location and thus may be considered a complete loss to the eroding area. The soil moved in saltation and surface creep, on the other hand, usually remains within the eroding area, especially when the erosive winds are from different directions. The maximum diameter of soil particles carried in suspension is on an average slightly greater than 0.1 mm.

The proportion of the three types of movement varies widely for different soils. In the cases examined, between 55 and 72 per cent of the weight of the soil was carried in saltation, 3 to 38 per cent in suspension, and 7 to 25 per cent in surface creep. Coarsely granulated soils erode mainly in saltation, and finely pulverized soils, in saltation and suspension.

The trapping capacity of stubble or ridged strips depends on the relative receptiveness of the surface and the length of jump of particles in saltation. The data presented supply information on the approximate percentage of soil grains that may be caught by trap strips of different widths.

The effectiveness of a particular type of trap depends also on the height and density of the obstructions and on the resistance of these obstructions to the abrasive action of wind-borne grains. Soils vary greatly in resistance to abrasion, but grain stubble is virtually unaffected.

---

1 R. A. Bagnold, "The Physics of Blown Sand and Desert Dunes," Methuen & Co. Ltd., London (1941).
2 Canada Department of Agriculture, Soil Research Laboratory 1943 Report of Investigations, Swift Current, Sask.
3 W. S. Chepil and R. A. Milne, "Comparative study of soil drifting in the field and in a wind tunnel," Sci. Agr., 19 (1939), pp. 249-57.
4 W. S. Chepil and R. A. Milne, "Wind erosion of soil in relation to roughness of surface," Soil Sci., 52 (1941), pp. 417-31.
5 W. S. Chepil, "Relation of wind erosion to the dry aggregate structure of a soil," Sci. Agr., 21 (1941), pp. 488-507.
6 F. E. Free, "The movement of soil material by the wind," U.S. Dept. Agr. Bur. Soils Bul., 68 (1911).
7 J. A. Udden, "Erosion, transportation, and sedimentation performed by the atmosphere," Jour. Geol., 2 (1894), pp. 318-31.

21

Aeolian Processes in Arctic Canada

J. B. Bird

The importance of wind as an erosive agent under periglacial conditions, when compared with other geomorphic processes, has not been established. Antevs (1928)[1] in a discussion of Iceland geomorphology considers that wind is the most important periglacial process, and Tricart (1955)[2] credits deflation with a greater transporting role than running water in periglacial areas. A more conservative view is held by Peltier (1950)[3] who believes that wind has a secondary role that increases in the later part of the periglacial cycle.

Erosional Landforms: Wind abrasion has been reported from many circumpolar lands. It is common in Iceland where there are wind-faceted basaltic boulders, and where fluting and smoothing on volcanic breccias result from wind action.[1] Similar forms are found in east Greenland where ventifacts and rock bosses have been etched by sand blast (Flint *in* Boyd, 1948)[4]. Near the mouth of the Kolyma River, northeast of Siberia, Sverdrup (1938)[5] found granite pillars that had been drastically reduced in diameter 4 m above the base by wind erosion. A rather similar gneiss mush-

Reprinted from *The Physiography of Arctic Canada*, by J. B. Bird. Johns Hopkins Press (1967), pp. 237-41.

[1] E. Antevs, "Wind deserts in Iceland," *Geog. Rev.* Vol. XVIII (1928), pp. 675-76.
[2] J. Tricant, "Géomorphologie climatique," Vol. II, Le modèle des pays froids: a modèle péreglaciaire, n.d. (1955), 267 pp. in *Cours de Géomorphologie*, Paris, Vol. III.
[3] L. C. Peltier, "The geographic cycle in pereglacial regions as it is related to climatic geomorphology," *Am. Assoc. Am. Geog.*, Vol. XL (1950), pp. 214-36.
[4] R. F. Flint, "Studies in glacial geology and geomorphology in Boyd, L. A., 'The coast of North East Greenland'," *Am. Geog. Soc. Sp. Pub.*, 30 (1948), pp. 91-210.
[5] H. V. Sverdrug, "Notes on erosion by drifting snow and transport of solid material by sea ice," *Am. J. Sci.*, Vol. XXXV (1938), pp. 370-3.

room pillar is depicted by James Ross (1835)[6] near Middle Lake on Boothia Isthmus, although in this case wind abrasion is unlikely to be the cause.

In arctic environments which have heavy snowfalls and moderate temperatures there is a continuous winter snow cover and the ground may be protected from wind action for more than half the year. In contrast, under arid continental conditions ridges and larger boulders are exposed for much of the winter, although they may gain some protection from a coating of ice. Sand particles, the most usual abrasive material, are rarely exposed on the ground in winter, and for abrasion to occur an alternative material must be present. A few minutes in strong snowdrift at low temperatures is sufficient to convince any geomorphologist that wind-driven snow is abrasive material; the hardness of ice particles increases as the temperature decrease (Teichert, 1939, 1948,[7] Blackwelder, 1940)[8] Ice has a Moh hardness of 2 at freezing point, 4 at —44°C., and about 6 between —50 to —80°C.; at temperatures in the lower range orthoclase feldspars are attacked, and such temperatures are not unknown in many parts of the Arctic. In northern Canada, particularly west of Hudson Bay, the winter winds are remarkably strong and constant for long periods; at Chesterfield Inlet northwest winds blow for two-fifths of the time in January, with a mean velocity of 8.7 m/sec, and exceptionally may reach 36 m/sec.

Detailed observations on wind erosion have been made in Peary Land where wind-borne particles polish dolerite and pitted sandstone; maximum erosion often occurs during the winter on the upper part of boulders exposed above the snow. Small boulders in tent rings, probably about 1,000 years old, have been reduced to half their diameter by wind action (Fristup, 1952-53)[9]. In the Frønlunds Fjord area of Peary Land, however, a controlled investigation failed to find abrasion in rocks of Moh hardness 3 after six months: softer diatomaceous brick earth was rather strongly eroded in the same period (Troelsen, 1952).[10]

6 J. Ross "Narrative of a second voyage in search of a North-West Passage," London (1835), p. 740.

7 C. Teichert, "Corrasion by wind-blown snow in polar regions," *Am. J. Sci.*, Vol. 237 (1939), pp. 146-48; "Corrasion by drifting snow," *J. Glac.*, Vol. I (1948), p. 145.

8 E. Blackwilder, "The hardness of ice," *Am. J. Sci.*, Vol. 238, (1940), pp. 61-62.

9 B. Fristrup, "Wind erosion within the arctic deserts," *Geog. Tidsster*, no. 52 (1952-53), pp. 51-65.

10 J. C. Troelsen, "An experiment on the nature of wind erosion," *Dansk. Geol. Foren. Medd.*, Vol. XII, pt. 2 (1952), pp. 221-22.

Evidence for active wind abrasion in northern Canada is not great. At most places where observations have been made, notably on Victoria Island, south of Coronation Gulf, on Ellesmere Island, and in northern Quebec, wind abrasion is slight and often non-existent (Jenness, 1952a;[11] Washburn, 1947;[12] Robitaille, 1959).[13] The wind has undoubtedly some erosive power, as may be seen when the lichen-free exposed sides of boulders are contrasted with the opposite face that is encrusted with lichens. Few studies have been made of lichen development on boulders, and it is not clear whether these differences are due solely to abrasion or whether the dehydrating action of the wind contributes. In spite of the generally negative evidence, close attention to the work of the wind discloses some signs of erosion. Ventifacts have been found in sandstone boulders north of the Gifford River, northwest Baffin Island (Marsden, pers. comm.),[14] in southern Bylot Island (Drury, 1962),[15] and in the western Arctic in the vicinity of Franklin and Darnley bays, where abrasion is active in relatively reisistant rocks, such a gabbro, as well as in softer rocks including limestone. Boulders have lost several inches from one side, although the process must be slow as no changes have occurred in boulders over a thirty-year period (Mackay, 1958b).[16] Farther west the sandstones forming the Richardson Range front have been sculptured by wind action. The small number of ventifacts is remarkable when it is recalled how widely they have been reported from Pleistocene periglacial areas. More detailed field studies may reveal more. In general, however, it must be concluded that wind erosional features are of minor significance in the scenery of northern Canada.

Deflation and Deposition

Deflation in polar regions probably attains a maximum in the vicinity of permanent ice sheets, especially if the ice margin is in

[11] J. L. Jenness, "Erosive forces in the physiography of western Arctic Canada, *Geog. Rev.*, Vol. XLII (1952), pp. 238-52.
[12] A. L. Washburn, "Reconnaissance geology of portions of Victoria Island and Adjacent Regions, Arctic Canada, *Geol. Soc. Am. Mem.*, no. 22 (1947), p. 142.
[13] B. Robitaille "Aperçu géomorphologique de la rive Québécoise du Detroit d'Hudson", *Rev. Can. Géog.* Vol. XIII, (1959), pp. 147-54.
[14] Marsden, personal communication.
[15] W. H. Drury, "Patterned ground and vegetation on Southern Bylot Island, N.W.T. Canada," *Harvard V. Gray Herb. Contributions*, no. 190 (1962), 111 pp.
[16] J. R. Mackay, "The Anderson River Map Area N.W.T.," *Geog. Br. Mem.*, 5 (1958), p. 137.

retreat; outwash is widespread and vegetation is sparse due to soil instability, glacial meltwaters, and climate.

In arctic North America there is little recently deglaciated land and active valley trains are restricted to parts of Alaska and small valleys in northeastern Canada. Deflation from these areas is necessarily restricted. Throughout the Arctic the density of the plant cover is the determining factor in deflation and there is little doubt that the wind transports quantities of fine material when the ground is unprotected by vegetation. Transportation is not confined to silts, and sands and coarse material may be moved, particularly when the ground is frozen. In Darnley Bay, wind-rolled pebbles have been found up to three miles offshore on the sea ice (Kindle, 1924).[17] Washburn (1947)[18] was told by Eskimos that mud balls were formed and rolled by the wind in the same area.

Several terrains in northern Canada are sources of fine wind-blown particles. The most widely distributed are the clay and silt-rich till terrains in Mackenzie District and the marine silts of the western arctic lowlands. Deflation is normally restricted because of the dense cover of hillock tundra, but where there is active erosion, such as in gullies, and in areas where frost heaving is powerful, it becomes important. There is deflation from moraines and outwash deposits that have a partial vegetation cover. The great sand sheet of central Victoria Island, the sand plains north of middle Back River, and the moraines between Great Bear Lake and Amundsen Gulf are in this category.

In the coastal areas of the western Arctic, deltas (including some that are elevated) of the larger rivers westward from Pelly Bay are a major source of fine-grained material. These include the deltas of Arrowsmith, Hayes, Western, Burnside, Hood, Coppermine, Rae, Hornaday, and Horton rivers. On many of these deltas winds of half gale force are sufficient to raise clouds of dust 200 m. (650 ft.), or more, into the air. Dust is particularly prominent around Burnside delta where strong dry winds, blowing off the plateau into the Bathurst trench, are common in spring. By the end of May the river may have flooded, swept the delta clear of

---

17 E. M. Kindle, "Observations on ice-borne sediments by the Canadian and other Arctic expeditions," *Am. J. Sci.* Vol. VII (1924), pp. 251-86.

18 A. L. Washburn, "Reconnaissance geology of portions of Victoria Island and Adjacent Regions, Arctic Canada," *Geol. Soc. Am. Mem.*, no. 22 (1947), p. 142.

snow and ice, and subsided again. At this time the delta surface is exceptionally vulnerable to wind action, and after a strong wind the snow and sea ice for many miles around are brown with a layer of fine sand and silt.

Other potential sources of wind-blown sediments include alluvial terraces and weathered rock outcrops. Even in areas where there are few sources of wind-blown sediments there is still some blown dust; late summer snowbanks in all parts of the Canadian Arctic contain visible mineral particles that have been blown onto the snow. When the snow melts a thin muddy film is left on the rocks.

Small gullies, particularly along terrace edges, trap wind-blown snow and mineral particles. If there is no longer a stream in the gully it often fills with beds of perennial "ice" separated by layers of sand. An extreme example was seen by the author at Mathiassen Brook on Southampton Island where the middle reaches are in a deep valley partly filled with outwash sands and terraces. Vegetation is not continuous and sand dunes are numerous. Blown sand and snow accumulate in gullies and at the foot of terrace bluffs; as the buried snow melts a hummocky small-scale thermokarst develops. Other examples have been noted on Somerset Island, in the Thelon Basin and on the Contwoyto upland.

When deposits of mixed particle-size are subjected to wind action there is a selective process which removes the finer particles and leaves a residual pebble and gravel concentrate that acts as a protective armour. This is common on elevated deltas, stream terraces, and on the crests of eskers in many parts of the Arctic.

Landforms produced entirely by deflation are few. The most widespread are shallow depressions, rarely more than three meters across, from which about 1 cm., and never more than a few cms., of fines have been blown away to leave a stony, and generally vegetation-free, residue. These patches are found throughout the Canadian Arctic. Although rarely studied they may be important, as without a plant cover the surface is left unprotected from other geomorphic processes. Conspicuous deflation forms have been noted from other arctic areas although not as yet from Canada. A typical case is found near Centrum Sø, Greenland, where a delta terrace of gravelly sands is covered with large polygons that initially had raised centers and depressed margins. Deflation has subsequently lowered the polygon centers so that today they are depressed between raised rims (Davies, 1960).[19]

Sand Dunes

Isolated patches of sand dunes are widely distributed in northern Canada, but the main dune areas are restricted to the western Arctic where a combination of suitable surficial deposits, and drier, warm summers results in a sparse tundra cover on sandy soils.

The largest area of aeolian deposits in northern Canada is between the middle Thelon and middle Back rivers. The dunes in this area are irregular in form and rarely show any preferred orientation. They are normally fixed by heath vegetation (and in more sheltered parts by low shrubs—principally birch and willow) with blow-outs evenly distributed through them. The ultimate source of the sand is the Dubawnt sandstone that underlies most of the area; the dunes develop from the surface deposits lying on the sandstone. Most widely distributed is the deep sandy till which frequently has blow-outs but rarely develops well-formed dunes. Larger but more restricted areas are lake shore dunes, rarely more than 8 m. (26 ft.) high, formed from the exposed sandy beaches as the lake level drops in the summer; such dunes are common on the shores of Aberdeen, Garry, and Pelly lakes. Other dunes accumulate along esker systems. Outwash sands form important sand dune areas, particularly, north of Aberdeen Lake, near Sandy Lake, and north of Pelly Lake. The beds and lower terraces of many of the rivers are an additional source of sand. The Tibielik River, entering the Thelon near the mouth of the Dubawnt, is typical of the valley areas. Roughly 40 per cent of this section is covered with dunes, and close to the bed of the stream in the lower part of the valley are isolated active dunes up to 8 m. (26 ft.) high. To the west of the river, where the source of sand is the abandoned channels of the Tibielik, there is a chaotic dune area with isolated dunes reaching 25 m. (82 ft.) in height.

Animals contribute to dune development where sandy areas are favored by colonies of Parry's ground squirrel (*Citellus parryi*), and less usually by the arctic fox (*Alopex lagopus*). The former may be directly responsible for blow-outs by destroying the vegetation and burrowing in the sand. They also attract the

---

19 W. E. Davies, "Surface fixtures of permafrost in arid areas," *19th Symp.* Greenland (1960), MS. p. 5. Same title in *Geology of the Arctic*, Toronto (1961).

Barren Ground grizzly bear (*Ursus richardsoni*) which tears up the vegetation over a wide area looking for squirrels.

In central Victoria Island featureless sand plains are widely scattered. Dunes have developed in some areas, but they are not high. There is also considerable bare sand on either side of Adelaide Peninsula, with an extension northward on the east side of King William Island near Matheson Point. Comparatively few dunes have formed and the underlying fluvioglacial sand is covered with a shallow pavement of gravel and pebbles, perhaps a result of wind erosion of the finer material, or possibly due to primary sorting when the area was submerged (Fraser and Hennoch, 1959).[20] Many rivers in the western Canadian Arctic have sand dunes, both on the present delta and on remnants of higher deltas, constructed during the postglacial marine transgression. Dunes are found at the mouth of Hayes River and on deltas farther west, including the Burnside, Coppermine, and the coastal zone of the Pleistocene delta of the Mackenzie. The majority of dunes have parabolic or rounded forms; they are invariably low, heights greater than 4 m. being uncommon, and are generally stable with only occasional blow-outs.

---

[20] J. K. Fraser and W. E. S. Henoch, "Notes on the glaciation of King William Island and Adelaide Peninsula, N.W.T.," *Geog. Br. Pup.*, no. 22 (1959), p. 39.

22

The Importance of Wind in the Geomorphology of the Mould Bay Area, Prince Patrick Island, N.W.T.

A. Pissart

1. Introduction

J. Tricart and A. Cailleux (1961, p. 214) stressed the importance of wind in many arctic periglacial regions. They state that this action is considerable in these regions because of:

1. The absence of vegetation
2. The frequence of violent winds
3. The paucity of precipitation
4. The concentration of water by freezing which has as an effect the drying of the soil surface
5. The availability of large quantities of fine debris produced by frost action.

Prince Patrick Island, being in the middle of the Canadian arctic, would appear at first sight to satisfy the conditions stated above; however, as we shall demonstrate, wind action is relatively unimportant.

2. Climatological Background and the Discontinuity of the Snow-cover

Prince Patrick Island is a part of the Queen Elizabeth archipelago and reaches the boundary of the frozen Arctic Ocean between lat. 75° 50′ and 77° 33′ N. and long. 115° 20′ and 122° 55′ W. A meteoroligical station has been in existence (Mould Bay Station) since 1948 and we therefore have access to climatological data over a 17-year period.

The climate of the island is very rigorous, as is illustrated by an average annual temperature of—18° C. and an average monthly temperature that only rises above 0° C. for the two months of

First published in French as "Le Rôle Geomorphologique du vent dans la Région de Mould Bay", *Zeitschrift für Geomorphologique*, N.F. Bd. 10, Heft 3 (1966), pp. 226-36. (Translated by D. R. W. Jones.)

July and August each year. Snow persists here for at least ten months. There are years, moreover, when melting is incomplete. Snow depth is never great, for of an annual average of 8.0 cm. (water equivalent) there is only some 50 mm. (water equivalent) of snow, producing a thickness of some 30 to 50 cm. This last figure represents nothing more than an average, for the snow, subjected to wind action, is spread in a highly irregular fashion. Some parts of the ground are thus always bare and contract with important snow accumulations that are found mainly on the upper parts of steep slopes facing away from the prevailing wind. A reconnaissance made from the air at the beginning of June, before melt had commenced, showed that topography plays the major role in the distribution of this snow. Extensive flat surfaces, such as the plain that occupies the entire western part of the island, are entirely snow-covered. Only a few rounded projections (notably pingos), and the sharp edges of the plateau immediately overlooking the river valleys, are bare.

Where the relief is more accentuated, wind turbulence increases, and the surface area totally cleared of snow increases considerably. However on this island, at the end of winter and under optimum conditions, it does not appear to exceed 30% of the total area.[1] The snow-free areas are essentially made up of rounded summits and steep slopes, although the upper portions of the lee side of these are locations of accumulation.

3. Wind Deflation

On projections that are bare throughout the year, traces of wind deflation appear in rare instances, and more often than not these are composed of sandy formations incompletely covered with pebbles. Surfaces where wind deflation is active to the point of removing the base from beneath the pebble formation are rarely found.

At first sight then, wind action appears to have little effect, and this impression is quite rapidly confirmed by the rare occurrence of wind-scoured pebbles. Wind-facetted pebbles are not found on Prince Patrick Island, and features showing wind effects are quite uncommon. There exist, of course, regional differences on the island according to relief and lithology. However, whatever the variations in the surface, the action of wind erosion only shows indirectly in the presence of niveo-aeolian deposits. These deposits permit us to estimate the importance of deflation and generally

---

[1] It is difficult to estimate the extent of the surface always exposed to deflation, for this varies throughout the year, and from year to year.

confirm that its role is a negligible part of the whole. We shall return to this further on.

In one isolated location, however, 9 km. to the east of Mould Bay, deflation appears to play a more important role. It occurs on broken structural surfaces of alternating beds of gently inclined sandstones and sands belonging to the Mould Bay formation (Jurassic). These surfaces are deeply incised by a network of valleys leading towards the S.-E. The succession of steep slopes encourages turbulence and the wind assists in sweeping clean the magnificent structural surfaces that are found here.[2] The snow accumulation downhill from these surfaces is particularly dirty, being laden with sand. In order to measure the importance of this deflation, we collected all the mineral particles on the snow accumulated on a square surface of 25 cm. of slope. Estimating that the surface affected by deflation has an approximately equal area, we calculate that erosion by deflation is, on the surface of the summits, of the order of 0.05 mm. per year.[3] However feeble, a deflation of this amount is altogether exceptional on Prince Patrick Island.

4. Wind-borne Deposits

We are able to distinguish two sorts of aeolian deposits according to their origins: those which derive from the denuded defiles which were discussed above, and those which have their origin in the alluvial plains. The two kinds are distinguishable not only by their topographical distribution, but also because the former, which are laid down during the winter, appear at the beginning of the thaw, whereas the latter, laid down before the snowfalls that completely cover the alluvial plains, only become visible when the thaw is more advanced. We shall consider each in turn.

A.   Aeolian deposits deriving from denuded slopes

The majority of dust grains carried away by wind deflation accumulate where the snow is gathered by wind action. It is therefore in the nivation hollows, usually near to the denuded slopes from which it was taken, that we find the greatest quantity of these dusts. In these places, the snow presents a very obvious stratification produced by layers loaded with mineral particles and layers of pure snow. Even in the most favourable cases, that is to say when loose sand showers down from above the snow-filled hollow, we have not been able to find layers composed solely of

---

[2] These structural surfaces have at times a slope of less than 2°.
[3] This estimate is difficult to make. Over and above the questions as to the surface, there are also problems that relate to the collection of the sample.

mineral particles; in every case we are dealing with strata containing large quantities of snow. Nevertheless, during the thaw these dusts collect on top of the melting snow so that the accumulation appears to be greater than it really is.

These niveo-aeolian deposits are without geomorphological importance, for the mineral matter involved is rather limited and it is deposited in the nivation hollows that develop rapidly through erosion. This aeolian material does not accumulate from year to year, but is removed before the beginning of the snows.

Before the thaw, as one moves further from these nivation hollows, the snow does not appear to have been dirtied by dust. Nor do cuttings made in the snow show mineral particles, at least to the naked eye. However, after several sunny days in June, at a temperature of less than 0° C., spots appeared here and there. By absorbing solar radiation more efficiently, mineral particles too small to be seen by the naked eye give rise to premature melting, and rapidly create small enclosed depressions, equivalent to cryonconite holes on glaciers. Usually, the melting takes place in depth beneath a remaining surface layer of ice 1 to 2 mm. thick, which serves as a hot-house pane accelerating the subjacent melting. In this fashion depressions with a depth of 10 to 25 cm. are sunk into the snow, and appear darker according to the quantity of dust particles that collect in them. In effect, melt-water intrudes into the subjacent snow while the mineral particles remain on the surface, concentrating and accelerating the process.[4]

The dark patches, which appear in June on, for example, the ice of Mould Bay bear witness to an accumulation of wind-borne dust. We should note, first of all, that there is no continuous dust-covered surface. Instead there are spots, or isolated areas. These deposits are clearly made in the shelter of micro-relief features that affect the snow-cover.

These dust spots occupy an area that varies according to location. Two traverses of Mould Bay, 6 km. in length and at intervals of 5 km., showed that the dust spots at one end covered 1% of the area, and at the other 0.1%. The causes of this distribution appeared clearly from the aircraft; they are related to the proximity of the coast, to the availability of fine-grain material, and also to the relief which determines local eddies of wind. It is for this

---

[4] This meltwater penetrates into the snow, where, because outside temperatures are always below 0°, it rapidly turns into ice. As a result, at the time of the general thaw these bodies of ice often survive longer than the surrounding snow.

reason that quite important trains of fine material extend for some 400 metres to the S.-E. of the Wind-Gap that lies between the main island and the peninsula situated on the other side of the bay to the S.-W. of Mould Bay.

Samples collected on the sea-ice of this bay were too small in volume to permit the establishment of a granulometric curve for these sediments. At more than one kilometre from the land there appeared to be hardly any granulometric variation from one place to another; the average size of sand grains appeared through a microscope to be 60μ, with larger elements rarely exceeding 100-200μ and the smallest measuring less than 1μ. We should add that numerous fragments of vegetation had been carried along with these particles of sand and may have contributed to the movement of the largest grains. In any event, these sand grains are too small to show wind-shaping. They are all angular (unmodified).

In sum, and after a complete flight over the island on the 18th June, it appeared clear that the deposition of mineral particles on the sea ice is a phenomenon of little importance, for the quantity of dust is always minimal, and the phenomenon is localized very close to the vicinity of coasts devoid of vegetal cover.

B. Wind deposits arising from deflation on the alluvial plains

Aeolian deflation is appreciable at the end of summer and the beginning of winter in the wide beds of rivers with braided channels. Mould Bay River offers a good example. For a distance of 3.5 km. upstream from its outflow into Mould Bay this water-course has a bed of about 100 m. in width. It is scoured temporarily by pulsations in flow resulting from snow melt, and is occupied later by a flow of water in the braided channels that diminishes progressively during the course of the summer. At the bottom of the bed, the alluviums are basically sandy and aeolian deflation can activate them easily. At the same time, the localization of these deposits in the valley floor, which is a place where snow accumulates during the winter, limits this wind action to the period preceding the first significant snowfalls. Near to these water-courses, there is no dust to accelerate melting at the beginning of spring, as the niveo-aeolian deposits are buried beneath the most recent layers of snow. It is not until later, when melting is more advanced, that layers of mineral particles appear, and their distribution clearly shows that the particles originate in the river bed. After the thaw, these sandy deposits remain in sufficient quantity on the banks to be locally quite visible, covering the very meagre vegetation that grows there.

This wind action is limited to the section of the river which appears temporarily as a flow of water in braided channels. Higher upstream the bed is much narrower and is also largely occupied by pebble which discourage all wind deflation.

The distribution of the aeolian layers is very irregular and depends on the micro-relief of the surface where deposition takes place. Very briefly, wind sedimentation is more important in the vicinity of the river banks and diminishes the further one travels from the river. These wind deposits, remaining after the thaw, have a thickness that varies and may reach 1 cm. in certain places. In these conditions, it is very difficult to estimate the importance of this form of wind accumulation. In our opinion, it appears reasonable to suggest a value that ranges between 0.5 and 0.01 mm. per year.

These wind-displaced sands were studied through a microscope. Of those measuring 0.7 mm., 20-30% of the grains had a round and dull appearance[5] acquired as a result of wind transportation. The percentage of these dull round particles in the river bed is very similar, which shows that this last wind transportation has left practically no record in the sediment itself.

Sand samples taken from well below the surface, for example on the flanks of pingos and in pockets of cryoturbation, showed a localized existence of 40-50% of round dull grains of the same 0.7 mm. dimension. Could wind action have been more intense in the past? It would require a study of numerous samples to verify this.

5. The Geomorphological Action of Wind

As we have seen, wind action is limited on Prince Patrick Island. The quantity of material taken from the snow-free upper slopes is, on the whole, rather small and has little influence on the evolution of the morphology. Niveo-aeolian deposits accumulate mainly in the nivation hollows, from where they are quite easily removed during the summer thaw.

The quantity of sediment carried further is minimal, as we were able to establish from the small quantities of dust existing on the sea ice.

The material carried from the river beds is deposited on flat surfaces nearby (frequently lower terraces), where it constitutes the most important and continuous wind accumulation on the island. This wind accumulation is indirectly apparent in the Mould Bay area thanks to the morphology of the tundra polygons.

[5] These grains are not perfectly rounded here, and there picotis aeolian is not visible.

The river bed in this region is bordered in places by low-centre type tundra polygons. The central parts of these structures are water-covered, with thermokarstic lakes of up to 1 m. in depth developing from the central basins.

These polygons are reduced and even disappear completely at a distance of approximately 100 m. from the steep banks, where wind accumulation is most intense. It is the filling up of these polygon centres by wind-borne sand that has prevented the thermokarst from developing here and, among other things, has almost completely obscured the polygonic forms.[6] We should note here that a little further upstream, above the zone of wind accumulation, the polygons exist with the usual characteristics right up to the immediate proximity of the river, and this observation proves quite clearly that it is not the river itself that determines the appearence of the polygons developed in the vicinity.

We should also add that in certain locations, for example near Satellite Bay, lines of sand several mm. in thickness show clearly in the landscape, as we noted from the air.

Geomorphological wind action is nonetheless of importance on Prince Patrick Island, but only through the intermediary of snow. The accumulation of snow in preferred locations determined by the wind affects the shaping of benches, cirques, and nivation niches which are the most striking features of the landscapes. These forms of snow action, which are so important in this climate, were studied by D. St. Onge in 1963 on Ellef Ringness Island, to which work we would recommend the reader.

6. Conclusion

The direct role of wind in the geomorphology of Prince Patrick Island is small. Deflationary forms are practically non-existent, and facetted pebbles were not found. Only a few very shallow aeolian deposits were observed. These niveo-aeolian deposits can hardly be compared with those in the Antarctic described by A. Cailleux (1962). All in all, the snow disappears almost completely each year in the region studied, and the wind deposits are of insufficient importance to noticeably retard melting. The reason for the limited importance of aeolian forms in this periglacial climate appear to be simple and of climatic origin; violent winds are rare. Winds with a velocity of over 47 m.p.h. (approximately 75 km/hour) have only been observed on the average once per year between 1948 and 1953.[7]

---

6 A profile taken in August 1966 at this place showed a thickness of windborne deposits greater than 2 m.

7 Observations are made every four hours.

Winds of over 32 m.p.h. (52 km/hour) last for a total of less than 6 days per year. It should also be noted that the most violent winds are always found in winter, when the snow cover seriously limits their geomorphological action.

Another aspect is that, in spite of the rigour of the climate, the vegetation cover is not always negligible; plants which knit the soil to outstanding buttes contribute considerably to fixing the soil.

True, dust storms, which would displace large quantities of sand and silt and would give rise to loss, do not appear to be found here. For this to take place, to have very violent winds would be required at the end of the summer period when the snow had disappeared or when the soil would be sufficiently dry to release such material. D. St. Onge (1965, p. 13) noted, nevertheless, that he had observed such phenomena in 1959 on Ellef Ringness Island. An examination of the climatological data of Isachsen would appear to indicate that this was an altogether exceptional occurrence.

J. Corbel (1958), who studied near Alert (the north point of Ellesmere), wrote: "A study of the terrain showed us only extremely weak aeolian action". The winds there, furthermore, are a little more violent than at Mould Bay. It appears certain, therefore, that this weakness of wind action is common to the whole Queen Elizabeth achipelago.

It is not without interest to note that A. Cailleux (1948, p. 120), through a study of sandy formations, arrived at the conclusion that quaternary aeolian action was less important in America than in Europe. He saw in this a proof of notable climatic differences probably due to the differing configuration of the continental masses. Similar climatic variations exist today within the most extreme regions of the globe.

References

Cailleux, A., "Les actions éoliennes périglaciaires en Europe," *Mém. Soc. Géol. France*, 28, rue Serpente, Paris, T. XXI (1942); "Etudes de géologie au détroit de McMurdo (Antarctique)," CNFRA 1 (1962), pp. 1-37.

Climatological summary, "Mould Bay—N.W.T. Canada, May 1948—December 1953," Department of transport, Meteorological branch, Toronto.

Corbel, J., "La neige dans les régions hautement polaires (Canada, Groenland) au delà du 80° latitude Nord.," *Revue de Geographie Alpine*, 46 (1958), pp. 343-66.

Heywood, W. W., "Isachsen area, Ellef Ringness island, District of Franklin, Northwest Territories," *Geological Survey of Canada*, paper 56-58 (1957), pp. 1-36.

St. Ongle, D., "La géomorphologie de l'ile Ellef Ringness, Territoire du Nord-Ouest, Canada," Etude géographique no. 38, Direction de la géographie, Ministrèré des Mines et des Relevés Techniques, Ottawa (1965), 58 pp.

Tricart, J., and A. Cailleux, "Le modelé périglaciaire," *Cours de géomorphologie*, C.D.U. Paris (1961), 350 pp.

23
Limestone Terrains in Southern Arctic Canada

J. B. Bird

Limestone is the most common sedimentary rock in arctic Canada south of Parry Channel, where it covers about 300,000 sq. km., or more than 12% of the land surface. Limestone terrains are the sites of most settlements, airfields, and military installations because of the reputed favorable engineering characteristics of terrains where there is continuous permafrost. Studies during the past 15 years have confirmed this observation generally for terrains close to sea level, but have shown also that there are great local variations.

Recognition of arctic limestone terrains from airphotos and their economic utilization for settlement purposes raises questions concerning their fundamental physical properties. In most parts of the world limestone is associated with special scenery and hydrological conditions due largely to the solubility of carbonate rocks in ground water. Geomorphologists have long recognized unique landforms in mid-latitudes where karst landscapes and their evolution were described first by Cvijic and Grund, and in the tropics where *kugel* and *turm* karst are developed [1, 2, 3, 4]. There are few geomorphological accounts of arctic limestone scenery, and it is uncertain to what extent cold climates influence the normal development of limestone landforms.

Distribution of Limestone

The limestones of arctic Canada are of three main geological ages. The oldest occurs in narrow bands and belts of crystalline limestone in Archean metasediments in several areas of the northern Canadian Shield. They are locally conspicuous in southern Baffin Island where in some parts, particularly around Lake Harbour, they are the dominant surface rock; even here few limestone outcrops are more than 100 m. wide.

First published in "Proceedings Permafrost International Conference", Purdue (1963), pp. 115-21.

The second group includes limestones and dolomites of early and late Proterozoic age. They are commonly thick bedded and gray or white in color. In the western Arctic, they occur in the Coppermine series where they are interlayered with trap rocks and diabase; in the northeast they form an important component of the rocks in northern Somerset Island and on Borden Peninsula, Baffin Island. The beds are often steeply dipping and occasionally vertical as they are in eastern Prince of Wales Island. In these circumstances, they are associated with striking hogback landforms.

Paleozoic limestones and dolomites form the third group. They were deposited in Ordovician, and in the north, Silurian seas. At one time they covered a large part, if not all the northern Canadian Shield, but today are preserved only in basins and troughs in the Shield and have been removed from intervening swells and arches. The Paleozoic limestones vary from thin bedded, flaggy, and often argillaceous formations to massive beds several meters thick. Cherty limestones, dolomite, and less commonly, shale beds, are present.

Six major basins containing Paleozoic limestones have been recognized including Hudson Bay, Foxe, Melville, Victoria Strait, Wollaston, and Jones-Lancaster Sound basins. Typically the sediments are about 1 km. deep in the centre of the basins. Around the margins the transition to the Shield rocks may extend over several miles, particularly where there is deep drift and limestone outliers; elsewhere the boundary is abrupt with limestone escarpments overlooking the Shield (Fig. 1).

Limestone landscapes in arctic Canada, like those developed on most other rocks, evolved their major elements prior to the final Pleistocene glaciation. They were further modified during glaciation and have experienced other changes in the brief postglacial period. Limestones have been denuded more rapidly than the majority of rocks in the Arctic and today are primarily lowland rocks. For several geomorphic reasons, limestone uplands have survived around Barrow Strait and Lancaster Sound where there are extensive plateaus between 370 to 600 m. above sea level [5]. Lower plateaus, 30 to 150 m., are found in the interior of many of the limestone plains.

Comparisons between the geomorphic and permafrost features of the limestone plains and plateaus, and between areas that experienced continental glaciation and others that escaped, make it possible to deduce the main characteristics of the arctic continental limestone terrains.

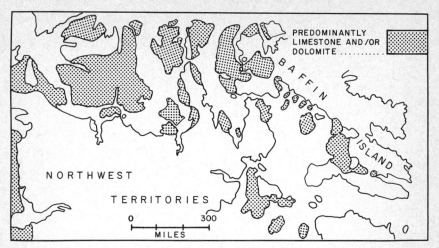

Figure 1. Distribution of carbonate rocks in the Southern Canadian Arctic; clastic and volcanic rocks occur in a few areas; Precambrian limestone is not shown.

Chemical Weathering

Limestone surfaces in most parts of the world are strongly affected by solution from ground and surface waters containing atmospheric carbon dioxide, soil bacteria, and decayed organic matter. The saturation equilibrium of carbon dioxide increases with decreased temperature so that at 0°C it is twice as great as at 25°C. Consequently, chemical weathering is greater in polar areas than in the tropics [6, 7, 8]. Bögli shows that the temperature-solution relationship is complex under natural conditions; the rates of limestone solution and limestone removal are not necessarily related and he believes that surface karst will develop more quickly under tropical conditions [9].

Surface karst is rare in northern Canada. Occasionally limited solution effects may be observed along joints and fissures; but solution has not proceeded quickly enough in the postglacial period of subaerial exposure, that varies from 6,000 to 12,000 years above the limit of marine transgression, for the results to be conspicuous.

Fissure solution is confined to massive limestone beds where they outcrop on horizontal surfaces and form pavements. On Somerset Island these are crossed by solution-enlarged joints 7 to 15 cm. deep. A more extensive pavement occurs on a low limestone plateau west of Wellington Bay, Victoria Island, that is covered with a rectangular fissure pattern. Solution fissures are not found on thin bedded limestones where mechanical weathering is the dominant process.

Large fissures occur at other points where they are associated with cambering; chemical weathering is, however, negligible. Such is the case along the limestone escarpment near Quillian Bay, Melville Peninsula, and in northeast Somerset Island.

Large trenches 5 to 20 m. wide, 0.5 to 1.5 m. deep and in some cases several hundred meters long, are found in limestone in the Victoria Strait Basin (Fig. 2). These trenches are partly filled with

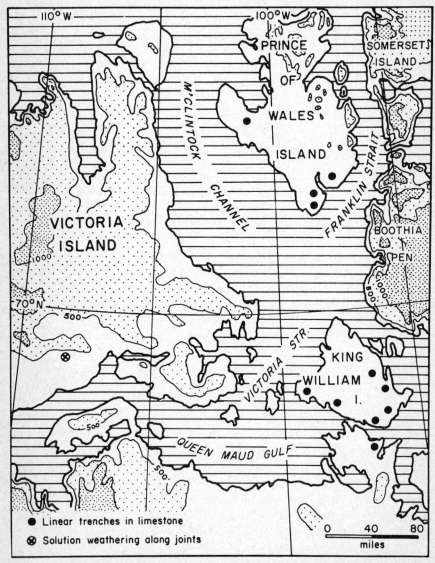

Figure 2. Solution fissures and trenches in the Victoria Strait Basin.

till and have been identified as preglacial solution fissures [10]. They may also be formed, however, by mechanical processes. Low ridges of shattered limestone have been observed at several localities in the Arctic including Melville Peninsula, Victoria Island, the Arctic mainland between Cape Kendall and Cape Krusenstern [11] and in North Greenland [12, 13], where they apparently develop from frost wedging along joints. If glacier ice subsequently removes the debris, the residual trenches would be comparable in dimensions to those in the Victoria Strait Basin.

Although solution by surface water in the Canadian Arctic is restricted, solution beneath snow is locally effective. Kauko has shown that the solubility of carbon dioxide in snow may be as much as 20 times as great as in water [14]. Maximum concentration is found after freezing and thawing at the snow surface [15]. Concentrated solution occurs on limestone surfaces that are snow covered during the thaw period, and this produces rough, etched rock with sharp ridges and pinnacles a few centimeters high.

Solution from snow lying on bedrock is rarely a significant geomorphic process primarily because of the small number of limestone pavements; where they do occur, the results are often conspicuous. An exceptionally large area has developed near the Prince Regent coast of Brodeur Peninsula north of Port Neill where there is a corroded, naked rock zone several miles wide. Subnival solution occurs on limestone fragments but the effect is generally less evident, although close examination often shows pitting and etching. Moisture that passes through frost-riven limestone plates dissolves calcite; this is redeposited on the underside as travertine crystals. These crystals are widely distributed in the central and northern Canadian Arctic, but are uncommon in the south.

Under suitable conditions subnival solution may be concentrated beneath perennial snowbanks. There are indications that the nivation-type limestone hollows on Cornwallis Island [16] and elsewhere around Parry Channel may have formed in this way.

The rate of subnival solution on limestone has not been determined. Observations of strong solution effects are more numerous from the northern Canadian Arctic (where deglaciation occurred relatively early) than from the southern Arctic. The time factor should not be given undue prominence as solution effects are apparent on limestone pebbles of elevated storm ridges on both sides of Bellot Strait to within a few feet of the sea, suggesting (from the continuing crustal emergence) that under favorable conditions they will develop in one or two centuries.

The extent of surface solution from snow and, to a lesser degree, from other forms of surface moisture varies with the local environment. In southwest Southampton Island, pebbles on beach swales invariably show more solution effects than on the ridges. In lagoons the solution on pebbles is greater than either. In many cases, solution is greater with height—reflecting the longer period of exposure since emergence from the sea; but greater differences result from lithological variations in the limestone.

Many features in karst regions develop from solution by ground water. Permafrost restricts underground drainage and consequently modifies karst formation. In the subarctic where permafrost is discontinuous there may be large scale underground drainage when the limestone is jointed. Caves, large springs, and *naledi* have been described from the central Lena Basin under these conditions [17]. There are few observations from northern Canada. Brandon notes that in the Mackenzie Lowlands region of discontinuous permafrost, there is considerable ground water flow, some of which occurs along solution channels [18]. Karst features are rare. They are believed to occur in northern Alberta and may exist locally south of Great Bear Lake. No karst development is found in the Hudson Bay Lowlands, the other subarctic limestone region in Canada, probably because of the low altitude.

Underground flow under arctic conditions of continuous permafrost is rare. Corbel considers that it occurs in continental arctic limestone area [19], but little geomorphological evidence in northern Canada supports this view. In the limestone plateaus of northwest Baffin, Somerset, and Devon Islands, no springs have been observed issuing from the sides of gorges. As some of the gorges are more than 400 m. deep, it is improbable that, if there is any quantity of ground water, none of it escapes as springs.

Springs are found occasionally elsewhere in limestone areas in the Canadian Arctic. On Southampton Island, several streams, the largest of which are Bursting and Unhealing Brooks in the south of the island, rise in springs that flow at least intermittently during the winter. The source of the water is probably talik in fluvioglacial deposits and in deeply weathered limestone rather than in bedrock. Most streams are undoubtedly fed solely by surface water and dry up as soon as the spring runoff is complete; streams in limestone areas are consequently often intermittent. In the southern Arctic where vegetation on limestone is locally continuous and where lakes are numerous, there may be some surface storage contributing to stream flow; in the far north perennial snow banks similarly may provide water throughout the summer, but in many limestone areas the smaller stream channels are dry, except in the spring and after rain.

Underground drainage is commonly marked by sink holes in karst regions. The only sink holes in stream beds (swallow holes) reported from the Canadian Arctic were found on Akpatok Island where they are up to 5 m. wide [20, 21]. Disappearing streams from other causes are not unknown in northern limestone areas and have been observed on Victoria, Somerset, and Southampton Islands. They result from small streams passing onto deeply weathered limestone and flowing at the bedrock-mantle interface instead of on the surface. The fact that identical phenomena are observed on felsenmeer developed or noncalcareous rocks in sufficient evidence that it is not a karst feature.

Solution hollows are not found although ponds in many limestone areas resemble them. Field inspection shows that they are forms of thermokarst: Smooth-sided ponds and lakes develop when local thawing of ground ice is followed by wave action on unconsolidated sediments.

Limestone caves are rare in the Canadian Arctic. They are found in modern cliffs near Leyson Point, Southampton Island, where they are developing by marine action. In northwest Somerset Island, rivers flowing in gorges are eroding caves on the outside of bends. In neither area is solution visible.

The conclusion is inescapable that under continental arctic conditions, in areas that were glaciated, and where there is continuous, deep permafrost, karst landscape due to chemical weathering has not developed. Where karst is found in northern subarctic areas, it may be contemporary but at least in unglacierized areas, it may also be a relic feature.

Mechanical Weathering

In dry arctic environments special landforms in limestone and other sedimentary rocks are primarily a consequence of mechanical weathering believed to be associated with frost riving. The process may commence evenly over a horizontal rock surface or may be localized initially along joints. In the latter case it leads occasionally to long, low ridges of shattered rock. The size of the fragments is linked closely with the lithology of the limestone; felsenmeer develop extremely rapidly when the rock is thin bedded and flaggy, such as is common in the central Canadian Arctic.

Development is slower on thicker bedded rock and is non-existent or is restricted to flaking in massive limestone. The variety of fragments in size and shape are practically endless. Near The Points, Southampton Island, dolomitic limestone plates formed postglacially average 1 m. in the long diameter; in contrast, limestone particles south of Amadjuak Lake, Baffin Island are initially 2 to 3 cm. in diameter and quickly disintegrate further, and at Batty Bay in east Somerset Island, limestone plates are locally paper thin.

The extensive distribution of limestone felsenmeer in areas that experienced Pleistocene continental glaciation suggests rapid postglacial weathering, but this conclusion is not always easy to substantiate. Preliminary measurements on emerged pavements on the west coast of Southampton Island show that mechanical weathering on thin-bedded limestone has attained depths of 15 to 30 cm. in 1000 years. At Mount Oliver, in southeast Somerset Island, weathered limestone 1.5 to 3.0 m. deep has developed in postglacial time, estimated to be roughly 9000 years.

If mechanical weathering results from frost scattering as the rock temperature fluctuates near freezing point, it will not progress indefinitely under arctic conditions as the annual temperature cycle is wholly below 0°C beneath the permafrost table. On a rock surface experiencing active weathering the permafrost table is at first in the bedrock below the frost-riven plates. With growth of the rock mantle, the permafrost table tends to sink as air and particularly ground water circulate in the shattered rock; an equilibrium state will be reached when the base of the weathered mantle coincides with the permafrost table. Investigations to test this deduction by determining the thermal condition of the near-surface were unsuccessful on Southampton Island in 1956, because of difficulties experienced in inserting instruments in deep limestone felsenmeer [22].

Once a deep limestone felsenmeer has formed, subsequent changes are restricted to progressive reduction of fragment size. Laboratory and field studies [23, 24] have shown that rock particles continue to disintegrate until a minimum size varying from 1 to 6μ, depending on the rock type, is reached. The process is accelerated by frost churning that raises the larger particles to the surface. In general it follows that the older the felsenmeer the smaller the mean particle size. In the brief postglacial period reduction of the initial frost-shattered limestone has not proceeded very far. Differences in particle size distribution between limestone felsenmeer close to sea level, that formed recently as the land emerged from the sea, and those that were exposed earlier at higher levels as the ice retreated, are due to differences of lithology rather than age. Residual surface deposits on limestone that survived the last continental glaciation unmodified are not known with certainty from the Canadian Arctic.

On the limestone plateaus on both sides of Parry Channel, the surface deposit is a silt rubble containing fragments up to boulders in size. The area was glacierized, but mainly by local rather than continental ice. The preglacial weathered mantle apparently survives with little change except for the addition of occasional erratic blocks. In these plateau areas, the landscape is in striking contrast to lowland limestone areas. Slopes are long, smooth, and gentle, the bedrock being buried except for occasional rock ledges. Mass transportation is considerable but results more from rill wash on the vegetation-free surface than from solifluction. In spring the ground frequently has the consistency of semiliquid mud and it may be impossible to cross on foot; in the summer, except after rain, it is brick-hard. The permafrost table is rarely more than 25 cm. below the surface.

Limestone felsenmeer have consequently very different terrain and permafrost properties depending on the stage of development from early forms with dry surface conditions, soil stability and lowering permafrost table to a late stage of alternating wet and dry conditions, instability, and high permafrost table.

A Periglacial Limestone Cycle of Erosion

Many arctic limestone landscapes are readily assigned to various stages of a theoretical erosion cycle. One group develops from limited uplift of an initial surface by a few tens of meters. Although this type is widely distributed in the Canadian Arctic, landscape elements resulting from glaciation and marine transgression dominate these areas. A second group has evolved on an initial surface about 400 m. above sea level. This includes those areas where the scenery has developed on the Barrow Upland Surcace on Devon, eastern Cornwallis, northwest Baffin, Somerset, and eastern Prince of Wales Islands [5]. Far from the sea, where rejuvenation has not occurred, the initial streams flow in wide, shallow valleys inherited from the previous cycle in which the valley sides are buried with debris and the stream beds are choked with boulders.

Rapid incision begins along the coast and spreads quickly inland leading to three distinct sectors in the landscape. The upper section and the wider interfluves are in the unchanged initial stage. In the middle section the rivers have numerous falls and rapids controlled locally by bedding planes in the limestone; the valley floor is littered with shattered and fallen rock, and the valley sides are deeply weathered with some scree. The lower section of the valley has a flat, often braided floor; the sides are steep but are rapidly modified by weathering and may be so rotten that they are impossible to climb, and it is dangerous to walk beneath them because of falling blocks. Valley sides formed partly of rock faces and partly of scree decline until the composite slope is replaced by a talus slope. Once the rock face is buried, retreat of the valley side is less rapid; without a supply of fresh scree, talus fragments weather, and transportation by rock falls and talus creep is replaced by rock creep in which interstitial ice may play a significant part.

Ultimately, when the silt fraction increases, solifluction appears. Slopes continue to decline and by the solifluction stage are concave-convex with active weathering restricted to the upper slopes where the mantle is thin or absent.

Across the middle of the slope transportation is restricted to the active layer that is normally above the rock surface, and there is no denudation. In the concave sector, accumulation by rill wash and solifluction prevails. Areas with gentle to moderate slopes of this stage are found in the low hills of northern Prince of Wales Island and on the southern islands of the Queen Elizabeth Archipelago. In some parts of the Canadian Arctic, steep and even vertical slopes are retained late into the cycle. The Buttes area at the base of Borden Peninsula is a good example.

This brief examination of limestone terrains, particularly around Parry Channel where glacial influence has been small, suggests that it is possible to analyze polar limestone terrains as part of a continuing spectrum of changing slope, particle size, drainage, and permafrost conditions. Early in the sequence are ephermeral rock pavements of little relief but considerable surface runoff; later is a stage of maximum relief, coarse debris, and little surface water, and, finally, is a stage of low relief, gentle slopes, high clay and silt content in the mantle, high surface runoff and, from the engineering point of view, deteriorating permafrost conditions.

The theoretical, orderly sequence of landform evolution was disturbed in northern Canada by the Pleistocene glaciations and the associated lowland marine transgression. The former has affected all Arctic Canada although on plateaus, particularly around Parry Channel, and in the lowlands of the Queen Elizabeth Islands, modifications have been slight. The marine transgression was confined to areas below about 100 m. in the northwest and 125 to 200 m. elsewhere.

Limestone Terrains

There are three main categories of arctic limestone terrains. In the first group are terrains found at all elevations that in the cyclic sense are youthful. The second group is restricted to areas invaded by the postglacial sea; the third group includes terrains developed on glacial drift.

Bare rock surfaces and felsenmeer form the first group. Extensive areas of the former are rare. In most uplands, outcrops are restricted to low limestone knobs several meters across that in many cases are shattered. Outcrops also occur in scarps and ledges and on the crests of low hills where the surficial mantle thins out. Occasionally, naked rock surfaces are more widely distributed as on the west coast of Brodeur Peninsula. In these situations the surface is often extremely rough due to chemical weathering.

True limestone felsenmeer are not common. Above the marine limit this is primarily because glacial drift buries the rock surface and restricts physical weathering. At all elevations, but particularly below the marine limit, block fields that have developed from bedrock are difficult to differentiate from pseudofelsenmeer. The latter have evolved through frost shattering of boulders concentrated by frost or wave action at the surface from till. Limestone felsenmeer that contain patches of fines, often in the form of patterned ground, are rarely if ever true felsenmeer but have developed over glacial or marine clays that contain limestone boulders.

The greatest variety of terrains is found in the limestone lowlands below about 100 m. where marine, glacial, and postglacial residual deposits are intermingled (Fig. 3). The most characteristic features are marine ridges and bars constructed during the emergence of the land from the postglacial sea. On exposed slopes greater than about 3° elevated storm ridges are normally present. A favoured site is on the sides of low plateaus that are widely distributed in limestone lowlands. In the majority of localities they have developed over bedrock that is rarely more than a metre below the surface. In some areas the limestone shingle has developed from weathered boulders concentrated on the surface of till or marine clays. From the surface indications, this situation is difficult to confirm unless clay plugs break through to the surface to form mud circles [25], or a natural section exists in a gorge. Whether the storm ridges are underlain by rock or clay, groundwater is often present at the base of the shingle during the summer. It may migrate downslope and can produce difficult icing conditions in underground structures.

Scarp............... ┬┬┬┬┬┬┬

Strandline......... — — — —

Fluvioglacial forms.. ─────────

Figure 3. Detail of geomorphology (left) and terrain (right) in the Coral Harbour Lowlands, Southampton Island. The whole area was transgressed by the postglacial sea except for two small sections in the north of the map (after Bronhofer).

Bedrock............

Unsorted drift.....

Ground moraine
(surface reworked)

Gravel and pebbles...

Sand and silt........

Water at flood stage..

Where the coastal slope is gentle, gravel and pebble bars form either along the shore or, in some cases, offshore. As the land rose during the postglacial period, a succession of ridges was formed, separated by depressions filled in summer by water or peat. The flatter the ground the more widely spaced are the ridges; in the extreme case where the ground is horizontal, they disappear entirely and a peat-covered plain overlying in some cases silt, and more generally limestone plates, is found. Ground ice forms at the base of the peat; arctic-type palsen with a clear-ice core and ice-wedge fissures are not uncommon. In the Koukdjuak Plain, western Baffin Island, the largest example in northern Canada, large thaw lakes with weakly developed orientation have formed. There are many smaller limestone plains of this type including the Cape Kendall area of Southampton Island, Saputing Plain near Berlinguet Inlet, Baffin Island, and the east side of Rasmussen Basin.

In addition to marine landforms and peat-covered plains, the limestone lowlands contain more restricted areas of rock outcrops that occur in low cliffs and pavements and in some glacial deposits. When the latter are deep, they may have survived essentially unmodified through the marine phase; subsequent solifluction leads to smooth, rounded slopes. Glacial landforms become numerous far inland and toward the upper limit of marine submergence. In extreme cases limestone landscapes are dominated by glacial landforms far below the marine limit—as occurs in the drumlin fields of eastern Victoria Island and the interior of southern Prince of Wales Island.

Close to the marine limit there is a transition of terrains toward the main types that are characteristic of the zone above it. The properties and distribution of the rubble terrain, broken occasionally by scree and rock-sided gorges, have already been described. Till terrains above the marine limit often resemble rubble terrains (Fig. 4). Locally the coarser fragments, often with some weathering, are concentrated on the surface and associated with sorting patterned ground develops. The latter are typified by the Amadjuak and Nettilling Plateau surfaces in western Baffin Island with their long gentle solifluction slopes. Vegetation is absent except for scattered plant clumps. The permafrost table is high and practically all precipitation runs off the surface. The main differences from rubble terrains lie in the faint drumlinization, occasional fluvioglacial deposits, and frequency of erratic boulders.

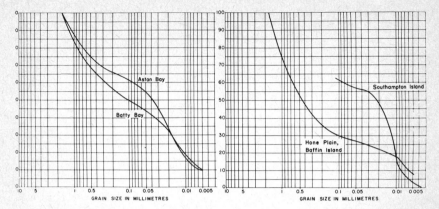

Figure 4. Analysis of the grain size of upland limestone rubble on Somerset Island (left) and unsorted limestone till from above the marine limit (right). In all four samples the size-fraction larger than sieve 10 was removed in the field. The Hone Plain is believed to have been washed by glacial melt water and this may explain the smaller silt fraction. Southampton Island till and Somerset Island rubble are similar.

The main limestone terrains have, therefore, distinct characteristics that enable them to be distinguished readily on the ground and—once their complexity is recognized—on aerial photographs. Ultimately, their properties derive from the lithology, geomorphic history, and climate. They are essentially different from limestone terrains in other parts of the world, and nowhere does karst in the accepted sense of the term develop.

References

[1] J. Cvijic, "Hydrographie Souterraine et Evolution Morphologique du Karst", *Rec. Trav. Inst. Géogr. Alpine,* Vol. VI (1918), pp. 375-426.
[2] A. Grund, "Der geographische Zyklus im Karst", *Z. Ges. Erdk. Berlin* (1914).
[3] H. V. Wissman, "Der Karst der humiden heissen und sommerheissen gebiete Ostasiens", *Erdkunde,* Vol VIII (1954), pp. 122-30.
[4] H. Lehmann, K. Krömmelbein, W. Lötschert, "Karstmorphologie, geologische und botanische Studien in der Sierra de los Organos auf Cuba", *ibid.,* Vol. X, 1956, pp. 185-204.
[5] J. B. Bird, "Recent Contributions to the Physiography of Northern Canada", *Z. Geomorph.,* Vol. III, 1959, pp. 151-74.
[6] J. Corbel, "Vitesse de l'erosion", *ibid.,* pp. 1-28.
[7] J. Corbel, "Morphologie périglaciaire dans l'Arctique", *Ann. Géogr.,* Vol. LXX (1961), pp. 1-24.
[8] J. Corbel, "Erosion en terrain calcaire: vitesse d'erosion morphologie", *ibid.,* Vol. LXVIII (1959), pp. 97-120.
[9] A. Bögli, "Kalklösung und Karrenbildung", *Z. Geomorph. Supp. Bd. 2* (1960), pp. 4-21.
[10] J. K. Fraser, W. E. S. Hennoch, "Notes on the Glaciation of King William Island and Adelaide Peninsula, NWT.", *Geog. Branch Paper 22* (1959).

[11] A. L. Washburn, "Classification of Patterned Ground and Review of Suggested Origins", *Geol. Coc. Am. Bull.*, Vol. LXVII (1956) pp. 823-65.

[12] W. E. Davies, "Polygonal Features on Bedrock, North Greenland", *U.S. Geol. Surv. Prog. Paper 424-D* (1961), pp. 218-19.

[13] W. E. Davies, "Surface Features of Permafrost in Arid Areas", *Geology of the Arctic*, Toronto (1961), pp. 981-87.

[14] Y. Kauko, L. Laitinen, "Die Kohlensäure-Sorption des natürlichen Schnees", *Suomen Kemistilehti*, Vol. VIIB (1935), p. 12.

[15] J. E. Williams, "Chemical Weathering at Low Temperature", *Geog. Rev.*, Vol XXXIX (1949), pp. 129-35.

[16] F. A. Cook, V. G. Raiche, "Simple Transverse Hollows at Resolute, NWT.", *Geog. Bull.*, No. 18, 1962, pp. 79-85.

[17] S. S. Korzhuyev, S. S. Nikolayev, "Karst Types in Permanfrost Regions and Features of Their Occurrence", (n Russian), *Izv. Akad. Nauk SSSR Ser. Geograf.*, no. 6 (1957), pp. 33-46.

[18] L. V. Brandon, "Groundwater in the Permafrost Regions of the Yukon, North Cordillera and Mackenzie District", *Proc. vst Can. Conf. Perm.* (1963), pp. 131-39.

[19] J. Corbel, "Hydrologie et morphologie des Nord-Ouest America", *Rev. Géom. Dyn.*, Vol. VIII (1957), pp. 97-112.

[20] I. H. Cox, "The Physical Geography of Akpatok Island", in Clutterbuck, H. Akpatok Island, *Geog. J.*, Vol. LXXX (1932), pp. 224-27.

[21] N. Polunin, "The Vegetation of Akpatok Island", Pt. I, *J. Ecol.*, Vol. XXII (1934), pp. 337-95.

[22] M. Bronhofer, "Field Investigations on Southampton Island and Around Wager Bay, NWT., Canada", *RAND Research Mem. 1936* (1957).

[23] I. C. McDowall, "Particle Size Reduction of Clay Minerals by Freezing and Thawing", *N.Z. J. Geol. Geoph.*, Vol. III (1960), pp. 337-43.

[24] J. Dylik, T. Klatka, "Recherches microscopiques sur la désintégration périglaciaire", *Bull. Soc. Sci. Lett. Lodz Class III*, Vol. III(no. 4 (1952), pp. 1-12.

[25] F. A. Cook, "Additional Notes on Mud Circles at Resolute Bay, NWT.", *Can. Geogr.*, no. 8 (1956), pp. 9-17.

24

Age and Origin of the Cypress Hills Plateau Surface in Alberta

P. D. Jungerius

Introduction

The Cypress Hills straddle the border between Alberta and Saskatchewan from 49° 35′ N. to 49° 40′ N., forming an elongated plateau about 130 km. long from east to west. This paper deals with the geomorphology of the Alberta section only (Figure 1).

Over a distance of approximately 25 km., the plateau summit slopes gradually eastward from an elevation of 1,440 metres southeast of Elkwater Lake, to 1,345 metres at the Saskatchewan border. It stands more than 500 metres above the general surface of the surrounding Great Plains from which it is separated by steep marginal slopes and a series of broad pediment-like features sloping down from the plateau. The surface is smooth, with only the edges deeply incised by gullies.

The plateau receives a mean annual precipitation of about 460 mm., 100 to 150 mm. more than the surrounding Great Plains which support a semiarid prairie flora. The northern slopes and part of the southern slopes are covered with forest comprising lodgepole pine, aspen poplar, spruce and black poplar (Wyatt et al., 1941). Grass is the dominant vegetation of the top of the plateau.

The upland is underlain by gravels of Oligocene age (Russell and Landes, 1940) composed primarily of quartzite, argillite and chert, with volcanic porphyry and granite as minor constituents. McConnell (1885, p. 69c) was of the opinion that these are alluvial deposits formed by a vigorous stream heading in the Rocky Mountains, and were originally distributed along a valley bordered by higher land.

First published in *Geographical Bulletin*, Vol. VIII, no. 4 (1966), pp. 307-18. Reprinted by permission of the Department of Energy, Mines and Resources.

According to Alden (1932) the present topographic position of these gravels indicates that a reversal of relief has taken place. Due to its high permeability, the gravel is considered to have retarded erosion of the valley bottom, which is now left as a gravel-capped erosion remnant, while extensive erosion removed hundreds of metres of Tertiary and Cretaceous deposits from the surrounding regions. This erosion was active from the middle of the Tertiary period to the onset of glaciation during the Pleistocene epoch (Gravenor and Bayrock, 1961). The plateau was not glaciated but its surface has been modified by periglacial processes, and it is now covered with thin aeolian deposits (Westgate, 1965). Alden gave the name Cypress Plains to an extensive erosion surface, of which the Cypress Hills form a part and assigned it to the Oligocene-early Miocene period. Although Alden's interpretations were based on inconclusive evidence they have not, until now, been challenged by subsequent workers (Broscoe, 1965).

Reconnaissance studies made in 1964 by M. J. J. Bik produced evidence requiring a different interpretation of the geomorphology of the Cypress Hills area from that given by previous workers. The present study forms part of the field program subsequently carried out under Bik's leadership. Grain-size distribution, clay

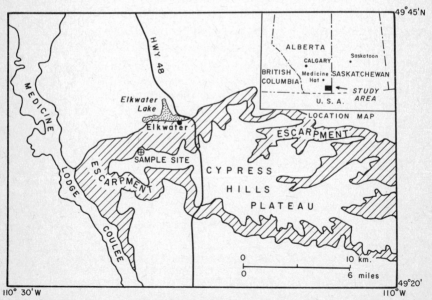

Figure 1. The Alberta section of the Cypress Hills plateau.

minerals and heavy minerals were analyzed in the laboratory of Physical Geography, University of Amsterdam, through the courtesy of Professor Dr. J. P. Bakker and Dr. H. J. Müller, Clay mineral data were provided by Dr. Th. W. M. Levelt, also of Amsterdam.

From a number of soil profiles studied and analyzed, a representative example (Figure 2) was selected for discussion in this paper. This profile is exposed in a quarry near the northern edge of the plateau, about 2 km. southwest of Elkwater Lake (Figure 1).

Materials

Description of the deposits shown in Figure 2

The profile includes, from the base upward:

Cypress Hills formation — conglomerate containing abundant well-rounded pebbles and cobbles of quartzite and argillite. The matrix is non-cemented and consists of a mixture of sand and clay. In scattered localities, remnants of a red palaeosol are found in the upper part. The top of the conglomerate is generally level but with shallow depressions about 10 metres wide and 30 to 40 cm. deep. It represents an erosion surface because the red

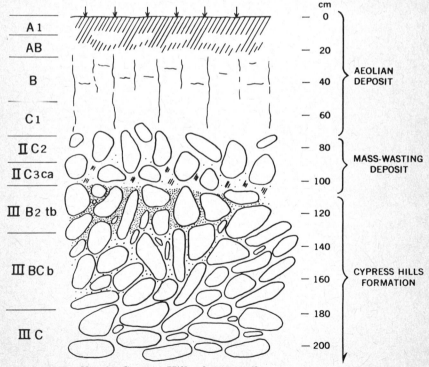

Figure 2. Profile of a Cypress Hills plateau soil.

soil which once was continuous over wide areas is now either absent or truncated. Although the section shown includes only about 1 metre of the conglomerate, this does not represent the total thickness of the formation.

Mass-wasting deposit — about 30 cm. of clayey material with frequent pebbles and cobbles derived from the underlying conglomerate. Where this deposit fills depressions in the Cypress Hills formation, it is as much as 60 cm. thick. Elsewhere it may be reduced to a mere stone-line. The top of this deposit forms a level surface extending over most of the plateau.

Aeolian deposit — 60 to 80 cm. of loam without coarse-grained clastics. This material overlies the level surface of the mass-wasting deposit as a continuous mantle.

Description of the soil profile

All morphological descriptions in the profile on the following page are based on the definitions and nomenclature of the Soil Survey Manual (United States Soil Survey Staff, 1962). The colour symbols refer to the Munsell Color Charts, and are given for dry colours. For pH figures see the section dealing with methods.

Methods

Representative samples were taken of all horizons. Mechanical analysis of the material to 2 mm. was carried out by sedimentation and sieving following H_2O_2 and HCl treatment to destroy the organic matter and carbonates, and following dispersion with $Na_4P_2O_7$ and Na_2CO_3. Organic carbon was determined by dichromate oxidation, and total carbonates by the gasiometric method. The pH was measured using a 2:5 soil:water, and 2:5 soil:1 N KCl solution ratio, respectively. The grain-size distribution is shown in Table I.

Heavy mineral analysis was carried out on the 30 to 500 micron fraction. This fraction was isolated by decantation and sieving. Heavy minerals were separated in bromoform (sp. gr. 2.89). One hundred translucent grains were identified in each slide, using the ribbon counting method (Van Harten, 1965), with 150 or 300 micron band-width depending on the number of grains on the slide. No determination was made on opaque, altered, coated, and other grains for which the optical properties could not be established. The results are recorded in Table II.

Powder specimens of the fraction less than 1 micron were prepared for X-ray analysis. X-ray photographs were obtained with a Guinier-De Wolff camera, using FeK radiation. Quantities were estimated by visual means, and by comparing the photographs with

TABLE I. MECHANICAL COMPOSITION OF A CYPRESS HILLS PLATEAU SOIL PROFILE, PERCENTAGE DISTRIBUTION.

Horizon	Depth (cm.)	Parent material	1400–2000 micron	1000–1400	1000–600	600–420	420–300	300–210	210–150	150–105	105–75	75–50	50–32	32–16	16–8	8–4	4–2	< 2 micron	Humus	C_aCO_3
A_1	0–10	Aeolian deposit	0.1	0.1	0.1	0.3	0.6	2.0	5.0	9.5	10.5	7.5	13.5	11.0	7.0	7.0	3.0	22.5	12.4	
AB	10–24				0.1	0.1	0.3	2.0	5.5	11.5	11.5	9.0	13.0	9.0	7.5	5.5	2.5	22.5	3.1	
B	24–51				0.1	0.1	0.2	1.5	6.0	12.5	13.5	10.0	10.5	7.0	5.0	7.0	1.5	25.5	1.2	
C_1	51–74						0.2	1.5	5.5	11.5	14.0	10.5	9.0	10.5	6.0	5.0	2.0	24.0		
IIC_2	74–86	Mass-wasting deposit	0.2	0.2	0.2	0.3	0.5	2.5	5.5	9.0	9.5	8.0	11.0	8.0	6.5	3.5	3.5	32.0		
IIC_{3ca}	86–103		0.5	0.4	0.4	0.5	0.8	2.0	3.0	4.5	4.5	4.5	12.5	10.0	8.5	4.0	2.5	41.0		5.1
$IIIB_{2tb}$	103–130	Cypress Hills formation	2.0	0.7	0.5	0.6	2.5	5.5	4.5	4.0	4.0	3.0	4.0	6.0	2.0	3.5	0.2	57.5		3.9
$IIIBC_b$	130–180		1.5	0.8	0.9	1.5	4.5	8.5	6.5	6.0	6.0	4.5	7.0	3.0	5.0	3.0	0.5	40.5		0.6
IIIC	180–200		0.9	0.6	0.7	1.0	4.0	11.0	8.0	7.5	7.0	6.0	6.5	4.0	4.5	2.0	2.0	34.5		2

TABLE II. HEAVY MINERAL COMPOSITION OF A CYPRESS HILLS PLATEAU SOIL PROFILE, PERCENTAGE DISTRIBUTION

Horizon	Depth (cm)	Parent material	Not determined opaque	Not determined others	Tourmaline	Zircon	Garnet	Rutile	Anatase	Sphene	Staurolite	Andalusite	Epidote group	Amphibole	Orthopyroxene	Clinopyroxene
A_1	0-10	Aeolian deposit	39	39	5	15	13	5	1	13		3	15	27		3
AB	10-24		35	28	8	15	12	3	1	9		3	22	24		3
B	24-51		40	23	5	35	12	4	4	7		1	22	10		
C_1	51-74		42	28	5	27	19	3	7	14		2	13	6		4
IIC_2	74-86	Mass-wasting deposit.	27	28	2	34	23	3		2		2	15	16	2	1
IIC_{2ca}	86-103		37	42	5	28	19	7		8	4	2	10	14		
$IIIB_{2tb}$	103-130	Cypress Hills formation	45	31	1	53	21	5		5			11	4		
$IIIBC_b$	130-180		29	59	2	45	16	6	1	8			15	6	1	
IIIC	180-200		20	52		34	8	2	2	16	3		28	7		

diffractograms from similar profiles. The results are listed in Table 3. The symbols indicate the intensity pattern of the reflection of each mineral in the successive horizons. The scale ranges from "trace" and (+) to ++++ according to intensity of the reflections.

TABLE III. CLAY MINERAL COMPOSITION OF A CYPRESS HILLS PLATEAU SOIL PROFILE.

Horizon	Depth (cm)	Parent material	Montmorillonite	Chlorite	Illite	Kaolinite	Quartz	Other Reflections
A1	0–10	Aeolian deposit	+(+)		+(+)	(+)	+(+)	
AB	10–24		+(+)		+(+)	(+)	+(+)	
B	24–51		++		+	trace	+(+)	weak 12.3 A
C1	51–74		+++(+)	trace	+	(+)	+(+)	
IIC2	74–86	Mass-wasting deposit	+++	trace	(+)	trace	+(+)	
IIC3ca	86–103		+	trace	trace	trace	+(+)	weak 13.4 A
IIIB2tb	103–130	Cypress Hills formation	++		+	trace	+(+)	
IIIBCb	130–180		+(+)	trace	(+)	trace	+(+)	
IIIC	180–200		++		+	trace	+(+)	

Discussion

Aeolian deposit

The histograms of the surface loam shown in Figure 3 have maxima in the clay, silt and fine sand fractions. The average dimensions of the silt fall between 16 and 50 microns and those of the sand are in the range 75-150 microns although surface loams with different composition have been found on other parts of the plateau There is at present no evidence that this material could have been deposited by any agent other than wind, but it differs in many respects from most aeolian deposits in which the sorting has generally reached a much higher degree of perfection.

The silt could have been acquired from the bordering Great Plains by transport in suspension, but the origin of the sand remains obscure as grains of this diameter are unlikely to be carried far or high in suspension by air currents (Zeuner, 1949). According to Westgate (1965), all the aeolian material here is derived from the glacial drift of the Plains, because the 62-177 micron heavy mineral fraction contains up to 17 per cent amphibole, a mineral which he failed to find in the underlying Cypress Hills formation. This is not substantiated by the present observations which found that amphibole does form part of the Cypress Hills conglomerate although in lesser quantities (Table 2). This finding is not due to the much wider fraction used, as the size of several amphibole grains of the

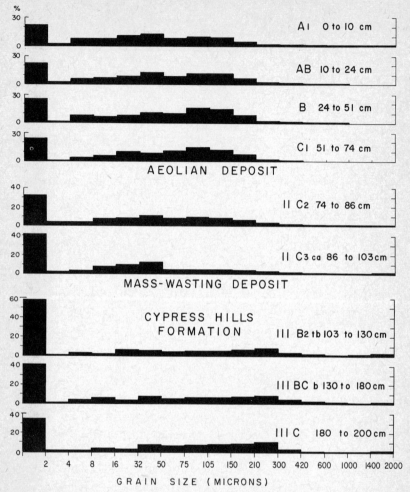

Figure 3. Histograms comparing grain-size distribution in a Cypress Hills plateau soil profile.

same fresh blue-green type as in the aeolian deposit, ranges from 62 to 177 microns. It is possible, of course, that they were washed from overlying materials although no such illuviation is evident from the grain-size analysis.

Westgate also found the aeolian material at the northern edge of the plateau to be coarse-grained and sandy. From the textural variation, he suggested strong northerly prevailing winds during the period of aeolian deposition. However, it is unlikely that winds strong enough to lift sand grains up the steep escarpment slope of the Cypress Hills would be unable to spread them over the horizontal summit surface. More probably the sand was derived from the mass-wasting deposits or from the Cypress Hills formation.

The soil profile in the surface loam conforms to the present surface, and is a thin Orthic Black, Chernozemic soil (National Soil Survey Committee of Canada, 1963). The dark surface horizon is formed by the decomposition of organic material derived from the present native grass vegetation. The lower or *AB* part of this horizon is now in what was formerly the upper part of the *B* horizon. This mixing is presumably the result of anthropogenic activities, as evidence of Indian settlement in this area is substantial. The dark tongues form part of an extensive polygonal pattern probably caused by filling up of desiccation cracks with humic material. The *B* horizon is characterized by a change in colour and structure, and seems to meet the requirements of *Bm* horizons as defined by the National Soil Survey Committee of Canada (1963).

The dominant clay materials in this soil are shown in Table III. There is a decrease of montmorillonite and an increase of illite toward the upper horizons. This phenomenon, often observed in Chernozemic soils has been attributed to the translocation of montmorillonite in preference to illite (Beavers *et al.*, 1955) or to fixation of potassium by expanding lattice clays (St. Arnaud and Mortland, 1963).

Mass-wasting deposit

In the mass-wasting deposit, the < 2 mm. fraction comprises about 50 per cent of the total. The remaining part is made up of pebbles and cobbles derived from the underlying Cypress Hills formation. The grain-size distribution of the matrix shows marked differences in this layer. The texture of the upper part is similar to that of the surface loam. Below 86 cm. the aeolian sand is absent and in this layer the fraction 50 to 2,000 microns reflects the properties of the matrix of the underlying conglomerate.

No separate mass-wasting deposit was recognized by Westgate (1965) who suggests that the pebbles were elevated into the aeolian cover by frost action. Although frost heaving is a common feature of many Cypress Hills profiles, it could not explain the morphological characteristics of the present layer. The erosive effect of this material on the underlying conglomerate, its level surface, and its occurrence over non-pebbly material elsewhere on the plateau indicate that it has been transported. The poor sorting and the many pebbles lying in vertical position are typical of solifluction under periglacial conditions, by which process the surface material flows over a frozen subsoil.

From the differences in grain-size distribution and clay mineral analysis in layers IIC2 and IIC3a (Tables 1 and 3), it appears that there are two solifluction deposits. A similar stratification has been found in other Cypress Hills soil profiles. In some of these, two solifluction layers were separated by an older aeolian deposit, indicating that the two layers are of different age.

No remnants of palaeosols were found in the mass-wasting deposit, and there is no way of establishing differences in age between the upper and lower material. The upper layer and the overlaying loam are presumably of the same age. The accumulation of $CaCO_3$ in the lower layer is a result of soil formation in the aeolian cover. This calcic horizon is wavy and may also occur in C or $IIIB$ horizons.

Cypress Hills formation

The buried soil derived from the Cypress Hills conglomerate is of particular interest to the geomorphology of the Cypress Hills plateau. It has a reddish coloured upper horizon, providing a marked contrast to the yellowish brown and greyish brown colours of the overlying soil. It is apparent from examination of the data (Table 1) that a striking increase in the percentage of clay has occurred in this horizon. This high clay content is interpreted as a result of illuviation from a former A horizon which has now been removed by erosion.

The translocated clay has formed distinct coatings adjacent to the pebbles, although in this particular profile the clay skins are neither as red nor as thick as in some other soil profiles from the plateau. Occasionally calcium carbonate is present between the red clay skin and the pebble surface, indicating that red soil was present before $CaCO_3$ accumulation occurred. The original chemical nature of the soil is not known as it is completely saturated with cations added from the overlying calcareous cover. Due to the same process the values of pH exceed 7.5 throughout the buried profile.

The profile morphology would appear at first glance to have been, in its original state, gradational toward the Red-Yellow Podzolic soils of the southeastern United States. This cannot, however, be so for its mineral composition is quite different. Red-Yellow Podzolic soils are restricted to relatively warm and moist climatic zones (McCaleb, 1959). Due to intensive leaching, they normally contain only small amounts of weatherable minerals in the sand and silt fraction, while the clay fraction has a low content of 2:1 lattice minerals. The reddish palaeosol in the Cypress Hills conglomerate, on the other hand, has considerable amounts of weatherable minerals, and of montmorillonitic clay (Tables II and III). This soil has much in common with the red and yellow mediterranean soils found in semihumid and semiarid climates under a deciduous forest or shrub vegetation (Bennema, 1963). Such soils also have a textural B, and normally contain substantial amounts of 2:1 lattice clay minerals and primary weatherable minerals.

The clay minerals of the palaeosol appear to be inherited directly from the parent material, with hardly any modification due to weathering. No iron oxides were found in this fraction and the red colour suggests that most of the iron is present in the form of coatings around the grains. X-ray analysis revealed the presence of feldspar not only in the matrix of the red horizon, but also in pieces of apparently weathered, soft rock fragments that were occasionally encountered in this horizon.

The conclusion which can be drawn from examination of the present evidence is that the surface of the Cypress Hills plateau is not a remnant of a Tertiary erosion level, but, rather an altiplanation surface probably dating from the Wisconsin glacial stage, and covered by material of aeolian origin. The surface has been produced by mass-wasting processes that were operative independent of base level control, and effective even on slopes of less than one degree. These processes have obliterated any drainage patterns that may have existed on the plateau.

The term "altiplanation" was used by Eakin (1916, pp. 67-82) to designate periglacial processes producing terraces and flattened summits in Alaska. Without being specific about the type and character of the processes involved, Bakker and Le Heux (1952) associated altiplanation with the formation of flattened upper slopes and summits by central rectilinear slope recession. Basing their arguments on mathematical considerations, they showed that high-level erosion surfaces of wide extent are not necessarily the result of polycyclic development. Hermans (1955, pp. 27-38) elaborating these theories, concluded that so-called Tertiary peneplain remnants in the southern Ardennes (Luxembourg) are, in fact, altiplanation surfaces produced by Pleistocene periglacial processes. The term "altiplanation" is now used to include all more or less horizontal degradational surfaces developed at high altitudes (Schieferdecker, 1959). In the case of the Cypress Hills altiplanation surface, it is the continuously active landslide movements along the edges which prevented the surface from being graded to the drainage system of the surrounding pediments.

At one stage in its development as an erosion surface, the plateau had an undulating relief. The elevations of this terrain were gradually lowered by downslope movement of material derived from the surface horizons. This is evident from the scattered remnants of the reddish palaesol which formerly occupied a wider area. These remnants indicate the position of the depressions which remained relatively unaffected, although even here the red soil lost its surface horizon due to the corrosive action of the material transported by the mass-wasting processes. A thin veneer of the material which was translocated by these processes is generally found below the aeolian loam, but many of the pebbles removed now cover the lower pediment-like surfaces surrounding the plateau.

The grain-size distribution of the disturbed material on the plateau suggests that the last time the mass-wasting processes were active was during the first stages of the aeolian accumulation which produced the present surface loam. The age of the aeolian deposit has yet to be established but evidence obtained so far indicates a late-Wisconsin age. The Cypress Hills plateau surface was then within the domain of the periglacial conditions which promoted the solifluction of the surface layer of the conglomerate. The ease with which this material flowed on the very gentle slopes is probably due to its high montmorillonite content.

As the climate became drier, solifluction decreased, and further deposits were no longer disturbed. The present stability results from the absence of slopes of sufficient inclination to be affected by the erosive processes still active on the adjacent pediment-like surfaces and which are operative on slopes exceeding one degree.

The red soil represents remnants of the oldest erosion surface recognizable in the Cypress Hills area. The formation of the soil required conditions of illuviation and oxidation. To allow for the liberation of iron, the climate must have been warm and subhumid, the rainfall having a marked seasonal distribution. The well-developed B horizon and the absence of a calcium carbonate accumulation zone belonging to this profile also demonstrate significant amounts of precipitation. That chemical weathering, on the other hand, was restricted and of limited duration is shown by the composition of the clay fraction.

Palaeosols have been extensively studied in the United States for correlation purposes and for the characterizing of the environmental conditions prevailing during their formation. A review of the literature (Ruhe, 1965) shows that the effect of interglacial climates on soil conditions varied considerably depending on other soil-forming factors. Where interglacial palaeosols have redder hues than modern soils, they are generally associated with warmer or moister climatic conditions.

It is generally assumed that semiarid and subhumid conditions have prevailed in the Plains area since the Oligocene epoch (Howard, 1960) because of its position in the rain shadow of the Rocky Mountains which gradually rose across the path of the moisture-bearing winds. On the other hand, inferences from Pleistocene fauna indicate the occurrence of interglacial intervals somewhat warmer and moister than at present (Taylor, 1965).

Climatic conditions required for the formation of the reddish soil were not necessarily very different from those prevailing at the present time. As is shown in Romania (Cernescu, 1964, pp. 57-58) an increase in the average rainfall is sufficient to change the soils along a given isotherm from the continental chernozem to the mediterranean red forest soil, provided the minimum annual precipitation is appreciably raised, say from 160 mm. to 260 mm. During the time when the red soil of the Cypress Hills was being formed, a similar range of climatic conditions may have prevailed with deciduous forest on the plateau, and grass covering the surrounding more arid Plains.

On the basis of present evidence the palaeosol should be correlated with one of the early interglacial stages of the Pleistocene epoch. This interpretation would also explain the vertical position of many pebbles in some of the red soil profiles. They were shifted from their original horizontal position, presumably by frost action, before the clay skins typical for the red soil were formed.

There are further indications that the undulating surface marked by the reddish soil is probably not much older than early-Pleistocene. The colour of the soil remnants suggests well-drained conditions even in the lowest portions of the terrain. This would seem to be possible only when the plateau was already in existence, with the groundwater level lowered by the formation of the surrounding pediment-like surfaces. The lowering of the groundwater level was necessarily most drastic near the edges of the plateau, and it is indeed here that most remnants of the reddish palaeosol are found. Although it has not been possible to reconstruct the topography of this older erosion surface from the scattered occurrences of the palaeosol, it is obvious that it had not much coincidence with the present surface, and that late Pleistocene degradational forces have done more than merely modify an existing relief.

Summary and Conclusions

The profile described in this paper reflects only a part of the history of the Cypress Hills plateau. Erosion has effectively destroyed any record of events between the formation of the red soil and the solifluction deposit. Evidence from other locations suggests at least one more period of soil formation which produced a brown soil. Occasionally older Pleistocene deposits, one of which is probably aeolian, are preserved beneath the solifluction layers. In general, however, the plateau has been a surface of erosion rather than of deposition, and the historical record preserved is, therefore, relatively short and far from complete.

It has too readily been assumed that the reversal of relief resulting in the formation of the Cypress Hills plateau indicates absence of erosion on the plateau surface, whereas, in fact, erosion has only proceeded more slowly here than on the surrounding Plains. It is not known how much Oligocene or even younger sediments may have been removed from the plateau because the present surface does not represent the original surface of the Tertiary beds. There are no traces of a middle-Tertiary erosion level found in this area, and speculations on its possible form and position are of little value.

References

Alden, W. C., "Physiography and glacial geology of eastern Montana and adjacent areas," *U.S. Geol. Surv. Prof.*, paper, 174, (1932), 133 pp.
Bakker, J. P. and Le Heux, J. W. N., "A remarkable new geomorphological law, I," *Proc. Kon. Ned. Akad. v. Wetensch*, Vol. LV, no. 4, series B (1952), pp. 399-410; "A remarkable new geomorphological law, II and III," *Proc. Kon. Ned. Akad. v. Wetensch*, Vol. XV, no. 5, series B (1952), pp. 554-71.
Beavers, A. H., Johns, W. D., Grim, R. E., and Odell, R. T., "Clay minerals in some Illinois soils developed from loess and till under grass vegetation, in 'Clays and Clay Minerals'," *Nat. Acad. Sci.*, N.R.C., pub. no. 395 (1955), pp. 356-72.
Bennema, J., "The red and yellow soils of the tropical and subtropical uplands," *Soil Sci.*, Vol. XCV (1963), pp. 250-57.
Broscoe, A. J., "The geomorphology of the Cypress Hills—Milk River Canyon area, Alberta," *Alta. Soc. Petr. Geol.*, 15th Ann. Field Conf. Guidebook, pt. 1 (1965), pp. 74-84.
Cernescu, N., "Guide-book of excursions, pt. 1, VIII," *Int. Congr. Soil Sci.*, Romania.
Eakin, H. M., "The Yukon-Koyukuk region, Alaska," *U.S. Geol. Surv. Bull.*, no. 631 (1916), 88 pp. and plate.
Gravenor, C. P. and Bayrock, L. A., "Glacial deposits of Alberta," Soils in Canada, *Roy. Soc. Can.*, spec. pub. no. 3 (1961), pp. 33-50.
Hermans, W. F., "Description et génèse des dépôts meubles de surface et du relief de l'Oesling," *Serv. Géol. Luxembourg*, (1955).
Howard, A. D., "Cenozoic history of northeastern Montana and northwestern Dakota with emphasis on the Pleistocene," *U.S. Geol. Surv. Prof.*, paper, 326, (196), 107 pp. and 8 plates.

McCaleb, S. B., "The genesis of red-yellow Podzolic soils," *Soil Sci. Soc. Amer. Proc.*, no. 23 (1959), pp. 164-68.

McConnell, R. G., "Report on the Cypress Hills, Wood Mountain, and adjacent country," *Can. Geol. Surv.*, Ann. Rept., (new series) Vol. I, Rept. C (1885), 85 pp.

National Soil Survey Committee of Canada, "Report on the fifth meeting of the National Soil Survey Committee held at Winnipeg, Manitoba," (1963), 92 pp.

Ruhe, R. V., "Quaternary paleopedology," *The Quaternary of the United States*, Wright, H. E. and Frey, D. G. (editors), Princeton Univ. Press, Princeton, N.J., (1965), pp. 755-64.

Russell, L. S. and Landes, R. W., "Geology of the southern Alberta plains," *Geol. Surv. Can.*, mem. 221 (1940), 223 pp. and maps.

St. Arnaud, R. J. and Mortland, M. M., "Characteristics of the clay fractions in a Chernozemic to Podzolic sequence of soil profiles in Saskatchewan," *Can. J. Soil Sci.*, Vol. XLIII (1963), pp. 336-49.

Schieferdecker, A. A. G., "Geological nomenclature," *Roy. Geol. and Min. Soc. of the Netherlands*, (1959).

Taylor, D. W., "The study of Pleistocene nonmarine mollusks in North America," *The Quarternary of the United States*, Princeton Univ. Press, Princeton, N.J., (1965), pp. 597-611.

United States Soil Survey Staff, "Identification and nomenclature of soil horizons," Supplement to U.S. Dept. Agric., Handbook no. 18, *Soil survey manual*, 1951, (1962).

Van Harten, D., "On the estimation of relative grain frequencies in heavy mineral slides," *Geol. en Mijnb.*, Vol. XLIV, no. 10 (1965), pp. 357-63.

Westgate, J. A., "The pleistocene stratigraphy of the Foremost—Cypress Hills area, Alberta," *Alta. Soc. Petr. Geol., 15th Ann. Field Conf. Guidebook*, pt. 1 (1965), pp. 85-111.

Wyatt, F. A., Newton, J. D., Bowser, W. E., and Odynsky, W., "Soil Survey of Milk River sheet," *Univ. Alta., Coll. Agric. Bull. no. 36*, (1941), 105 pp., maps and plates.

Zeuner, F. E., "Frost soils on Mount Kenya, and the relation of frost soils to aeolian deposits," *J. Soil Sci.*, Vol. I, no. 1 (1949), pp. 20-30 and plates.

25
Organic Terrain and Geomorphology

N. W. Radforth

Those who specialize in geomorphology do not accept landform as a static phenomenon and recognize that land conformations prevailing at any one time are transient symbols of reference for purposes of classification. The dynamic concept is favoured and it is evolution of form that inspires inquiry. No doubt this is why the Arctic and subarctic make an attractive panorama. The ice cap retreated and left new configurations that rested on the old. Nothing is more fascinating than a journey from the fresh landforms of the north to the old ones of the south to observe how time has consorted with forces to wear the land into senescence.

But wear and tear are not inevitable, for in vast areas of the land, new depth is added. I refer not to that mineral increment that climate translocated from one place to another. I mean organic deposition that has the inherent power to grow—a new landscape crowned with living plants growing on their ancestors beneath. This combination creates new aspects of form.

Organic Land
Surficial features

Those whose business it is to read the character of the land want to know how to distinguish organic terrain, or muskeg as it is frequently called in Canada, from mineral terrain. This has never been my concern because mineral terrain has been outside my ivory tower. Recently I have given the matter some thought, for I have been impressed by those who without formal training can pick out organic from mineral terrain, especially when they are airborne.

Their first symbol of recognition is hidden in the vegetation. To find it they must fly at somewhat less than 5000 ft., if they are novices. They discover, if mineral terrain is in line of sight and is tree-covered, that the tree crowns will vary usually markedly in

First published in *Canadian Geographer*, Vol. VI, nos. 3-4 (1962), pp. 166-71.

level. Also, if the trees are deciduous there will be characteristic admixtures of form (Figure 1). When organic terrain comes into view there are no convoluted admixtures (Figure 2) and tree crowns form no longer in multiple level but in a single plane. This plane may be either sloping, curved, or horizontal.

Where there is treeless mineral terrain the cover is changeable as to proportions of tall shrubs, low shrubs, and herbaceous plants except where the landform is constant over a large area, for example, prairie. On treeless organic terrain the cover is also changeable but less heterogeneous. Configurations of over-topping tall or dwarf shrubs are delineated against a background of herbaceous cover, which is mainly grass-like.

To attempt further separation of mineral and organic terrain by reference to cover requires partial reference to land conformation. Thus, characteristically, grass-like cover for the prairie is on a gently rolling surface whereas for organic terrain it is on a flat expanse.

Figure 1.

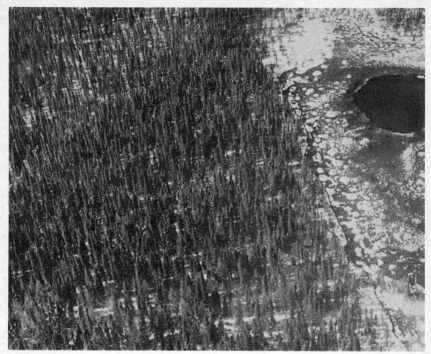

Figure 2.

The principle of cover-form heterogeneity for mineral and homogeneity for organic terrain is also better understood when basic land conformation is included in the consideration. When muskeg is confined, there is zonation of homogeneous cover-form but often there is local over-growing at the zone interfaces and the total effect simulates heterogeneity on mineral terrain. Usually, however, there is no difficulty in detecting the organic terrain because the delineated margin of the muskeg is observed to coincide with the edge of a land depression.

It has been reported elsewhere[1] that those natural attributes of muskeg cover which lend themselves to ready identification are stature, habit, and presence or absence of woodiness. Use of them results in the designation of nine classes of cover, which usually occur in combinations. On application of the rule that if a given class of cover does not exceed 25 per cent it is not significant in symbolizing prevailing character, no more than three classes will appear in a combination and often fewer.

1 N. W. Radforth, "Suggested classification of muskeg for the engineer". *Eng. Jour.*, 35 (1952), pp. 1199-1210.

Combinations of classes are known as cover formulae and are portrayed by grouping appropriate letters chosen from A to I, the nine symbols of cover class. Typical examples of cover formulae are AEI (trees over 15 ft. tall overtopping low shrubs less than 3 ft. occurring with mosses—not lichens), FI (non-woody sedge to grass-like growth with mosses), and HE (lichen-like habit with low shrubs less than 3 ft. tall), and so on. The predominating cover class (tallest overtopping layer having greatest coverage) is always on the left of the formula with those on its right in descending order of pre-dominance.

The number of formulae that occurs is happily few by compari-son with the number of mathematical possibilities. This, coupled with the fact that there is much muskeg, means that in nature there must be frequent recurrence of formulae. This is indeed the case and furthermore some show a much higher frequency of occurrence than others; for example, EI and FI occur much more frequently than do ADF or BDE.

Since cover formulae integrate to provide three dimensional air-form patterns,[2] the latter once identified can be extrapolated back to cover; for example, Marbloid, a 30,000-ft.-altitude air-form pat-tern is constituted large of HE, EH, and EI and not FI, DFI, which are basic to Dermatoid, another 30,000-ft. air-form pattern, and not AEI or ADE, which are basic to Stipploid.

Topographic features

Surficial cover for organic terrain has greater significance than it does for mineral landscape. The reason is that muskeg cover forms the culminating element to a whole series of buried elements that collectively convey ordered post-Pleistocene vegetal history. The history explains the structural composition in the vertical dimension of the peat, which facilitates comprehension of difference in organic terrain.[3] There is no equivalent for this to assist in the interpretation of muskeg-free terrain.

Botanical organization of organic land masses plays such a dominant role in imparting character that examination of it in rela-tion to geomorphic manifestations seems reasonable. Like botanical phonomena, the topographic ones are repetitive and not at all for-tuitous in their occurrence. Also, evidence of correlation between

---

[2] N. W. Radforth, "Organic Terrain Organization from the Air (altitudes 1,000-5,000 feet)," *Defence Research Board,* Handbook no. 2, DR no. 124, (1958), Ottawa.

[3] H. R. Eydt, J. M. Stewart, and N. W. Radforth, "The structural aspect of peat," *Proc. Seventh Muskeg Research Conference.* NRC Tech. Memo. no. 71, (1961), Ottawa, pp. 12-23.

cover and conformation is inescapable. For instance, in one family of cover formulae characterized by EI and EH, closely applied mounds occur. Their bases are usually much wetter than their peaks, which rise to one or two feet in height. With another family of cover classes, that represented by FI, the mounds become very widely scattered or if they coalesce they do so to form ridges in tortuous or crescentic parallel arrangements. Whenever D cover class enters the cover formula, intermittent traps or depressions with vertical sides intervene with high frequency. Usually this designates a drainage course and almost invariably the lower end of the gradient terminates in terrain with FI cover in which not mounds but hummocks characteristically appear (Figure 3).

Figure 3. A drainage course in DFI background terminating in FI hummocks in the foreground, about five miles north of Kapuskasing, Ontario.

Superimposed upon this micro-topography is a macro-pattern. Thus, where mounds occur, the terrain is at higher elevation than is ridged or hummocked terrain. Sometimes the elevation reaches a height of 20 to 30 ft., and this is a common feature in semi-marine climates where the condition is known as "raised bog land." This macro-feature is common in Labrador and Newfoundland. On the other hand the topographic feature sometimes defines the air-form pattern. The anastomosing ridges of Reticuloid air-form pattern,[4] the lace bogs referred to by Allington[5] and Sjörs[6] are usually of FI or FBI cover unless the water table is low or raised bog effect becomes superimposed and then BDE may form the framework of the reticulum.

Because of the relationships already suggested between topography and vegetal cover, presence of topographic features can be interpreted once air-form pattern is known.

Ice effect

So far, topographic differences have been expressed as suggesting relationship to cover which in turn reflects structural relationship established in the vertical dimension of peat. These relationships can be either accentuated or sometimes altered by an ice factor.

There are three conditions where freezing temperatures are influential. First, there is permafrost, which the writer rightly or wrongly regards as terrain that is frozen indefinitely. In this circumstance, the elevations are accentuated and peat plateaus occur. Probably because of the ice that becomes incorporated into peat year in and year out, the plateaus rise to a height of 10 or 12 feet above the surrounding terrain.

Another condition of frozen ground also occurs perennially but from two to several years, not indefinitely. Temperatures arranged on this order of time produce the condition designated by the author as "climafrost." Obviously climafrost and permafrost can occur together but it is significant that the former may occur in permafrost-free country, a situation which has been found and studied in the organic terrain south of Waboden, Manitoba. It is climafrost that is largely responsible for ice polygons, common features in Marbloid country. It is also important where perforations in subsurface ice typify FI-covered muskeg.[7]

---

4 N. W. Radforth, "The application of aerial survey over organic terrain," *Roads and Engineering Construction Magazine.* NRC Tech. Memo no. 42, (1956).
5 K. R. Allington, "The bogs of central Labrador Ungava: an examination of their physical characteristics," *McGill Sub-Arctic Res.* Montreal, paper, 7, (1959), (thesis).
6 J. Sjörs, "Surface patterns in boreal peatlands." *Endeavour*, 20, (1961), London, pp. 217-224.
7 N. W. Radforth, "The ice factor in muskeg," Presented at the First Canadian Conference on Permafrost, Ottawa, April 17-18, 1962.

Finally, there is active frost, perhaps better defined as "seasonal frost," for it gives rise to winter ice. Perhaps the best example of it is ice-knolling, which is associated with the organic factor to produce mounds. The process that forms the sub-surface condition known as ice-knolling in mounds and the vertical growth of ice-blocks in peat plateaus likens to the ice dynamics in pingos.

The ice factor in muskeg contributes to its own special kind of topography in organic terrain, but it is a hidden topography in that it is beneath the surface. The link between the ice and the organic factors is always real. One very significant example of relationship involves not the peat but the living cover—the H, or lichenaceous cover. Aerial inspection on flights from south to north reveal EI changing to EH and finally HE. The latitude at which H makes its appearance is now known to bear some relationship to the southern limit of permafrost.[8]

The Water Factor

In recent work, the study of hidden topography has taken on another aspect which does not concern ice. Structurally weak peats have been discovered beneath FIE cover in small masses averaging about eight feet in diameter, and surrounded by stronger peat with FEI cover. Analysis suggests that the weak areas were once open water—small ponds in an open expanse of floating peat (Figure 4). Thus paleo-topographic features may in some cases have occurred only to become obliterated in the course of time.

In the case described, the organic factor has taken over from the water factor but often the reverse may be the cause of topographic differential. Thus, in northern shallow peat plateaus erosion caused by local drainage produces irregularities. Fringe irregularities caused by seasonal melt water at the edges of peat plateaus are not as common or as accenuated in the south as they are in the north. Melt water may also encourage shift in position of ponds of the type shown in Figure 5, and of water courses that arise from time to time, particularly where FI and DFI abound.

The mechanical factor of water erosion and the biotic factor combined do more than climate in the control of vegetal succession. Where drainage change does not exist, usually there is no change in the botanical history of the peat, which means that the culminating flora is the same in composition as the ancestral ones. There is therefore no climax in the classical sense that the expression "climax" conveys.

8 N. W. Radforth, "Distribution of organic terrain in northern Canada," Presented at the Seventh Muskeg Research Conference, Ottawa, April 18-19, 1961.

Figure 4. Aerial view of a confined muskeg near Parry Sound, Ontario, show-ing secondary areas of open water interrupting the muskeg at its fringe.

Mineral Land

There is of course topographic control resulting from conforma-tion of the mineral terrain lying beneath the organic.

In these circumstances, the water factor, if it arises as a result of organic differentials, is secondary and the primary effect of water is a function of contour of the mineral sub-layer. In heavily forested muskeg where the peat is usually relatively shallow, drain-age is always towards DFI cover. Where the formula is ADE (in contrast to AEI) there will be intermittent shallow depressions and the mineral sub-layer will be gently rolling. Seasonal flooding in ADE country will be common. Where ADF occurs the depressions become contiguous, and extensive flooding can be expected for a period lasting well into the summer.

At the macro-topographic level one might expect controls im-posed by mineral terrain on muskeg formation to be impressive. There is evidence to support such expectation but discriminatory relationships are hard to appreciate because every landform en-countered across North America may support muskeg so far as the

Figure 5. Aerial view of FI-DFI country showing ponds in their temporary position in organic terrain between Hay River and Fort Smith, N.W.T.

author can confirm. Whatever their genesis, hillsides or mountainsides may be muskeg-covered (Figure 6). Whatever the age or form of a major drainage system (Figure 7), it too can support muskeg. Sedimentary or glaciated flats serve equally well as foundation for muskeg[9] (Figure 8). Glacial features, beach-lines, Precambrian exposures, or land surface depressions may all become muskeg-covered.

It is usually easy to designate the mineral landform occurring beneath organic overburden. It is the determination of mineral soil type that is difficult. There are as yet no objectively established rules which enable an observer to predict mineral sub-layer. There is a useful hypothesis in which it is proposed that contemporary cover of the muskeg bears relationship to type of mineral aggregate beneath the peat. Although there is now much evidence to support this claim, it has been derived by empirical procedure and is not

---

9 N. W. Radforth, "Organic terrain," in "Soils in Canada," *Roy. Soc. Can. Special Publications*, no. 3, ed. R. F. Legget, (University of Toronto Press), 1961.

Figure 6.

yet conclusive. At this time the statement will be appreciated that it is almost invariably the case that the A and the B families (cf. reference 9) occur over course aggregate mixed with silt and sand. If the formula is AH, it is safe to claim that the mineral matter beneath will be sand. Commonly the H and E families prescribe for sandy silt except for HE in which case the constitution of the mineral matter is usually gravelly sand. The D family relates to silty terrain and the F family to clay. An explanation for these phenomena has yet to be supplied.

Confined Muskeg

Kettle holes, and depressions that simulate kettle holes, examples of which are found on the Canadian Shield, require special study for organic terrain interpretation.

Analyses and distribution of cover formulae support the classical claim that zonation, oriented with respect to the margin of the water body, characterizes the depression. On the other hand the writer finds that the successional concepts conveyed in most text-

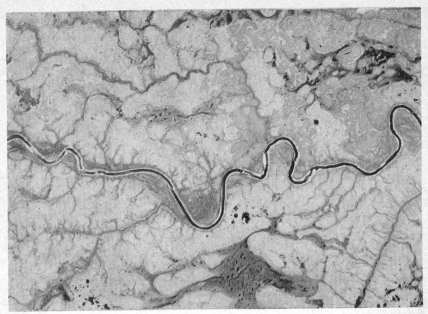

Figure 7.

books on geomorphology are very much in need of qualification. For instance, when peat commences to form at the edge of a glacial lake, it is not always the case that the initial cover is sedge-like (Figure 9). The so-called xeromorphic condition which is constituted of cricoids in company with mosses may usurp the privilege of the sedge-like plants as the initial colonizers. There are other cases where exceptions become the rule, and certainly evidence suggests that confined muskeg cover does not conform to the climax formation theory for it appears at the time of writing that depth of organic terrain and possibly depth of depression will turn out to be very influential in controlling cover type. Certainly this, combined with other factors not as yet adequately designated or understood will prove to outweigh climate in cover control and peat development.

Application of Organic Terrain Studies

Forestry and agricultural industries, together with engineering development in one aspect or another, stand to benefit from the results of organic terrain studies. Devices for terrain interpretation can be fashioned to afford prediction pertinent to application of almost any kind. Prediction procedures based on cover and air-form pattern have been tested on a circumpolar basis and the results

Figure 8.

are most encouraging. For Canada, off-the-road access is the most significant requirement for northern development, and it is most encouraging to realize that for any given vehicle not only can routes across organic terrain be selected for which it can be calculated that the operation contemplated will succeed but also the operational costs can be estimated.[10] This is achieved in one preliminary aerial survey that is attractively inexpensive.

But it is not the application of the studies that encouraged the writer to present this paper at this time; it is the realization that a sister science of paleoecology can throw some light on geomorphic principles and provide an extension for geomorphic study.

---

[10] J. G. Thompson, "Vehicle design from field test data," Presented at the First International Conference on the Mechanics of Soil-Vehicle Systems, Torino-Saint Vincent, Italy, June 1961.

Figure 9.

Index